MATHEMATICAL MODELING

Staff of Research and Education Association

Research and Education Association
505 Eighth Avenue
New York, N. Y. 10018

MATHEMATICAL MODELING

Printed in the United States of America

Library of Congress Catalog Card Number 82-80745

International Standard Book Number 0-87891-538-9

PREFACE

This book is a valuable reference source for applied mathematicians, scientists, and engineers in formulating models to illustrate and study various phenomena.

The techniques described in this book are aimed at simplifying the formulation of mathematical models by reducing the modeling process to a series of routine operations, which can be performed either manually or by computer. As a result, the techniques that are described, are very useful to those working in the fields of simulation sciences and experimentation.

To illustrate the principles involved, applications and examples are chosen from fields in which mathematical modeling is a large part of scientific research efforts. From these examples, the underlying principles and analytic techniques can be extended to other applications which may be encountered, such as marketing, finance, sociological and psychological studies, and biomedical applications.

The information in this book was originated and sponsored by the National Aeronautics and Space Administration and edited by James C. Howard.

CONTENTS

SCALARS, VECTORS, AND TENSORS

1.1 SUMMATION CONVENTION

Many of the advantages inherent in the tensor method derive from the simplifying nature of the tensor notation, in general, and the summation convention in particular. This convention, which lends itself to the design of computational algorithms, is well suited to computer applications. For example, consider the following set of equations:

$$y^1 = a_{11}x^1 + a_{12}x^2 + a_{13}x^3$$

$$y^2 = a_{21}x^1 + a_{22}x^2 + a_{23}x^3$$

$$y^3 = a_{31}x^1 + a_{32}x^2 + a_{33}x^3$$

These equations can be written very compactly as follows:

$$y^i = \sum_{j=1}^{j=3} a_{ij}x^j , \quad i = 1,2,3 \tag{1.1.1}$$

A further simplification is possible by adopting the summation convention (ref 5). This convention permits the removal of the summation sign on condition that the occurrence of two like indices in a given expression denotes summation on the appropriate indices. Hence, since j occurs twice in the expression on the right-hand side of equation (1.1.1), this equation can be written simply as

$$y^i = a_{ij}x^j \tag{1.1.2}$$

The advantages of the summation convention are more evident if one considers matrix multiplication. For example, the product of two matrices A and B, requires that the elements of the A matrix be combined with the elements of the B matrix according to well-established rules. Because these rules have to be memorized and may be forgotten if not frequently employed, utilization of the summation convention acts as a spur to the memory and suggests the order of multiplication if this has been forgotten. A simple example will illustrate this point. Consider the product of two matrices A and B, where each is a three by three matrix. In this case

$$A = \begin{pmatrix} a_{11} & a_{12} & a_{13} \\ a_{21} & a_{22} & a_{23} \\ a_{31} & a_{32} & a_{33} \end{pmatrix} \tag{1.1.3}$$

and

$$B = \begin{pmatrix} b_{11} & b_{12} & b_{13} \\ b_{21} & b_{22} & b_{23} \\ b_{31} & b_{32} & b_{33} \end{pmatrix} \tag{1.1.4}$$

Therefore

$$AB = \begin{pmatrix} a_{11} & a_{12} & a_{13} \\ a_{21} & a_{22} & a_{23} \\ a_{31} & a_{32} & a_{33} \end{pmatrix} \begin{pmatrix} b_{11} & b_{12} & b_{13} \\ b_{21} & b_{22} & b_{23} \\ b_{31} & b_{32} & b_{33} \end{pmatrix} = E$$

where

$$E = \begin{pmatrix} a_{11}b_{11} + a_{12}b_{21} + a_{13}b_{31} & a_{11}b_{12} + a_{12}b_{22} + a_{13}b_{32} & a_{11}b_{13} + a_{12}b_{32} + a_{13}b_{33} \\ a_{21}b_{11} + a_{22}b_{21} + a_{23}b_{31} & a_{21}b_{12} + a_{22}b_{22} + a_{23}b_{32} & a_{21}b_{13} + a_{22}b_{23} + a_{23}b_{33} \\ a_{31}b_{11} + a_{32}b_{21} + a_{33}b_{31} & a_{31}b_{12} + a_{32}b_{22} + a_{33}b_{32} & a_{31}b_{13} + a_{32}b_{23} + a_{33}b_{33} \end{pmatrix}$$

If the summation convention is employed, there is no need to write out the matrices in this manner in order to obtain the product. In terms of this convention, the elements of the product matrix E are given by

$$e_{ij} = a_{ik} b_{kj} \qquad (1.1.5)$$

Since the index k occurs twice in the expression on the right-hand side of equation (1.1.5), this expression must be summed on k, for all admissible values of k. Hence, for the three by three matrices being considered

$$e_{ij} = a_{i1} b_{1j} + a_{i2} b_{2j} + a_{i3} b_{3j} \qquad (1.1.6)$$

and by permitting i and j to assume the values

$$i = 1,2,3$$

$$j = 1,2,3$$

the fully expanded form of AB is obtained. Likewise, if any element of the product is required, it can be obtained by assigning specific values to i and j. A more complicated example involving the product of three matrices will show that the summation convention is a convenient shorthand, a compact and well-adapted code for expressing complicated relationships. Again, let the matrices A and B be as given in equations (1.1.3) and (1.1.4), respectively, and let

$$C = \begin{pmatrix} c_{11} & c_{12} & c_{13} \\ c_{21} & c_{22} & c_{23} \\ c_{31} & c_{32} & c_{33} \end{pmatrix} \qquad (1.1.7)$$

then the product

$$ABC = \begin{pmatrix} a_{11} & a_{12} & a_{13} \\ a_{21} & a_{22} & a_{23} \\ a_{31} & a_{32} & a_{33} \end{pmatrix} \begin{pmatrix} b_{11} & b_{12} & b_{13} \\ b_{21} & b_{22} & b_{23} \\ b_{31} & b_{32} & b_{33} \end{pmatrix} \begin{pmatrix} c_{11} & c_{12} & c_{13} \\ c_{21} & c_{22} & c_{23} \\ c_{31} & c_{32} & c_{33} \end{pmatrix} = D$$

can be replaced by the more compact equation

$$ABC = a_{ih}b_{hk}c_{kj} = D \qquad (1.1.8)$$

where the appearance of the repeated subscripts h and k implies that the summation convention is to be observed. For example, the ijth element of the matrix D can be obtained by summing first on k and then on h or vice versa, that is

$$d_{ij} = a_{ih}b_{h1}c_{1j} + a_{ih}b_{h2}c_{2j} + a_{ih}b_{h3}c_{3j}$$

If the expression is now summed on h, the ijth element of the required product matrix assumes the following form

$$d_{ij} = a_{i1}b_{11}c_{1j} + a_{i2}b_{21}c_{1j} + a_{i3}b_{31}c_{1j}$$

$$+ a_{i1}b_{12}c_{2j} + a_{i2}b_{22}c_{2j} + a_{i3}b_{32}c_{2j}$$

$$+ a_{i1}b_{13}c_{3j} + a_{i2}b_{23}c_{3j} + a_{i3}b_{33}c_{3j}$$

Although the cases considered so far have demonstrated the convenience of the summation convention, later applications will be dictated by necessity.

1.2 TENSORS

Physical entities that can be adequately characterized by the specification of their magnitudes are referred to as scalar quantities. Examples of scalar quantities are: temperature, volume, mass, and energy. Other quantities, however, such as forces and velocities, need for their complete specification not only magnitude but also a direction in space. Such quantities are termed vectors (ref. 6). Although a single quantity is not sufficient to completely specify them, vectors should be considered as single entities. Treating vectors in this manner greatly facilitates the processing of vector algebra and the derivation of formulas.

Although a scalar quantity has magnitude only, and a vector or tensor of rank one has both magnitude and direction, tensors of rank greater than one belong to a class of entities that depends on more than one vector. The chief aim of tensor calculus is the investigation of relations that remain valid in all coordinate systems. The condition of invariance with respect to coordinate transformations leads to the transformation laws that the tensor components must obey. Most books on tensor calculus define tensors in terms of these transformation laws. However, for present purposes, it is more satisfactory to define a tensor in terms of a linear form in the base vectors. When this is done, a vector or a tensor of rank one is defined as

$$\bar{A} = A_i \bar{a}^i = A^j \bar{a}_j \tag{1.2.1}$$

where A_i and A^j are the covariant and contravariant tensor components, respectively, and $\bar{a}^i$ and $\bar{a}_j$ are the corresponding base vectors which are, in general, functions of the coordinates. The meaning of these components will be explained presently. It should be remembered that the covariant and contravariant forms of the vector $\bar{A}$ in equation (1.2.1) must be expanded in accordance with the summation convention, since the indices "i" and "j" occur twice, that is,

$$A = A_1 \bar{a}^1 + A_2 \bar{a}^2 + A_3 \bar{a}^3 = A^1 \bar{a}_1 + A^2 \bar{a}_2 + A^3 \bar{a}_3$$

A point to be observed here and in all subsequent equations, is that each step is so formulated that it is amenable to mechanization.

Since the concepts of covariance and contravariance are not encountered in the study of elementary vector analysis, the meaning of these terms and the need for them in the present context will be explained. At the outset, it should be emphasized that the covariance or contravariance of vector or tensor components is not an intrinsic property of the entity under consideration. The distinction is due to the way in which the entity is related to its environment, the coordinate system to which it is referred (ref. 1). The two sets of quantities A_i and A^i represent the same vector $\bar{A}$ referred to two different base systems. The vectors $\bar{a}^i$ and $\bar{a}_i$ that constitute the systems of base vectors, to which the covariant and contravariant components are referred, are said to be reciprocal systems of vectors. When reciprocal bases are subsequently defined, it will be seen that the system of unit base vectors specifying an orthogonal Cartesian reference frame is its own reciprocal. Hence, the distinction between covariant and contravariant vector components vanishes in this case. This explains why there is no preoccupation with these representations in the study of elementary vector analysis. However, when problems are formulated in curvilinear systems of coordinates, it is frequently useful in specifying vector and tensor components to employ a given base system and its reciprocal.

As one might expect, the base vectors $\bar{a}^1$, $\bar{a}^2$, $\bar{a}^3$ are called the reciprocal system to $\bar{a}_1$, $\bar{a}_2$, $\bar{a}_3$ when the following relations are satisfied:

$$\bar{a}^1 \cdot \bar{a}_1 = \bar{a}^2 \cdot \bar{a}_2 = \bar{a}^3 \cdot \bar{a}_3 = 1$$

These relations will be satisfied if the reciprocal system to $\bar{a}_1$, $\bar{a}_2$, $\bar{a}_3$ is defined as follows:

$$\bar{a}^1 = \frac{\bar{a}_2 \times \bar{a}_3}{[\bar{a}_1 \bar{a}_2 \bar{a}_3]}$$

$$\bar{a}^2 = \frac{\bar{a}_3 \times \bar{a}_1}{[\bar{a}_1 \bar{a}_2 \bar{a}_3]}$$

$$\bar{a}^3 = \frac{\bar{a}_1 \times \bar{a}_2}{[\bar{a}_1 \bar{a}_2 \bar{a}_3]}$$

It will be remembered that $[\bar{a}_1 \bar{a}_2 \bar{a}_3]$ is the familiar scalar triple product of elementary vector analysis (ref. 2). It is simply the scalar product of two vectors, one of which is itself the vector product of two vectors, that is,

$$[\bar{a}_1 \bar{a}_2 \bar{a}_3] = \bar{a}_1 \cdot (\bar{a}_2 \times \bar{a}_3)$$

The scalar triple product has the important cyclical property

$$\bar{a}_1 \cdot (\bar{a}_2 \times \bar{a}_3) = \bar{a}_2 \cdot (\bar{a}_3 \times \bar{a}_1) = \bar{a}_3 \cdot (\bar{a}_1 \times \bar{a}_2)$$

The proof of this relation follows from the fact that each of these expressions represents the volume of the parallelepiped whose edges are $\bar{a}_1$, $\bar{a}_2$, $\bar{a}_3$.

In terms of these relations and definitions, it is seen that

$$\bar{a}^1 \cdot \bar{a}_1 = \bar{a}^2 \cdot \bar{a}_2 = \bar{a}^3 \cdot \bar{a}_3 = 1 \qquad \cdot$$

The symmetry of these relations shows that if $\bar{a}^1$, $\bar{a}^2$, $\bar{a}^3$ is the reciprocal system to $\bar{a}_1$, $\bar{a}_2$, $\bar{a}_3$ then $\bar{a}_1$, $\bar{a}_2$, $\bar{a}_3$ is the reciprocal system to $\bar{a}^1$, $\bar{a}^2$, $\bar{a}^3$, that is,

$$\bar{a}_1 = \frac{\bar{a}^2 \times \bar{a}^3}{[\bar{a}^1 \bar{a}^2 \bar{a}^3]}$$

$$\bar{a}_2 = \frac{\bar{a}^3 \times \bar{a}^1}{[\bar{a}^1 \times \bar{a}^2 \times \bar{a}^3]}$$

$$\bar{a}_3 = \frac{\bar{a}^1 \times \bar{a}^2}{[\bar{a}^1 \bar{a}^2 \bar{a}^3]}$$

and

$$\bar{a}_1 \cdot \bar{a}^1 = \bar{a}_2 \cdot \bar{a}^2 = \bar{a}_3 \cdot \bar{a}^3 = 1$$

As previously indicated, the conventional system of unit vectors $\hat{i}, \hat{j}, \hat{k}$ that is used to specify an orthogonal Cartesian system of coordinates is seen to be its own reciprocal.

Let $\hat{i}^1, \hat{j}^1, \hat{k}^1$ be the reciprocal system to $\hat{i}, \hat{j}, \hat{k}$; then, because the scalar triple product $[\hat{i}\hat{j}\hat{k}]$ of three orthogonal unit vectors is clearly equal to unity

$$\hat{i}^1 = \frac{\hat{j} \times \hat{k}}{[\hat{i}\hat{j}\hat{k}]} = \hat{i}$$

$$\hat{j}^1 = \frac{\hat{k} \times \hat{i}}{[\hat{i}\hat{j}\hat{k}]} = \hat{j}$$

$$\hat{k}^1 = \frac{\hat{i} \times \hat{j}}{[\hat{i}\hat{j}\hat{k}]} = \hat{k}$$

It should be evident now that there is nothing mysterious or obscure about the concepts of covariance and contravariance. These are simply convenient terms for describing vector and tensor components which are referred to a given base system, on the one hand, and to the reciprocal of the given system on the other.

The vector $\bar{A}$ may also be expressed in terms of its physical components as follows:

$$\bar{A} = \mathcal{A}^\alpha \hat{a}_\alpha \tag{1.2.2}$$

where $\mathcal{A}^\alpha$ is the physical component and $\hat{a}_\alpha$ is a set of unit base vectors.

Since the index α occurs twice in the expression on the right-hand side of equation (1.2.2), the summation convention must be observed, that is

$$\overline{A} = \mathscr{A}^1 a_1 + \mathscr{A}^2 a_2 + \mathscr{A}^3 a_3$$

The transformation from physical to tensor components, and vice versa will be considered in a later section.

Before proceeding to more general tensor forms, it should be remarked that the term "tensor" was used by Einstein in connection with the sets of quantities transforming in accordance with the covariant and contravariant laws. The formulation of covariant and contravariant laws, as well as an outline of the essential features of the algebra and calculus of covariant and contravariant tensors, is due to G. Ricci (ref. 1). Because of the usefulness of covariant and contravariant laws of transformation in applications to geometry and physics, the term tensor is generally used in the sense contemplated by Einstein. In the present context, however, a tensor of rank "*r*" associated with a point *P* of an "*n*" dimensional space is defined as an *r*-linear form in the base vectors associated with the point whose coefficients are, in general, functions of the coordinates of the point, and which is invariant with respect to coordinate transformations. When the condition of invariance with respect to coordinate transformations is imposed on an *r*-linear form, the transformation law for the tensor components is obtained. These components will be seen to transform in accordance with the covariant and contravariant laws and to satisfy the definition of a tensor in the sense in which it was used by Einstein.

In terms of this definition, a tensor of rank two may assume the following alternative forms:

$$A^{ij}\bar{a}_i\bar{a}_j \tag{1.2.3}$$

$$A_j{}^i\bar{a}_i\bar{a}^j \tag{1.2.4}$$

$$A_{ij}\bar{a}^i\bar{a}^j \tag{1.2.5}$$

Form (1.2.3) represents a doubly contravariant tensor or dyadic. Form (1.2.4) is a mixed tensor or dyadic, having one index of covariance and one index of contravariance. Form (1.2.5) is a doubly covariant tensor or dyadic.

More generally, a tensor of rank *r* associated with a point in *N* dimensional space is an *r*-linear form in the base vectors associated with the point, and is invariant with respect to the choice of coordinate system. In terms of this definition, a trilinear form having a *y*-coordinate representation

$$B^{ijk} \bar{b}_i \bar{b}_j \bar{b}_k$$

and an *x*-coordinate representation

$$A^{\alpha\beta\gamma} \bar{a}_\alpha \bar{a}_\beta \bar{a}_\gamma$$

will represent a tensor of rank three if

$$B^{ijk} \bar{b}_i \bar{b}_j \bar{b}_k = A^{\alpha\beta\gamma} \bar{a}_\alpha \bar{a}_\beta \bar{a}_\gamma$$

1.3 PHYSICAL EXAMPLES

1.3.1 The Stress Tensor

Because the name tensor originated in the study of tensions or stresses, it is appropriate to use the stress tensor (ref. 7) to illustrate the physical meaning of a tensor of rank two.

In the study of elasticity, certain quantities are introduced that are more complex than vectors. The stresses or tensions in the interior of a deformed body are defined by a collection of six numbers which behave like the six components of a new quantity. It was W. Voigt, the crystal physicist, who first named these new quantities, tensors. The word clearly recalls their origin, since the first one identified was the system of tensions of a deformed solid. In this connection it should be remarked that an elastic stress is defined as the intensity of force acting at any point in a deformed body, that is, the force per unit area.

Consider, for example, a uniform bar having the dimensions shown in sketch (a), and acted on by an axial force of "*F*" pounds.

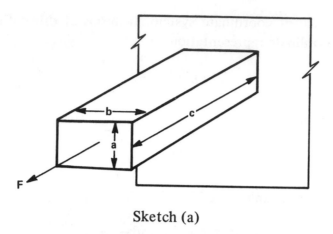

Sketch (a)

The stress across any cross section normal to the force vector is the force per unit area, that is,

$$\frac{F}{a \cdot b} \text{ lb/in.}^2$$

When an elastic body is subjected to a stressing agent, it is deformed. The extent of the deformation determines what is called the strain. In the case of the uniform bar subjected to an axial force, the strain is defined as the relative elongation, or the change in length per unit length.

Hooke's law states that strain is proportional to stress within the elastic limit (see sketch (b)).

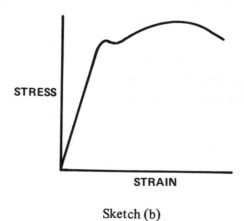

Sketch (b)

According to Hooke's law (ref. 8)

$$\frac{\text{stress}}{\text{strain}} = E$$

where E is Young's modulus of elasticity for the material. Knowing the stress and Young's modulus for a material, it is easy to compute the strain, which is simply

$$\frac{\text{stress}}{E}$$

The elongation or the extension of the bar is given by the product of the strain and the length of the bar, that is

$$\text{elongation} = \text{strain (length of the bar)}$$

$$= (\text{stress}/E)c$$

Similarly, volumetric strain is defined as the change in volume per unit volume. For example, consider a volume of elastic material bounded by the closed surface shown in sketch (c).

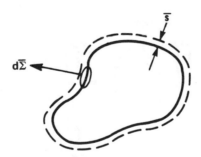

Sketch (c)

When this volume is subjected to a normal stress, it will assume the form shown by the dotted line. If the distance between the stressed and unstressed surfaces be denoted by the vector $\bar{s}$, the change in volume due to the stress will be given by the following integral

$$\text{change of volume} = \iint^{\Sigma} \bar{s} \cdot d\bar{\Sigma}$$

where $d\overline{\Sigma}$ is an element of area of the closed surface. For reasons that will become apparent as we proceed, it is expedient to convert this surface integral to a volume integral. This can be achieved by the use of Gauss' divergence theorem (ref. 9), which states that for any vector point function $\overline{A}$, which together with its derivative in any direction, is uniform, finite, and continuous

$$\iint^{\Sigma} \overline{A} \cdot d\overline{\Sigma} = \iiint^{\tau} \text{div } \overline{A} \, d\tau$$

where $d\tau$ is an element of volume, and div $\overline{A}$ is the divergence function.

If the vector $\overline{A}$ be expressed in component form as follows:

$$\overline{A} = X\hat{i} + Y\hat{j} + Z\hat{k}$$

where X, Y, Z are the Cartesian components of $\overline{A}$, and $\hat{i}, \hat{j}, \hat{k}$ are a triad of mutually orthogonal unit vectors, div $\overline{A}$ assumes the form

$$\text{div } \overline{A} = \frac{\partial X}{\partial x} + \frac{\partial Y}{\partial y} + \frac{\partial Z}{\partial z}$$

x, y, z being the Cartesian coordinates of an arbitrary point in the material.

Applying this theorem to the expression for the change of volume gives

$$\iint^{\Sigma} \overline{s} \cdot d\overline{\Sigma} = \iiint \text{div } \overline{s} \, d\tau$$

Therefore,

$$\text{change of volume} = \iiint \text{div } \overline{s} \, d\tau$$

If we consider an infinitesimal volume $d\tau$ of the material, then the change in this element of volume is

$$\text{div } \overline{s} \, d\tau$$

and the volumetric strain, which is the change in volume per unit volume, is given by the ratio

$$\frac{\text{div } \overline{s} \, d\tau}{d\tau} = \text{div } \overline{s}$$

Assume that $\bar{s}$ has Cartesian components u, v, w, that is

$$\bar{s} = u\hat{i} + v\hat{j} + w\hat{k}$$

then the volumetric strain (volume dilation) is

$$\epsilon_v = \frac{\partial u}{\partial x} + \frac{\partial v}{\partial y} + \frac{\partial w}{\partial z} = e_x + e_y + e_z$$

The quantities e_x, e_y, e_z denote relative changes in length (elongations) of an elementary volume in three coordinate directions caused by the normal stresses.

In discussing the stress across a given surface, we are obviously dealing with a situation that depends on two vectors as indicated in sketch (d); that is, the effect of a force on a surface depends not only on the force, but also on the size and orientation of the surface (ref. 10).

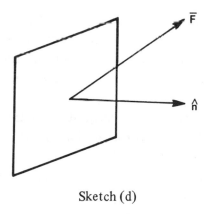

Sketch (d)

One vector $\bar{F}$ represents the force vector acting on the surface; the other vector $\hat{n}$, being the normal to the surface, uniquely characterizes the surface. Hence, the stress acting on the surface depends on the two vectors $\bar{F}$ and $\hat{n}$. In accordance with our definition of a tensor of rank "r" as an "r" linear form in the base vectors, the stress may be tentatively classified as a tensor of rank two. It should be remarked, however, that to qualify as a tensor the components must also transform in accordance with the covariant and contravariant laws to be defined.

After this brief discussion of the physical meaning of stress, the stress tensor will be derived.

Consider the element of volume enclosed by the infinitesimal tetrahedron shown in figure 1.3.1

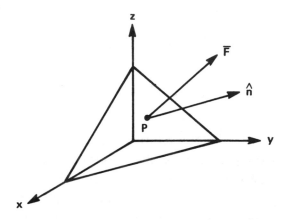

Figure 1.3.1.— Force and surface vectors in a Cartesian reference frame.

where $\hat{n}$ is a unit vector whose direction is normal to the element of area $d\bar{s}$ of a surface S passing through the point P of an elastic medium, and $\bar{F}$ is the resultant force. If the elastic medium is in a state of equilibrium, the resultant of all the forces acting on the element must vanish, and the resultant moment of these forces about any point must vanish also.

An examination of figure 1.3.1 again suggests that the tensor character of stress derives from the fact that it depends on the two vectors $\hat{n}$ and $\bar{F}$ rather than on a single vector. For the sake of clarity and simplicity, the stress tensor will be derived relative to a Cartesian system of axes. Relative to this system, $\bar{F}$ may be expressed in terms of its components as follows:

$$\bar{F} = F_X\hat{i} + F_Y\hat{j} + F_Z\hat{k} \tag{1.3.1}$$

The quantities F_X, F_Y, F_Z can be resolved into components perpendicular to each face and components parallel to each face. The components perpendicular to each face produce normal or direct stresses, and the components parallel to each face produce shearing stresses. Likewise, the area can be resolved into components relative to the three Cartesian axes (see sketch (e)).

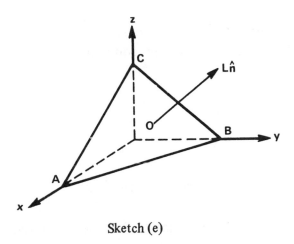

Sketch (e)

If $\overline{OL}$ perpendicular to ABC in the direction of the unit vector $\hat{n}$ is the vector that represents the size and orientation of the area $ABC = d\bar{s}$, than the x component of $\overline{OL}$, that is, $\hat{i} \cdot d\bar{s}$, represents in magnitude and direction the projection of $d\bar{s}$ on the yz plane. Similarly, the y and z components of $d\bar{s}$ are the projections of $d\bar{s}$ on OAC and OAB, respectively. These components are $\hat{j} \cdot d\bar{s}$ and $\hat{k} \cdot d\bar{s}$. Hence, the components of $\boldsymbol{F}$ may be rewritten in the following form:

$$F_X = f_{xx}\, ds_x + f_{xy}\, ds_y + f_{xz}\, ds_z \tag{1.3.2}$$

$$F_Y = f_{yx}\, ds_x + f_{yy}\, ds_y + f_{yz}\, ds_z \tag{1.3.3}$$

$$F_Z = f_{zx}\, ds_x + f_{zy}\, ds_y + f_{zz}\, ds_z \tag{1.3.4}$$

and

$$ds_x = \hat{i} \cdot d\bar{s} \tag{1.3.5}$$

$$ds_y = \hat{j} \cdot d\bar{s} \tag{1.3.6}$$

$$ds_z = \hat{k} \cdot d\bar{s} \tag{1.3.7}$$

The double subscript notation should be noted. The first subscript in f_{xx}, f_{xy}, and f_{xz} refers to the fact that these stresses all emanate from the component F_X, whereas the second subscript designates the projected area on which the stress acts. Thus, f_{xy} means the stress is due to F_X acting on the element of area ds_y.

When the components of area, equations (1.3.5) through (1.3.7), are substituted in equations (1.3.2) through (1.3.4), the components of the elastic force assume the following form:

$$F_X = (f_{xx}\hat{i} + f_{xy}\hat{j} + f_{xz}\hat{k}) \cdot d\bar{s}$$

$$F_Y = (f_{yx}\hat{i} + f_{yy}\hat{j} + f_{yz}\hat{k}) \cdot d\bar{s}$$

$$F_Z = (f_{zx}\hat{i} + f_{zy}\hat{j} + f_{zz}\hat{k}) \cdot d\bar{s}$$

Substitution of these values in equation (1.3.1) yields the required form

$$\bar{F} = (f_{xx}\hat{i}\hat{i} + f_{xy}\hat{i}\hat{j} + f_{xz}\hat{i}\hat{k} + f_{yx}\hat{j}\hat{i} + f_{yy}\hat{j}\hat{j} + f_{yz}\hat{j}\hat{k} + f_{zx}\hat{k}\hat{i} + f_{zy}\hat{k}\hat{j} + f_{zz}\hat{k}\hat{k}) \cdot d\bar{s}$$

But $d\bar{s} = \hat{n}\, ds$ and the resultant stress across $d\bar{s}$ at the point P is defined by $\bar{f}$, where

$$\bar{f} = \frac{\bar{F}}{ds} = \Phi \cdot n \tag{1.3.8}$$

and $\overline{\Phi}$ is the stress tensor which is defined as follows:

$$\overline{\Phi} = f_{xx}\hat{i}\hat{i} + f_{xy}\hat{i}\hat{j} + f_{xz}\hat{i}\hat{k} + f_{yx}\hat{j}\hat{i} + f_{yy}\hat{j}\hat{j} + f_{yz}\hat{j}\hat{k} + f_{zx}\hat{k}\hat{i} + f_{zy}\hat{k}\hat{j} + f_{zz}\hat{k}\hat{k}$$

$$\tag{1.3.9}$$

By using the condition that the resultant of the moments about any point of all the forces acting on the infinitesimal tetrahedron vanish, it can be shown that the stress tensor must be symmetric, that is

$$f_{xy} = f_{yx}$$

$$f_{xz} = f_{zx}$$

$$f_{yz} = f_{zy}$$

Although it is beyond the scope of the present illustration, it can be shown that the stress tensor, equation (1.3.9), is a tensor of rank two. That is, in addition to being a two linear form in the base vectors, it transforms in accordance with its variance. As indicated elsewhere, the distinction between covariance and contra-variance vanishes in orthogonal Cartesian coordinate systems, and in these coordinate systems it may be treated either as a covariant or a contravariant tensor. It will be shown that if a contravariant formulation is adopted, the components of the stress tensor in the "y" coordinate system will be related to its components in the "x" coordinate system as follows:

$$f^{ij}(y) = \frac{\partial y^i}{\partial x^\alpha} \frac{\partial y^j}{\partial x^\beta} f^{\alpha\beta}(x)$$

Similarly, if the stresses are treated as components of a covariant tensor, the components of stress in the "y" coordinate system will be shown to be related to the components in the "x" coordinate system by the following covariant transformation:

$$f_{ij}(y) = \frac{\partial x^\alpha}{\partial y^i} \frac{\partial x^\beta}{\partial y^j} f_{\alpha\beta}(x)$$

In these equations the indices i, j, α, β are used, for convenience, instead of x, y, z. Each index can assume the values 1, 2, 3. The choice of variance will depend on whether "y" is known as a function of "x," or "x" is known as a function of "y." If $y = y(x)$ is the form available, then the contravariant transformation would be the obvious choice, but if $x = x(y)$ is known, the covariant transformation would be simpler. If both $y = y(x)$ and $x = x(y)$ are known, it is immaterial which transformation is used.

Consider, for example, the contravariant transformation. The transformation equations are obtained by summing on the dummy indices α and β as follows:

$$f^{ij}(y) = \frac{\partial y^i}{\partial x^1} \frac{\partial y^j}{\partial x^1} f^{11}(x) + \frac{\partial y^i}{\partial x^1} \frac{\partial y^j}{\partial x^2} f^{12}(x) + \frac{\partial y^i}{\partial x^1} \frac{\partial y^j}{\partial x^3} f^{13}(x)$$

$$+ \frac{\partial y^i}{\partial x^2} \frac{\partial y^j}{\partial x^1} f^{21}(x) + \frac{\partial y^i}{\partial x^2} \frac{\partial y^j}{\partial x^2} f^{22}(x) + \frac{\partial y^i}{\partial x^2} \frac{\partial y^j}{\partial x^3} f^{23}(x)$$

$$+ \frac{\partial y^i}{\partial x^3} \frac{\partial y^j}{\partial x^1} f^{31}(x) + \frac{\partial y^i}{\partial x^3} \frac{\partial y^j}{\partial x^2} f^{32}(x) + \frac{\partial y^i}{\partial x^3} \frac{\partial y^j}{\partial x^3} f^{33}(x)$$

By assigning all possible values to the superscripts "*i*" and "*j*," the nine transformation equations are obtained. If the shear stress component $f^{23}(y)$ is required, then

$$f^{23}(y) = \frac{\partial y^2}{\partial x^1} \frac{\partial y^3}{\partial x^1} f^{11}(x) + \frac{\partial y^2}{\partial x^1} \frac{\partial y^3}{\partial x^2} f^{12}(x) + \frac{\partial y^2}{\partial x^1} \frac{\partial y^3}{\partial x^3} f^{13}(x)$$

$$+ \frac{\partial y^2}{\partial x^2} \frac{\partial y^3}{\partial x^1} f^{21}(x) + \frac{\partial y^2}{\partial x^2} \frac{\partial y^3}{\partial x^2} f^{22}(x) + \frac{\partial y^2}{\partial x^2} \frac{\partial y^3}{\partial x^3} f^{23}(x)$$

$$+ \frac{\partial y^2}{\partial x^3} \frac{\partial y^3}{\partial x^1} f^{31}(x) + \frac{\partial y^2}{\partial x^3} \frac{\partial y^3}{\partial x^2} f^{32}(x) + \frac{\partial y^2}{\partial x^3} \frac{\partial y^3}{\partial x^3} f^{33}(x)$$

Likewise, the covariant form would appear as follows:

$$f_{23}(y) = \frac{\partial x^1}{\partial y^2} \frac{\partial x^1}{\partial y^3} f_{11}(x) + \frac{\partial x^1}{\partial y^2} \frac{\partial x^2}{\partial y^3} f_{12}(x) + \frac{\partial x^1}{\partial y^2} \frac{\partial x^3}{\partial y^3} f_{13}(x)$$

$$+ \frac{\partial x^2}{\partial y^2} \frac{\partial x^1}{\partial y^3} f_{21}(x) + \frac{\partial x^2}{\partial y^2} \frac{\partial x^2}{\partial y^3} f_{22}(x) + \frac{\partial x^2}{\partial y^2} \frac{\partial x^3}{\partial y^3} f_{23}(x)$$

$$+ \frac{\partial x^3}{\partial y^2} \frac{\partial x^1}{\partial y^3} f_{31}(x) + \frac{\partial x^3}{\partial y^2} \frac{\partial x^2}{\partial y^3} f_{32}(x) + \frac{\partial x^3}{\partial y^2} \frac{\partial x^3}{\partial y^3} f_{33}(x)$$

where

$$f_{11}(x) = f_{xx} \; ; \qquad f_{12}(x) = f_{xy} \; ; \qquad f_{13}(x) = f_{xz}$$

$$f_{21}(x) = f_{yx} \; ; \qquad f_{22}(x) = f_{yy} \; ; \qquad f_{23}(x) = f_{yz}$$

$$f_{31}(x) = f_{zx} \; ; \qquad f_{32}(x) = f_{zy} \; ; \qquad f_{33}(x) = f_{zz}$$

Although these operations can be performed by human operators, it will be seen in subsequent sections that they can be executed with speed and efficiency by using a computational algorithm and a digital computer that exploits the advantages of the summation convention.

It is seen that a tensor of rank two as exemplified by the stress tensor, equation (1.3.9), has nine components in three-dimensional space, and that the covariant and contravariant transformations give rise to nine terms on the right-hand side. In two-dimensional space, the stress tensor would have four components and four transformation equations would be required, each having four terms on the right-hand side. In general, a tensor of rank two has n^2 components in "n" dimensional space, and a tensor of rank "r" has n^r components in "n" dimensional space.

1.3.2 Inertia Tensor

Another example of a tensor of rank two is the inertia tensor. Let "m" be the mass of a particle of a body at the point P, and let "$\bar{r}$" be the position vector of the particle relative to the fixed point 0 (see fig. 1.3.2).

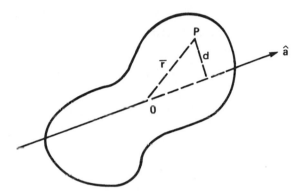

Figure 1.3.2.— Mass distribution relative to $\hat{a}$.

The moment of inertia of the body about an axis through the point 0 parallel to the unit vector $\hat{a}$ is given by the following sum:

$$M_I = \Sigma \, md^2$$

where "d" is the perpendicular distance of the point P from the axis $\hat{a}$ and

$$d^2 = (\bar{r} \times \hat{a})^2 = (\bar{r} \times \hat{a}) \cdot (\bar{r} \times \hat{a})$$

Therefore

$$M_I = \Sigma \, m(\bar{r} \times \hat{a}) \cdot (\bar{r} \times \hat{a})$$

The expression on the right-hand side of this equation may be treated as a triple scalar product and expanded accordingly,

$$M_I = \Sigma \; m\{\bar{r} \cdot [\hat{a} \times (\bar{r} \times \hat{a})]\}$$

$$M_I = \Sigma \; m\{\bar{r} \cdot [\bar{r} - (\hat{a} \cdot \bar{r})\hat{a}]\}$$

$$M_I = \Sigma \; m[r^2 - (\hat{a} \cdot \bar{r})^2]$$

This equation may be rewritten in the following form:

$$M_I = \hat{a} \cdot [\Sigma \; m(r^2\bar{I} - \bar{r}\bar{r})] \cdot \hat{a}$$

where $\bar{I}$ is the idemfactor or the identical dyadic. The idemfactor has the property that the scalar product of $\bar{I}$ and any vector $\bar{r}$ is always equal to $\bar{r}$; that is, if

$$\bar{I} \cdot \bar{r} = \bar{r} \cdot \bar{I} = \bar{r}$$

for all values of $\bar{r}$, then $\bar{I}$ is an idemfactor. In particular

$$\bar{I} = \hat{i}\hat{i} + \hat{j}\hat{j} + \hat{k}\hat{k}$$

is an idemfactor, since

$$\bar{r} \cdot \bar{I} = (x\hat{i} + y\hat{j} + z\hat{k}) \cdot (\hat{i}\hat{i} + \hat{j}\hat{j} + \hat{k}\hat{k}) = (x\hat{i} + y\hat{j} + z\hat{k}) = \bar{r}$$

and

$$\bar{I} \cdot \bar{r} = (\hat{i}\hat{i} + \hat{j}\hat{j} + \hat{k}\hat{k}) \cdot (x\hat{i} + y\hat{j} + z\hat{k}) = (\hat{i}x + \hat{j}y + \hat{k}z) = \bar{r}$$

In terms of this notation the moment of inertia assumes the form

$$M_I = a \cdot \overline{\overline{\Phi}} \cdot a$$

where $\overline{\overline{\Phi}}$ is the inertia tensor or, as it is sometimes called, the inertia dyadic, that is,

$$\overline{\overline{\Phi}} = m(r^2\bar{I} - \bar{r}\bar{r})$$

As in the case of the stress tensor, it is seen that the inertia tensor assumes the form of a dyadic, or a two linear form in the vector $\bar{r}$. This justifies its classification as a tensor of rank two since, in addition, it transforms in accordance with the covariant and contravariant laws to be defined in a subsequent section.

The inertia tensor occurs in the study of rotational motion. For example, consider the case of a rigid body rotating about the fixed point 0, and let $\bar{\omega}$ be the angular velocity of the body at any instant. If the angular momentum of the body be denoted by $\bar{H}$

$$\bar{H} = \Sigma \, \bar{r} \times m\bar{V}$$

but

$$\bar{V} = \bar{\omega} \times \bar{r}$$

Therefore

$$\bar{H} = \Sigma \, m\bar{r} \times (\bar{\omega} \times \bar{r})$$

or

$$\bar{H} = \Sigma \, m[r^2 \bar{\omega} - (\bar{r} \cdot \bar{\omega})\bar{r}]$$

This equation may be rewritten as follows:

$$\bar{H} = \bar{\omega} \cdot [\Sigma \, m(r^2 \bar{I} - \bar{r}\bar{r})]$$

or

$$\bar{H} = \bar{\omega} \cdot \bar{\Phi} = \bar{\Phi} \cdot \bar{\omega}$$

where $\bar{\Phi}$ is the inertia dyadic.

Similarly, if the kinetic energy of rotation of the body be denoted by T

$$2T = \Sigma \, mV^2 = \Sigma \, m(\bar{\omega} \times \bar{r})^2$$

or

$$2T = \Sigma \, m(\bar{\omega} \times \bar{r}) \cdot (\bar{\omega} \times \bar{r})$$

Again, the expression on the right-hand side of this equation can be treated as a triple scalar product. Hence

$$2T = \Sigma \, m\{\bar{\omega} \cdot [\bar{r} \times (\bar{\omega} \times \bar{r})]\}$$

or

$$2T = \Sigma \, m\{\bar{\omega} \cdot [r^2\bar{\omega} - (\bar{r} \cdot \bar{\omega})\bar{r}]\}$$

which may be rewritten as follows:

$$2T = \bar{\omega} \cdot [\Sigma \, m(r^2\bar{I} - \bar{r}\bar{r})] \cdot \bar{\omega}$$

or

$$2T = \bar{\omega} \cdot \bar{\Phi} \cdot \bar{\omega}$$

where $\bar{\Phi}$ is the inertia tensor again.

The components of the inertia tensor may be obtained by expanding $\bar{\Phi}$ as follows:

$$\bar{\Phi} = \Sigma \, m[(x^2 + y^2 + z^2)(\hat{i}\hat{i} + \hat{j}\hat{j} + \hat{k}\hat{k}) - (x\hat{i} + y\hat{j} + z\hat{k})(x\hat{i} + y\hat{j} + z\hat{k})]$$

or

$$\bar{\Phi} = [\Sigma \, m(y^2 + z^2)\hat{i}\hat{i} + \Sigma \, m(x^2 + z^2)\hat{j}\hat{j} + \Sigma \, m(x^2 + y^2)\hat{k}\hat{k}$$
$$- \Sigma \, mxy\hat{i}\hat{j} - \Sigma \, mxz\hat{i}\hat{k} - \Sigma \, myx\hat{j}\hat{i} - \Sigma \, myz\hat{j}\hat{k} - \Sigma \, mzx\hat{k}\hat{i} - \Sigma \, mzy\hat{k}\hat{j}]$$

The following definitions are required:

$$I_{xx} = \Sigma \, m(y^2 + z^2) \quad ; \quad I_{yy} = \Sigma \, m(x^2 + z^2) \quad ; \quad I_{zz} = \Sigma \, m(x^2 + y^2)$$

$$I_{xy} = I_{yx} = \Sigma \, mxy \quad ; \quad I_{xz} = I_{zx} = \Sigma \, mxz \quad ; \quad I_{yz} = I_{zy} = \Sigma \, myz$$

In terms of these definitions, the inertia tensor assumes the more familiar form

$$\Phi = (I_{xx}\hat{i}\hat{i} - I_{xy}\hat{i}\hat{j} - I_{xz}\hat{i}\hat{k} - I_{yx}\hat{j}\hat{i} + I_{yy}\hat{j}\hat{j} - I_{yz}\hat{j}\hat{k} - I_{zx}\hat{k}\hat{i} - I_{zy}\hat{k}\hat{j} + I_{zz}\hat{k}\hat{k})$$

It is interesting to note that

$$\hat{i}\cdot\overline{\Phi}\cdot\hat{i} = I_{xx} \quad ; \quad \hat{j}\cdot\overline{\Phi}\cdot\hat{j} = I_{yy} \quad ; \quad \hat{k}\cdot\overline{\Phi}\cdot\hat{k} = I_{zz}$$

$$\hat{i}\cdot\overline{\Phi}\cdot\hat{j} = -I_{xy} \quad ; \quad \hat{i}\cdot\overline{\Phi}\cdot\hat{k} = -I_{xz} \quad ; \quad \hat{j}\cdot\overline{\Phi}\cdot\hat{k} = -I_{yz}$$

The inertia tensor, like the stress tensor, is seen to have nine components. As previously indicated, this is characteristic of a tensor of rank two in three-dimensional space. Hence, the same law that was used to transform the components of the stress tensor can be used in this case also. The transformation of the inertia tensor will be considered in more detail in a later section.

1.4 TRANSFORMATION LAWS

1.4.1 Vector Components

To facilitate the computer processing of vectors and dyadics, all such entities should be expressed in terms of their tensor components and a corresponding set of base vectors, rather than in terms of their physical components and a set of unit base vectors. When referred to a general curvilinear coordinate system, a vector $\overline{A}$ may be expressed in the following alternative forms:

$$\overline{A} = A^i\bar{a}_i = A_j\bar{a}^j \tag{1.4.1}$$

As previously indicated, when a certain index occurs twice it means that the expression is to be summed with respect to that index for all admissible values of the index, that is

$$A^i\bar{a}_i = \sum_{i=1}^{n} A^i\bar{a}_i \tag{1.4.2}$$

$$A_j \bar{a}^j = \sum_{j=1}^{n} A_j \bar{a}^j \tag{1.4.3}$$

where A^i, A_j are the tensor components of the vector $\bar{A}$, and $\bar{a}_i$, $\bar{a}^j$ are the corresponding systems of base vectors. In accordance with established convention, contravariant components will be denoted by superscripts and covariant components by subscripts. It is necessary to keep in mind the distinction between contravariance and covariance because if general coordinate transformations are contemplated, the transformation law for the components of a contravariant vector denoted by superscripts differs from that for a covariant vector denoted by subscripts. It must be emphasized, however, that the covariance or contravariance of tensor components is not an intrinsic property of the entity under consideration. The distinction is due to the way in which the entity is related to its environment, or the coordinate system, to which it is referred. For a transformation from a coordinate system x to a coordinate system y given by

$$y^i = y^i(x^1, x^2, x^3) \tag{1.4.4}$$

the transformation law for the components of a contravariant vector A^i will be derived in the following section and will be shown to have the following form:

$$B^j(y) = \frac{\partial y^j}{\partial x^i} A^i(x) \tag{1.4.5}$$

where $A^i(x)$ are the contravariant components in the x coordinate system and $B^j(y)$ are the components when referred to the y coordinate system. For the same transformation of coordinates, other vectors, such as the gradient of a scalar point function, obey a different transformation law. These are the covariant vectors denoted by subscripts. Assuming that the coordinate transformation is reversible and one-to-one, the appropriate transformation law for these vector components is

$$B_j(y) = \frac{\partial x^i}{\partial y^j} A_i(x) \tag{1.4.6}$$

where $A_i(x)$ are the covariant components in the x coordinate frame and $B_j(y)$ are the covariant components when referred to the y coordinate frame. As the following argument shows, the distinction between these two transformation laws vanishes when the transformation is orthogonal Cartesian. Let x^i be the components of a position vector $\bar{r}$ when referred to the x coordinate system which is orthogonal Cartesian. Likewise, let y^j be components of the same vector when referred to another orthogonal Cartesian system. In this case, the transformation of coordinates is given by

$$y^i = \alpha_j{}^i x^j \tag{1.4.7}$$

where the $\alpha_j{}^i$ are constants. The position vector $\bar{r}$ is invariant with respect to coordinate transformations. Hence, the square of the vector is also invariant. Therefore,

$$x^j x^j = y^i y^i = \alpha_j{}^i \alpha_k{}^i x^j x^k = \delta_k{}^j x^j x^k$$

and

$$\alpha_j{}^i \alpha_k{}^i = \delta_k{}^j \tag{1.4.8}$$

where $\delta_k{}^j$ is the Kronecker delta, that is

$$\delta_k{}^j = \begin{array}{ll} 1 & \text{for } j = k \\ 0 & \text{for } j \neq k \end{array}$$

Equation (1.4.8) is the orthogonality condition which may be used to solve equation (1.4.7) for x^j. If both sides of equation (1.4.7) are multiplied by $\alpha_k{}^i$

$$\alpha_j{}^i \alpha_k{}^i x^j = \alpha_k{}^i y^i$$

and

$$\delta_k{}^j x^j = x^k = \alpha_k{}^i y^i$$

Therefore

$$x^j = \alpha_j{}^i y^i \tag{1.4.9}$$

From equation (1.4.7) it is seen that

$$\frac{\partial y^i}{\partial x^j} = \alpha_j{}^i \tag{1.4.10}$$

and from equation (1.4.9)

$$\frac{\partial x^j}{\partial y^i} = \alpha_j{}^i \tag{1.4.11}$$

It follows from equations (1.4.10) and (1.4.11) that

$$\frac{\partial y^i}{\partial x^j} = \frac{\partial x^j}{\partial y^i} \tag{1.4.12}$$

At this point, it is instructive to give an example of a covariant vector or a covariant tensor of rank one. Consider the components of a gradient vector and let ϕ be a uniform, continuous scalar point function. Let the gradient of this function with respect to the x^j coordinate in the X-reference frame be denoted by

$$\frac{\partial \phi}{\partial x^j}$$

Likewise, let the gradient of this function with respect to the y^i coordinate in the Y-coordinate reference frame be denoted by

$$\frac{\partial \phi}{\partial y^i}$$

These gradients are related as follows:

$$\frac{\partial \phi}{\partial y^i} = \frac{\partial \phi}{\partial x^j} \frac{\partial x^j}{\partial y^i}$$

Moreover, let

$$\frac{\partial \phi}{\partial y^i} = B_i(y)$$

and let

$$\frac{\partial \phi}{\partial x^j} = A_j(x)$$

then

$$B_i(y) = \frac{\partial x^j}{\partial y^i} A_j(x)$$

This is seen to satisfy the mathematical definition of a covariant vector given in equation (1.4.6).

1.5 BASE VECTORS

The transformation laws and, hence, the covariant and contravariant character of the base vectors and their reciprocals may be obtained as follows: Let the differential of a position vector be denoted by $d\bar{r}$. Then if $\bar{a}_i(x)$ are the base vectors in the x coordinate system, and $\bar{b}_j(y)$ are the base vectors in the y coordinate system, the differential $d\bar{r}$ may be expressed in the following alternative forms:

$$d\bar{r} = \bar{a}_i(x)dx^i = \bar{b}_j(y)dy^j = \bar{b}_j(y)\frac{\partial y^j}{\partial x^i} dx^i \qquad (1.5.1)$$

Therefore

$$\bar{a}_i(x) = \frac{\partial y^j}{\partial x^i} \bar{b}_j(y) \qquad (1.5.2)$$

Likewise

$$\bar{a}_i(x) \frac{\partial x^i}{\partial y^j} dy^j = \bar{b}_j(y)dy^j$$

and

$$\bar{b}_j(y) = \frac{\partial x^i}{\partial y^j} \bar{a}_i(x) \qquad (1.5.3)$$

It is seen from equations (1.5.2) and (1.5.3) that the base vectors $\bar{a}_i$ and $\bar{b}_j$ obey the covariant transformation law; consequently, the use of subscripts is justified.

1.5.1 Reciprocal Base Vectors

To each system of base vectors $\bar{a}_i$ there exists a reciprocal system of vectors $\bar{a}^j$ with the following property

$$\bar{a}_i \cdot \bar{a}^j = \delta_i{}^j = \bar{a}^j \cdot \bar{a}_i \tag{1.5.4}$$

where $\delta_i{}^j$ is the Kronecker delta; that is

$$\delta_i{}^j = \begin{array}{ll} 1 & \text{for } j = i \\ 0 & \text{for } j \neq i \end{array}$$

Scalar multiplication of each side of equation (1.5.2) by $\bar{b}^j(y)$ gives, on using (1.5.4)

$$\bar{b}^j(y) \cdot \bar{a}_i(x) = \frac{\partial y^j}{\partial x^i} \tag{1.5.5}$$

Similarly, from equations (1.5.3) and (1.5.4) it is seen that

$$\bar{a}^i(x) \cdot \bar{b}_j(y) = \frac{\partial x^i}{\partial y^j} \tag{1.5.6}$$

Equation (1.5.1), referred to the reciprocal system of base vectors, assumes the form

$$d\bar{r} = \bar{a}^i(x)dx_i = \bar{b}^j(y)dy_j \tag{1.5.7}$$

Therefore

$$dy_j = b_j(y) \cdot \bar{a}^i(x)dx_i$$

$$dy_j = \frac{\partial x^i}{\partial y^j} dx_i \tag{1.5.8}$$

and

$$dx_i = \bar{a}_i(x) \cdot \bar{b}^j(y)dy_j$$

Therefore

$$dx_i = \frac{\partial y^j}{\partial x^i}\, dy_j \tag{1.5.9}$$

From equations (1.5.7) and (1.5.8)

$$\bar{a}^i(x)dx_i = \bar{b}^j(y)\frac{\partial x^i}{\partial y^j}\, dx_i$$

Therefore

$$\bar{a}^i(x) = \frac{\partial x^i}{\partial y^j}\, \bar{b}^j(y) \tag{1.5.10}$$

Likewise, from equations (1.5.7) and (1.5.9)

$$\bar{b}^j(y) = \frac{\partial y^j}{\partial x^i}\, \bar{a}^i(x) \tag{1.5.11}$$

From equations (1.5.10) and (1.5.11), it is seen that the reciprocal base vectors obey the contravariant law of transformation; therefore, the superscript notation is justified.

1.6 VECTOR TRANSFORMATIONS

Equations (1.5.10) and (1.5.11) may be used to obtain the transformation law for a vector $\bar{A}$, where

$$\bar{A} = A^i\bar{a}_i = A_j\bar{a}^j \tag{1.6.1}$$

If $\bar{A} = A^i(x)\bar{a}_i(x)$ when the vector $\bar{A}$ is referred to the x coordinate system, and if $\bar{A} = B^j(y)\bar{b}_j(y)$ when referred to the y coordinate system, the invariance of $\bar{A}$ requires that

$$B^j(y)b_j(y) = A^i(x)a_i(x) \tag{1.6.2}$$

From equations (1.5.2) and (1.6.2), the appropriate transformation law is obtained as follows:

$$B^j(y) = \frac{\partial y^j}{\partial x^i} A^i(x)$$

(1.6.3)

Equation (1.6.3) is the contravariant transformation law for the components of the vector $\bar{A}$. When $\bar{A}$ is referred to the x coordinate system with base vectors $\bar{a}_i(x)$, which obey the covariant transformation law, the components $A^j(x)$ obey the contravariant transformation law; hence, the use of superscripts is justified. If $\bar{A}$ is referred to the reciprocal base system $\bar{a}^i$, then from equation (1.6.1):

$$\bar{A} = A_i\bar{a}^i$$

On a transformation of coordinates from the x coordinate system to the y coordinate system, invariance of $\bar{A}$ requires that

$$A_i(x)\bar{a}^i(x) = B_j(y)\bar{b}^j(y)$$

(1.6.4)

From equations (1.5.10) and (1.6.4), the appropriate transformation law is obtained as follows:

$$B_j(y) = \frac{\partial x^i}{\partial y^j} A_i(x)$$

(1.6.5)

It is seen that when a vector $\bar{A}$ is referred to a coordinate system with reciprocal base vectors, which obey the contravariant law, the corresponding components of $\bar{A}$ obey the covariant law, and the use of subscripts is therefore justified.

1.7 RAISING AND LOWERING OF INDICES

1.7.1 Lowering Indices

The vector $\bar{A}$ may be expressed in the alternative forms given in equation (1.6.1). Scalar multiplication of each side of equation (1.6.1) by a_j gives

$$(\bar{a}_i \cdot \bar{a}_j)A^i = A_j(\bar{a}^j \cdot \bar{a}_j)$$

(1.7.1)

By substitution from equation (1.5.4) in equation (1.7.1) the following result is obtained:

$$g_{ij}A^i = A_j \tag{1.7.2}$$

where

$$\bar{a}_i \cdot \bar{a}_j = g_{ij}$$

Again by substitution for A_j from equation (1.7.2) in equation (1.6.1)

$$\bar{a}_i = g_{ij}\bar{a}^j \tag{1.7.3}$$

Equation (1.7.2) gives the transformation from the contravariant components to the covariant components of a vector. The corresponding transformation of base vectors is given by equation (1.7.3). These operations are usually referred to as lowering the index (ref. 1).

1.7.2 Raising Indices

Scalar multiplication of each side of equation (1.6.1) by $\bar{a}^i$ gives

$$A^i(\bar{a}_i \cdot \bar{a}^i) = A_j(\bar{a}^j \cdot \bar{a}^i) \tag{1.7.4}$$

Substitution from equation (1.5.4) in equation (1.7.4) gives

$$A^i = g^{ij}A_j \tag{1.7.5}$$

where

$$\bar{a}^i \cdot \bar{a}^j = g^{ij}$$

When this expression for A^i is substituted in the left-hand side of equation (1.6.1) the following result is obtained

$$g^{ij}A_j\bar{a}_i = A_j\bar{a}^j$$

Therefore

$$\bar{a}^j = g^{ij}\bar{a}_i \qquad (1.7.6)$$

Equation (1.7.5) enables the contravariant components of a vector to be expressed in terms of its covariant components. Equation (1.7.6) gives the corresponding transformation of base vectors. These operations are usually referred to as raising the index.

1.8 BIVECTOR TRANSFORMATIONS

A second-order tensor is characterized by having two indices. Both indices can be superscripts, in which case the tensor is doubly contravariant. Tensors of this kind are sometimes referred to as the contravariant components of a bivector. When both indices are subscripts, the tensors are doubly covariant, or the components of a covariant bivector. It sometimes happens that one of the indices is a superscript and the other one a subscript. Entities of this kind are called mixed tensors or the components of a mixed bivector.

1.8.1 Contravariant Bivectors

As in the case of vectors or first-order tensors, bivectors are entities whose properties are independent of the reference frames used to describe them. Equations (1.6.2) and (1.6.4) are mathematical expressions of this statement, insofar as it applies to vectors. As might be expected, the invariance of a bivector, in going from a coordinate system x with base vectors $\bar{a}_j(x)$ to a coordinate system y with base vectors $\bar{b}_j(y)$, involves the equality of two dyadics. The coefficients of the individual dyads in these dyadics are the components of the bivectors. If in the x coordinate system with base vectors $\bar{a}_j$ the bivector is given by

$$A^{\alpha\beta}(x)\bar{a}_\alpha(x)\bar{a}_\beta(x)$$

and if in the y coordinate system with base vectors $\bar{b}_i$ this bivector assumes the form

$$B^{ij}(y)\bar{b}_i(y)\bar{b}_j(y)$$

invariance requires that

$$B^{ij}(y)\bar{b}_i(y)\bar{b}_j(y) = A^{\alpha\beta}(x)\bar{a}_\alpha(x)\bar{a}_\beta(x) \tag{1.8.1}$$

By substitution from equation (1.5.2) in equation (1.8.1),

$$B^{ij}(y)\bar{b}_i(y)\bar{b}_j(y) = A^{\alpha\beta}(x)\frac{\partial y^i}{\partial x^\alpha}\bar{b}_i(y)\frac{\partial y^j}{\partial x^\beta}\bar{b}_j(y) \tag{1.8.2}$$

Therefore, by equating coefficients of like dyads in equation (1.8.2), the required transformation law is obtained as follows:

$$B^{ij}(y) = \frac{\partial y^i}{\partial x^\alpha}\frac{\partial y^j}{\partial x^\beta}A^{\alpha\beta}(x) \tag{1.8.3}$$

This is the transformation law for the components of a contravariant bivector.

1.8.2 Covariant Bivectors

Since covariant bivectors are characterized by two subscripts, it follows that the formulation of the dyadics will be in terms of the reciprocal base vectors. That is, if $A_{\alpha\beta}(x)$ are the components of the covariant bivector in the x coordinate system, and $B_{ij}(y)$ are the corresponding components in the y coordinate system, invariance of the bivectors requires that

$$B_{ij}(y)b^i(y)b^j(y) = A_{\alpha\beta}(x)a^\alpha(x)a^\beta(x) \tag{1.8.4}$$

Substitution from equation (1.5.10) in equation (1.8.4) gives

$$B_{ij}(y)\bar{b}^i(y)\bar{b}^j(y) = A_{\alpha\beta}(x)\frac{\partial x^\alpha}{\partial y^i}\bar{b}^i(y)\frac{\partial x^\beta}{\partial y^j}\bar{b}^j(y) \tag{1.8.5}$$

Therefore

$$B_{ij}(y) = \frac{\partial x^\alpha}{\partial y^i}\frac{\partial x^\beta}{\partial y^j}A_{\alpha\beta}(x) \tag{1.8.6}$$

Equation (1.8.6) is the transformation law for the components of a covariant bivector.

1.8.3 Mixed Bivectors

A mixed bivector has one index of covariance and one index of contravariance. In this case, the bivectors consist of base vectors and reciprocal base vectors. The invariance requirements may be stated as follows:

$$B_j{}^i(y)\bar{b}_i(y)\bar{b}^j(y) = A_\beta{}^\alpha(x)\bar{a}_\alpha(x)a^\beta(x) \tag{1.8.7}$$

Substitution from equations (1.5.2) and (1.5.10) in equation (1.8.7) gives

$$B_j{}^i(y)\bar{b}_i(y)\bar{b}^j(y) = A_\beta{}^\alpha(x)\frac{\partial y^i}{\partial x^\alpha}\bar{b}_i(y)\frac{\partial x^\beta}{\partial y^j}\bar{b}^j(y)$$

Therefore

$$B_j{}^i(y) = \frac{\partial y^i}{\partial x^\alpha}\frac{\partial x^\beta}{\partial y^j}A_\beta{}^\alpha(x) \tag{1.8.8}$$

The components of mixed bivectors transform according to equation (1.8.8).

1.9 PHYSICAL COMPONENTS

The transformation from covariant to contravariant components and vice versa was discussed in preceding sections. This section is concerned with the transformation from covariant and contravariant components to physical components and vice versa.

It frequently happens that an analysis can be performed and the results obtained, without reference to physical components. However, sometimes a quantity, such as a force, is known only in terms of its physical components. In this case, the transformation from physical components to tensor components must be determined. The appropriate transformations may be obtained as follows:

$$\bar{a}_i \cdot \bar{a}_j = g_{ij}$$

Therefore

$$\bar{a}_i \cdot \bar{a}_i = g_{(ii)}$$

Let

$$\bar{a}_i = \alpha_i \hat{a}_i$$

where α_i is a scalar magnitude and a_i is a unit vector. With this notation

$$\bar{a}_i \cdot \bar{a}_i = (\alpha_i)^2 = g_{(ii)}$$

Therefore

$$\alpha_i = \sqrt{g_{(ii)}}$$

that is

$$\bar{a}_i = \sqrt{g_{(ii)}}\hat{a}_i \tag{1.9.1}$$

where the parentheses imply suspension of the summation convention. Hence, if A_i are the contravariant tensor components of a vector $\bar{A}$, and if $\mathscr{A}^i$ are the corresponding physical components, then

$$A^i \bar{a}_i = (\sqrt{g_{(ii)}}A^i)\hat{a}_i = \mathscr{A}^i \hat{a}_i$$

Therefore

$$\mathscr{A}^i = \sqrt{g_{(ii)}}A^i \tag{1.9.2}$$

Likewise, let

$$\bar{a}^i \cdot \bar{a}^j = g^{ij}$$

and let

$$\bar{a}^i = \beta^i \hat{a}^i$$

where β^i is a scalar magnitude and $\hat{a}^i$ is a unit reciprocal base vector. Therefore

$$\bar{a}^i \cdot \bar{a}^i = g^{(ii)} = (\beta^i)^2$$

$$\left.\begin{array}{l} \beta^i = \sqrt{g^{(ii)}} \\ \bar{a}^i = \sqrt{g^{(ii)}}\,\hat{a}^i \end{array}\right\} \tag{1.9.3}$$

Hence, if A_i are the covariant components of a vector $\bar{A}$, and if $\mathscr{A}_i$ are the corresponding physical components

$$A_i \bar{a}^i = \left(\sqrt{g^{(ii)}} A_i \right) \hat{a}^i = \mathscr{A}_i \hat{a}^i$$

Therefore

$$\mathscr{A}_i = \sqrt{g^{(ii)}} A_i$$

Moreover, if the coordinate system is orthogonal, the physical components can be expressed in the following alternative forms:

$$\mathscr{A}_i = \sqrt{g^{(ii)}} A_i = \frac{1}{\sqrt{g_{(ii)}}} A_i \qquad (1.9.4)$$

Equation (1.7.2) may be used to show that $\mathscr{A}_i = \mathscr{A}^i$ in orthogonal coordinate systems. From equations (1.9.4) and (1.7.2):

$$\mathscr{A}_i = \frac{1}{\sqrt{g_{(ii)}}} A_i = \frac{1}{\sqrt{g_{(ii)}}} \left[g_{(ii)} A^i \right] = \sqrt{g_{(ii)}} A^i = \mathscr{A}^i$$

that is

$$\mathscr{A}_i = \mathscr{A}^i$$

That $\mathscr{A}_i \neq \mathscr{A}^i$ in nonorthogonal coordinate systems may be seen as follows:

$$\mathscr{A}_i = \sqrt{g^{(ii)}} A_i = \sqrt{g^{(ii)}} g_{ij} A^j$$

Therefore, in this case

$$\mathscr{A}_i \neq \sqrt{g_{(ii)}} A^i = \mathscr{A}^i$$

The fact that $\mathscr{A}^i \neq \mathscr{A}_i$ in nonorthogonal coordinate systems is a consequence of the relation

$$A_i = g_{ij} A^j$$

1.10 TENSOR COMPONENTS

It should be noted that a vector, or a tensor of rank one, has three components in three dimensional space. For the case $i = 1,2,3$ and $j = 1,2,3$, equation (1.6.3) represents three equations in which the right-hand members each have three terms:

$$B^1 = \frac{\partial y^1}{\partial x^1} A^1 + \frac{\partial y^1}{\partial x^2} A^2 + \frac{\partial y^1}{\partial x^3} A^3$$

$$B^2 = \frac{\partial y^2}{\partial x^1} A^1 + \frac{\partial y^2}{\partial x^2} A^2 + \frac{\partial y^2}{\partial x^3} A^3$$

$$B^3 = \frac{\partial y^3}{\partial x^1} A^1 + \frac{\partial y^3}{\partial x^2} A^2 + \frac{\partial y^3}{\partial x^3} A^3$$

Although a vector, or a tensor of rank one, has n components in n dimensional space, with each transformation equation having n terms on the right-hand side, a tensor of rank two has n^2 components, with each transformation equation containing n^2 terms on the right-hand member. For example, for the case $i = 1, j = 2$, equation (1.8.3) assumes the following form:

$$B^{12} = \frac{\partial y^1}{\partial x^1} \frac{\partial y^2}{\partial x^1} A^{11} + \frac{\partial y^1}{\partial x^1} \frac{\partial y^2}{\partial x^2} A^{12} + \frac{\partial y^1}{\partial x^1} \frac{\partial y^2}{\partial x^3} A^{13}$$

$$+ \frac{\partial y^1}{\partial x^2} \frac{\partial y^2}{\partial x^1} A^{21} + \frac{\partial y^1}{\partial x^2} \frac{\partial y^2}{\partial x^2} A^{22} + \frac{\partial y^1}{\partial x^2} \frac{\partial y^2}{\partial x^3} A^{23}$$

$$+ \frac{\partial y^1}{\partial x^3} \frac{\partial y^2}{\partial x^1} A^{31} + \frac{\partial y^1}{\partial x^3} \frac{\partial y^2}{\partial x^2} A^{32} + \frac{\partial y^1}{\partial x^3} \frac{\partial y^2}{\partial x^3} A^{33}$$

It will be noted that α and β have each taken on their three possible values which resulted in nine terms on the right, whereas $i = 1$ and $j = 2$ have been retained throughout. Now since i and j may each have the three values 1, 2, 3, there will be nine such equations, each containing nine terms on the right. In relativistic mechanics there are four dimensions to be considered. In this case, equation (1.8.3) will represent 16 equations, each containing 16 terms on the right-hand side. Likewise, a contravariant tensor of rank three is defined by the following equation

$$B^{ijk}(y) = \frac{\partial y^i}{\partial x^\alpha} \frac{\partial y^j}{\partial x^\beta} \frac{\partial y^k}{\partial x^\gamma} A^{\alpha\beta\gamma}(x) \tag{1.10.1}$$

The number of equations represented by (1.10.1) and the number of terms on the right-hand side of each transformation equation depends on the dimensionality of the space. In n dimensional space, (1.10.1) will have n^3 components, with each transformation equation consisting of n^3 terms on the right-hand side. More specifically, there will be 8 components in two-dimensional space, 27 components in three-dimensional space, and 64 components in four-dimensional space. And, in general, if the components of a mixed tensor in the X-coordinate system are denoted by

$$A_{\alpha_1 \alpha_2, \ldots, \alpha_s}^{\beta_1 \beta_2, \ldots, \beta_r}(x)$$

its components in the Y-coordinate system will be

$$B_{i_1 i_2, \ldots, i_s}^{j_1 j_2, \ldots, j_r}(y)$$

where

$$B_{i_1 i_2, \ldots, i_s}^{j_1 j_2, \ldots, j_r}(y) = \left(\frac{\partial y^{j_1}}{\partial x^{\beta_1}} \frac{\partial y^{j_2}}{\partial x^{\beta_2}} \cdots \frac{\partial y^{j_r}}{\partial x^{\beta_r}} \frac{\partial x^{\alpha_2}}{\partial y^{i_1}} \frac{\partial x^{\alpha_2}}{\partial y^{i_2}} \cdots \frac{\partial x^{\alpha_s}}{\partial y^{i_s}} \right) A_{\alpha_1 \alpha_2, \ldots, \alpha_s}^{\beta_1 \beta_2, \ldots, \beta_r}$$

1.11 ALGEBRA OF TENSORS

The following results are stated without proof. For a rigorous derivation of these results, the reader is referred to standard texts on the subject.

THEOREM I. *The sum or difference of two tensors which have the same number of covariant indices and the same number of contravariant indices is a tensor of the same type and rank as the given tensors.*

For example, take a vector having components $A^i(x)$ when referred to the X-coordinate system and let $B^i(x)$ be another such vector referred to the same reference frame. Since $A^i(x)$ and $B^i(x)$ are contravariant tensors of rank one, they obey the corresponding transformation law. The components of these vectors when referred to the Y-coordinate system are

$$A^i(y) = \frac{\partial y^i}{\partial x^j} A^j(x) \qquad (1.11.1)$$

$$B^i(y) = \frac{\partial y^i}{\partial x^j} B^j(x) \qquad (1.11.2)$$

Therefore

$$A^i(y) + B^i(y) = \frac{\partial y^i}{\partial x^j} [A^j(x) + B^j(x)] \qquad (1.11.3)$$

or

$$C^i(y) = \frac{\partial y^i}{\partial x^j} C^j(x) \qquad (1.11.4)$$

where

$$C^i(y) = A^i(y) + B^i(y)$$

and

$$C^j(x) = A^j(x) + B^j(x) \qquad (1.11.5)$$

It is to be noted that (1.11.3) may be obtained by adding (1.11.1) and (1.11.2) as if each of these represented a single equation containing only a single term on the right, rather than a set of equations each containing several terms on the right. Thus the notation takes care that the corresponding components shall be added correctly.

THEROEM II. *The set of quantities consisting of the product of each component of a tensor having p indices of contravariance and q indices of covariance, by each component of a tensor having r components of contravariance and s components of covariance, defines a tensor called the outer product. The product tensor is contravariant of rank p + r and covariant rank q + s.*

Again, as in the case of addition and subtraction of tensor components, the tensor notation automatically assures that the outer product of

$$A^i(y) = \frac{\partial y^i}{\partial x^\alpha} A^\alpha(x) \qquad (1.11.6)$$

and

$$B^j(y) = \frac{\partial y^j}{\partial x^\beta} B^\beta(x) \qquad (1.11.7)$$

can be written immediately as

$$C^{ij}(y) = \frac{\partial y^i}{\partial x^\alpha} \frac{\partial y^j}{\partial x^\beta} C^{\alpha\beta}(x) \qquad (1.11.8)$$

where

$$C^{ij}(y) = A^i(y)B^j(y) \qquad (1.11.9)$$

and

$$C^{\alpha\beta}(x) = A^\alpha(x)B^\beta(x) \qquad (1.11.10)$$

By writing out the equations in full for the two-dimensional case, the reader can easily verify that this is a valid procedure. In the two-dimensional case

$$A^i(y) = \frac{\partial y^i}{\partial x^1} A^1(x) + \frac{\partial y^i}{\partial x^2} A^2(x) \qquad (1.11.11)$$

$$B^j(y) = \frac{\partial y^j}{\partial x^1} B^1(x) + \frac{\partial y^j}{\partial x^2} B^2(x) \qquad (1.11.12)$$

Therefore

$$A^i(y)B^j(y) = C^{ij}(y)$$

where

$$C^{ij}(y) = \left[\frac{\partial y^i}{\partial x^1} A^1(x) + \frac{\partial y^i}{\partial x^2} A^2(x)\right]\left[\frac{\partial y^j}{\partial x^1} B^1(x) + \frac{\partial y^j}{\partial x^2} B^2(x)\right] \qquad (1.11.13)$$

Hence

$$C^{ij}(y) = \left[\frac{\partial y^i}{\partial x^1}\frac{\partial y^j}{\partial x^1} A^1(x)B^1(x) + \frac{\partial y^i}{\partial x^1}\frac{\partial y^j}{\partial x^2} A^1(x)B^2(x) + \frac{\partial y^i}{\partial x^2}\frac{\partial y^j}{\partial x^1} A^2(x)B^1(x)\right.$$

$$\left. + \frac{\partial y^i}{\partial x^2}\frac{\partial y^j}{\partial x^2} A^2(x)B^2(x)\right] \qquad (1.11.14)$$

Summing equation (1.11.8) on α and β gives

$$C^{ij}(y) = \frac{\partial y^i}{\partial x^1}\frac{\partial y^j}{\partial x^1} C^{11}(x) + \frac{\partial y^i}{\partial x^1}\frac{\partial y^j}{\partial x^2} C^{12}(x)$$

$$+ \frac{\partial y^i}{\partial x^2}\frac{\partial y^j}{\partial x^1} C^{21}(x) + \frac{\partial y^i}{\partial x^2}\frac{\partial y^j}{\partial x^2} C^{22}(x) \qquad (1.11.15)$$

which, in view of equation (1.11.14), does represent completely the product of the two given equations.

Moreover, it is possible to multiply a covariant tensor by a contravariant one, thus obtaining a mixed tensor as follows. The outer product of

$$A^i(y) = \frac{\partial y^i}{\partial x^\alpha} A^\alpha(x) \qquad (1.11.16)$$

and

$$B_j(y) = \frac{\partial x^\beta}{\partial y^j} A_\beta(x) \qquad (1.11.17)$$

is

$$C_j{}^i(y) = \frac{\partial y^i}{\partial x^\alpha}\frac{\partial x^\beta}{\partial y^j} C_\beta{}^\alpha(x) \qquad (1.11.18)$$

Thus, the outer product of a contravariant tensor of rank one with a covariant tensor of rank one is a mixed tensor of rank two. The product tensor has one index of contravariance and one index of covariance. More generally, the outer product of

$$A^{ij}_{k}(y) = \frac{\partial y^i}{\partial x^\alpha} \frac{\partial y^j}{\partial x^\beta} \frac{\partial x^\gamma}{\partial y^k} A^{\alpha\beta}_{\gamma}(x) \qquad (1.11.19)$$

and

$$B_m{}^l(y) = \frac{\partial y^l}{\partial x^\delta} \frac{\partial x^\epsilon}{\partial y^m} A_\epsilon{}^\delta(x) \qquad (1.11.20)$$

is

$$C^{ijl}_{km}(y) = \frac{\partial y^i}{\partial x^\alpha} \frac{\partial y^j}{\partial x^\beta} \frac{\partial x^\gamma}{\partial y^k} \frac{\partial y^l}{\partial x^\delta} \frac{\partial x^\epsilon}{\partial y^m} C^{\alpha\beta\delta}_{\gamma\epsilon}(x) \qquad (1.11.21)$$

Hence, if any two tensors of ranks p and q are multiplied together to form their outer product, the result is a tensor of rank $p + q$. Moreover, if the tensor of rank p is a mixed tensor with p_1 indices of contravariance and p_2 indices of covariance, and if the tensor of rank q has q_1 indices of contravariance and q_2 indices of covariance, then their outer product will be a mixed tensor having $p_1 + q_1$ indices of contravariance and $p_2 + q_2$ indices of covariance.

Although the tensor calculus makes it easy to perform these operations, it should be emphasized that the operations represent complicated processes. Equation (1.11.19) is the transformation law for a mixed tensor of rank three and represents a whole set of equations. As previously indicated, the number of equations depends on the dimensionality of the space being considered. There will be 8 equations for two-dimensional space, 27 equations for three-dimensional space, and 64 equations for four-dimensional space. Each of these equations will have a corresponding number of terms on the right-hand side. And equation (1.11.20) is the transformation law for a mixed tensor of rank two. It also represents a set of equations that depends on the dimensionality of the space being considered, namely, 4 for two-dimensional space, 9 for three-dimensional space, and 16 for four-dimensional space — all with a corresponding number of terms on the right-hand side of each equation. The outer product of these two equations gives rise to the set of equations (1.11.21). This equation is the mathematical definition of a tensor of rank five, having three indices of contravariance and two indices of covariance. The set resulting from this outer product contains 32 equations for two-dimensional space; 243 for three-dimensional space, 1024 for four-dimensional space, and so on. And, of course, each equation will have a correspondingly large number of terms on the right-hand side (ref. 11).

In addition to the outer product of two tensors, which gives rise to a tensor of rank higher than the rank of the individual tensors, another kind of tensor product which gives rise to a tensor of lower rank than the individual tensors is defined by the following theorem (ref. 12):

THEOREM III. *If in a mixed tensor, contravariant rank p and covariant rank q, a contravariant index and a covariant index are equated, and the resulting tensor summed with respect to that index, the resulting set of p + q − 2 sums is a mixed tensor, contravariant of rank (p − 1), and covariant of rank (q − 1).*

Consider the mixed tensor

$$B_k^{ij}(y) = \frac{\partial y^i}{\partial x^\alpha} \frac{\partial y^j}{\partial x^\beta} \frac{\partial x^\tau}{\partial y^k} \, A_\tau^{\alpha\beta}(x) \qquad (1.11.22)$$

If the indices i and k are made equal, this tensor becomes

$$B_i^{ij}(y) = \frac{\partial y^i}{\partial x^\alpha} \frac{\partial y^j}{\partial x^\beta} \frac{\partial x^\tau}{\partial y^i} \, A_\tau^{\alpha\beta}(x) \qquad (1.11.23)$$

but

$$\frac{\partial x^\tau}{\partial y^i} \frac{\partial y^i}{\partial x^\alpha} = \delta_\alpha^{\ \tau} \qquad (1.11.24)$$

where $\delta_\alpha^{\ \tau}$ is the Kronecker delta, that is

$$\delta_\alpha^{\ \tau} = \begin{array}{ll} 1 & \text{for} \quad \tau = \alpha \\ 0 & \text{for} \quad \tau \neq \alpha \end{array}$$

Therefore

$$B_i^{ij}(y) = \frac{\partial y^j}{\partial x^\beta} \, \delta_\alpha^{\ \tau} A_\tau^{\alpha\beta}(x) \qquad (1.11.25)$$

$$B_i^{ij}(y) = \frac{\partial y^j}{\partial x^\beta} A_\alpha^{\alpha\beta}(x) \qquad (1.11.26)$$

By the summation convention, the left-hand side is to be summed on i, and the right-hand side summed on α. To clarify this operation assume that the space involved is two-dimensional. When equation (1.11.26) is written out explicitly, the following result is obtained:

$$B_1^{1\,1}(y) + B_2^{2\,1}(y) = \frac{\partial y^1}{\partial x^1}\left[A_1^{1\,1}(x) + A_2^{2\,1}(x)\right] + \frac{\partial y^1}{\partial x^2}\left[A_1^{1\,2}(x) + A_2^{2\,2}(x)\right]$$

$$B_1^{1\,2}(y) + B_2^{2\,2}(y) = \frac{\partial y^2}{\partial x^1}\left[A_1^{1\,1}(x) + A_2^{2\,1}(x)\right] + \frac{\partial y^2}{\partial x^2}\left[A_1^{1\,2}(x) + A_2^{2\,2}(x)\right]$$

This equation may be rewritten as follows:

$$C^j(y) = \frac{\partial y^j}{\partial x^\beta} A^\beta(x) \qquad (1.11.27)$$

where

$$C^1(y) = B_1^{1\,1}(y) + B_2^{2\,1}(y)$$

$$C^2(y) = B_1^{1\,2}(y) + B_2^{2\,2}(y)$$

and

$$A^1(x) = A_1^{1\,1}(x) + A_2^{2\,1}(x)$$

$$A^2(x) = A_1^{1\,2}(x) + A_2^{2\,2}(x)$$

Hence, by making one upper and one lower index equal, a tensor of rank three has been reduced to a tensor of rank one. The operation of equating one contravariant index to one covariant index is known as contraction. If it is possible to apply the operation of contraction to the outer product of two tensors, the result is a tensor called the inner product. It should be noted that when, as a result of contraction of one or more pairs of indices, there remain no free indices, the resulting quantity is a scalar.

At this point some readers may wish to skip the remaining tensor theory and proceed directly to section 2.1 of aeronautical applications, which only requires a knowledge of rank two tensor transformations. Having seen the utility of the more elementary tensor operations and the ease with which the summation convention can be utilized with a simple computational algorithm, the reader will wish to return to a study of the remaining theory which deals with the Christoffel symbols and their role in obtaining derivatives. Once the Christoffel symbols are understood and expressions for vector derivatives obtained, the way is clear to proceed with the formulation of models of diverse phenomena. It will be seen, however, that all formulations, from the simplest to the most complex, require operations involving only summation and differentiation. The simplicity of these operations is adequately demonstrated in sections 2.1 through 2.9 of the chapter on aeronautical applications.

1.12 VECTOR DERIVATIVES AND THE CHRISTOFFEL SYMBOLS

The scalar product of any two base vectors $\bar{a}_i$ and $\bar{a}_j$ may be defined as follows:

$$\bar{a}_i \cdot \bar{a}_j = \bar{a}_j \cdot \bar{a}_i = g_{ij} \qquad (1.12.1)$$

Likewise, the scalar product of the reciprocal base vectors $\bar{a}^i$ and $\bar{a}^j$ may be defined as

$$\bar{a}^i \cdot \bar{a}^j = \bar{a}^j \cdot \bar{a}^i = \bar{g}^{ij} \qquad (1.12.2)$$

The symmetry of g_{ij} and g^{ij} follows from the nature of the scalar product. Certain combinations of the partial derivatives of these scalar products with respect to the system coordinates are useful in obtaining the derivatives of a vector or formulating the equations of mathematical physics in a general curvilinear coordinate system. The definitions that follow are ascribed to Christoffel and are called Christoffel symbols. There are two of these symbols, the first of which is defined as

$$[ij,k] = \frac{1}{2}\left(\frac{\partial g_{ik}}{\partial x^j} + \frac{\partial g_{jk}}{\partial x^i} - \frac{\partial g_{ij}}{\partial x^k}\right) \qquad (1.12.3)$$

The Christoffel symbol of the second kind is

$$\begin{Bmatrix} k \\ ij \end{Bmatrix} = g^{kl}[ij,l] \tag{1.12.4}$$

1.12.1 *Derivatives of a Contravariant Vector*

The utility of the Christoffel symbols is immediately apparent when an attempt is made to find the partial derivatives of a base vector, or its reciprocal, with respect to any system coordinate. Any vector $\bar{A}$ may be expressed in the forms given in equation (1.2.1). Furthermore, since the base vectors are, in general, functions of the coordinates, it follows that the derivative of $\bar{A}$ with respect to any coordinate can involve the Christoffel symbols. From equation (1.2.1), the partial derivative of the contravariant form of the vector $\bar{A}$ with respect to the coordinate x^k is given by

$$\frac{\partial \bar{A}}{\partial x^k} = \frac{\partial A^i}{\partial x^k} \bar{a}_i + A^i \frac{\partial \bar{a}_i}{\partial x^k} \tag{1.12.5}$$

Since $a_i \cdot a_j = g_{ij}$

$$\frac{\partial g_{ij}}{\partial x^k} = \frac{\partial a_i}{\partial x^k} \cdot \bar{a}_j + \bar{a}_i \cdot \frac{\partial \bar{a}_j}{\partial x^k} \tag{1.12.6}$$

Likewise,

$$\frac{\partial g_{jk}}{\partial x^i} = \frac{\partial \bar{a}_j}{\partial x^i} \cdot \bar{a}_k + \bar{a}_j \cdot \frac{\partial \bar{a}_k}{\partial x^i} \tag{1.12.7}$$

and

$$\frac{\partial g_{ik}}{\partial x^j} = \frac{\partial \bar{a}_i}{\partial x^j} \cdot \bar{a}_k + \bar{a}_i \cdot \frac{\partial \bar{a}_k}{\partial x^j} \tag{1.12.8}$$

Since

$$\bar{a}_i = \frac{\partial \bar{r}}{\partial x^i} \tag{1.12.9}$$

it follows that

$$\frac{\partial \bar{a}_i}{\partial x^j} = \frac{\partial}{\partial x^j}\left(\frac{\partial \bar{r}}{\partial x^i}\right) = \frac{\partial}{\partial x^i}\left(\frac{\partial \bar{r}}{\partial x^j}\right) = \frac{\partial \bar{a}_j}{\partial x^i} \qquad (1.12.10)$$

From equations (1.12.6) through (1.12.10)

$$\frac{\partial \bar{a}_i}{\partial x^j} \cdot \bar{a}_k = [ij,k] \qquad (1.12.11)$$

Therefore, if equation (1.5.4) is used, the rate of change of the base vector a_i with respect to x_j assumes the form

$$\frac{\partial \bar{a}_i}{\partial x^j} = [ij,k]\bar{a}^k \qquad (1.12.12)$$

Equation (1.12.12) gives the required rate of change of the base vector a_i with respect to a system coordinate, in terms of the Christoffel symbol of the first kind and the reciprocal base vectors. A more convenient form is obtained if both sides of equation (1.12.12) are multiplied scalarly by the reciprocal base vector $\bar{a}^l$ to yield

$$\frac{\partial \bar{a}_i}{\partial x^j} \cdot \bar{a}^l = [ij,k]\bar{a}^k \cdot \bar{a}^l \qquad (1.12.13)$$

From equation (1.12.2), it is seen that

$$\bar{a}^k \cdot \bar{a}^l = g^{kl}$$

Therefore

$$\frac{\partial \bar{a}_i}{\partial x^j} \cdot \bar{a}^l = [ij,k]g^{kl} \qquad (1.12.14)$$

In terms of the defining formula (1.12.4), equation (1.12.14) may be rewritten as follows:

$$\frac{\partial \bar{a}_i}{\partial x^j} \cdot \bar{a}^l = \begin{Bmatrix} l \\ ij \end{Bmatrix} \qquad (1.12.15)$$

Therefore

$$\frac{\partial \bar{a}_i}{\partial x^j} = \begin{Bmatrix} l \\ ij \end{Bmatrix} \bar{a}_l \tag{1.12.16}$$

By substitution of equation (1.12.16) in equation (1.12.5) the partial derivative of a vector $\bar{A}$ with respect to the system coordinate x^k is

$$\frac{\partial \bar{A}}{\partial x^k} = \frac{\partial A^i}{\partial x^k} \bar{a}_i + A^i \begin{Bmatrix} l \\ ik \end{Bmatrix} \bar{a}_l \tag{1.12.17}$$

The indices i and l in the second term on the right side of equation (1.12.17) are dummy indices, and may therefore be replaced by any other convenient indices, except k. To have a common base vector a_i, equation (1.12.17) may be rewritten as follows

$$\frac{\partial \bar{A}}{\partial x^k} = \left(\frac{\partial A^i}{\partial x^k} + \begin{Bmatrix} i \\ jk \end{Bmatrix} A^j \right) \bar{a}_i = A^i_{,k} \bar{a}_i \tag{1.12.18}$$

Furthermore, since

$$\frac{\partial \bar{A}}{\partial x^k} \frac{dx^k}{dt} = \frac{d\bar{A}}{dt}$$

and

$$\frac{\partial A^i}{\partial x^k} \frac{dx^k}{dt} = \frac{dA^i}{dt}$$

the intrinsic derivative, or the total derivative with respect to the parameter, t, of the contravariant form of the vector $\bar{A}$, may be obtained from equation (1.12.18) in the following form:

$$\frac{d\bar{A}}{dt} = \left(\frac{dA^i}{dt} + \begin{Bmatrix} i \\ jk \end{Bmatrix} A^j \frac{dx^k}{dt} \right) \bar{a}_i = A^i_{,k} \frac{dx^k}{dt} \bar{a}_i \tag{1.12.19}$$

where $A^i{}_{,k}$ is the covariant derivative of the contravariant vector A^i with respect to x^k.

The notation $A^i{}_{,j}$ suggests that the covariant derivative of a contravariant vector is not a simple covariant or contravariant vector. As the notation implies, $A^i{}_{,j}$ is a mixed tensor, with one index of contravariance and one index of covariance. If a single-valued, reversible functional transformation of the form given in equation (1.4.4) is assumed, the transformation law for this type of entity is

$$B_k{}^i(y) = \frac{\partial y^i}{\partial x^\alpha} \frac{\partial x^\beta}{\partial y^k} A_\beta{}^\alpha(x) \qquad (1.12.20)$$

where $A_\beta{}^\alpha(x)$ are the components in the x coordinate system and $B_k{}^i(y)$ are the corresponding components in the y coordinate system.

In an orthogonal Cartesian reference frame

$$g_{ij} = \bar{a}_i \cdot \bar{a}_j = \delta_j{}^i = \bar{a}^j \cdot \bar{a}^j = g^{ij} \qquad (1.12.21)$$

Therefore, since all these scalar products are constants, it follows that the Christoffel symbols vanish. In this case, the covariant derivative of a contravariant vector reduces to the sum of the partial derivatives of its physical components along a set of fixed axes

$$\frac{\partial \bar{A}}{\partial x^k} = \frac{\partial A^i}{\partial x^k} \, \bar{a}_i \qquad i = 1,2,3$$

Likewise, the intrinsic derivative of a vector reduces to the ordinary time rates of change of the physical components along a set of fixed axes.

For a general space of three dimensions, equation (1.12.19) assumes the form

$$\frac{d\bar{A}}{dt} = \left(\frac{dA^1}{dt} + f^1\right) \bar{a}_1 + \left(\frac{dA^2}{dt} + f^2\right) \bar{a}_2 + \left(\frac{dA^3}{dt} + f^3\right) \bar{a}_3 \qquad (1.12.22)$$

$$f^1 = A^1 \left(\begin{Bmatrix} 1 \\ 11 \end{Bmatrix} \frac{dx^1}{dt} + \begin{Bmatrix} 1 \\ 12 \end{Bmatrix} \frac{dx^2}{dt} + \begin{Bmatrix} 1 \\ 13 \end{Bmatrix} \frac{dx^3}{dt} \right)$$

$$+ A^2 \left(\begin{Bmatrix} 1 \\ 21 \end{Bmatrix} \frac{dx^1}{dt} + \begin{Bmatrix} 1 \\ 22 \end{Bmatrix} \frac{dx^2}{dt} + \begin{Bmatrix} 1 \\ 23 \end{Bmatrix} \frac{dx^3}{dt} \right)$$

$$+ A^3 \left(\begin{Bmatrix} 1 \\ 31 \end{Bmatrix} \frac{dx^1}{dt} + \begin{Bmatrix} 1 \\ 32 \end{Bmatrix} \frac{dx^2}{dt} + \begin{Bmatrix} 1 \\ 33 \end{Bmatrix} \frac{dx^3}{dt} \right) \qquad (1.12.23)$$

$$f^2 = A^1 \left(\begin{Bmatrix} 2 \\ 11 \end{Bmatrix} \frac{dx^1}{dt} + \begin{Bmatrix} 2 \\ 12 \end{Bmatrix} \frac{dx^2}{dt} + \begin{Bmatrix} 2 \\ 13 \end{Bmatrix} \frac{dx^3}{dt} \right)$$

$$+ A^2 \left(\begin{Bmatrix} 2 \\ 21 \end{Bmatrix} \frac{dx^1}{dt} + \begin{Bmatrix} 2 \\ 22 \end{Bmatrix} \frac{dx^2}{dt} + \begin{Bmatrix} 2 \\ 23 \end{Bmatrix} \frac{dx^3}{dt} \right)$$

$$+ A^3 \left(\begin{Bmatrix} 2 \\ 31 \end{Bmatrix} \frac{dx^1}{dt} + \begin{Bmatrix} 2 \\ 32 \end{Bmatrix} \frac{dx^2}{dt} + \begin{Bmatrix} 2 \\ 33 \end{Bmatrix} \frac{dx^3}{dt} \right) \qquad (1.12.24)$$

$$f^3 = A^1 \left(\begin{Bmatrix} 3 \\ 11 \end{Bmatrix} \frac{dx^1}{dt} + \begin{Bmatrix} 3 \\ 12 \end{Bmatrix} \frac{dx^2}{dt} + \begin{Bmatrix} 3 \\ 13 \end{Bmatrix} \frac{dx^3}{dt} \right)$$

$$+ A^2 \left(\begin{Bmatrix} 3 \\ 21 \end{Bmatrix} \frac{dx^1}{dt} + \begin{Bmatrix} 3 \\ 22 \end{Bmatrix} \frac{dx^2}{dt} + \begin{Bmatrix} 3 \\ 23 \end{Bmatrix} \frac{dx^3}{dt} \right)$$

$$+ A^3 \left(\begin{Bmatrix} 3 \\ 31 \end{Bmatrix} \frac{dx^1}{dt} + \begin{Bmatrix} 3 \\ 32 \end{Bmatrix} \frac{dx^2}{dt} + \begin{Bmatrix} 3 \\ 33 \end{Bmatrix} \frac{dx^3}{dt} \right) \qquad (1.12.25)$$

The intrinsic derivative of a contravariant vector in a space of three dimensions contains 27 Christoffel symbols. However, because of the symmetry of the Christoffel symbols

$$\begin{Bmatrix} k \\ ij \end{Bmatrix} = \begin{Bmatrix} k \\ ji \end{Bmatrix} \qquad (1.12.26)$$

and the number of independent Christoffel symbols reduces to 18.

1.12.2 Derivatives of a Covariant Vector

The second alternative from equation (1.4.1) may be used to express the vector $\overline{A}$ in terms of its covariant tensor components and reciprocal base vectors, that is

$$\overline{A} = A_i \bar{a}^i \tag{1.12.27}$$

In this case, the partial derivative of the vector $\overline{A}$ with respect to x^k is given by

$$\frac{\partial \overline{A}}{\partial x^k} = \frac{\partial A_i}{\partial x^k} \bar{a}^i + A_i \frac{\partial \bar{a}^i}{\partial x^k} \tag{1.12.28}$$

From equation (1.5.4)

$$\frac{\partial \bar{a}^i}{\partial x^k} \cdot \bar{a}_j + \bar{a}^i \cdot \frac{\partial \bar{a}_j}{\partial x^k} = 0$$

Therefore

$$\frac{\partial \bar{a}^i}{\partial x^k} \cdot \bar{a}_j = -\bar{a}^i \cdot \frac{\partial \bar{a}_j}{\partial x^k} \tag{1.12.29}$$

Substituting equation (1.12.16) in equation (1.12.29) gives

$$\frac{\partial \bar{a}^i}{\partial x^k} \cdot \bar{a}_j = -\bar{a}^i \cdot \begin{Bmatrix} l \\ jk \end{Bmatrix} \bar{a}_l = -\begin{Bmatrix} i \\ jk \end{Bmatrix}$$

Therefore

$$\frac{\partial \bar{a}^i}{\partial x^k} = -\begin{Bmatrix} i \\ jk \end{Bmatrix} \bar{a}^j \tag{1.12.30}$$

Substituting equation (1.12.30) in equation (1.12.28) gives the partial derivative of the vector $\overline{A}$ with respect to x^k in the following form:

$$\frac{\partial \overline{A}}{\partial x^k} = \frac{\partial A_i}{\partial x^k} \bar{a}^i - \begin{Bmatrix} i \\ jk \end{Bmatrix} A_i \bar{a}^j \tag{1.12.31}$$

The indices i and j in the second term on the right side of equation (1.12.31) are dummy indices, and may therefore be replaced by any other indices, except k. In terms of the base vectors a^i, equation (1.12.31) may be rewritten as follows

$$\frac{\partial \bar{A}}{\partial x^k} = \left(\frac{\partial A_i}{\partial x^k} - \left\{ \begin{matrix} j \\ ik \end{matrix} \right\} A_j \right) \bar{a}^i = A_{i,k} \bar{a}^i \tag{1.12.32}$$

where $A_{i,k}$ defines the covariant derivative of the covariant vector A_i with respect to x^k.

It may be noted that the covariant derivative of a covariant vector is not a vector. As the notation implies, $A_{i,j}$ is a doubly covariant tensor, that is, a tensor with two indices of covariance. If a single-valued, reversible functional transformation of the form given in equation (1.4.4) is again assumed, the transformation law for entities of this kind is

$$B_{ij}(y) = \frac{\partial x^\alpha}{\partial y^i} \frac{\partial x^\beta}{\partial y^j} A_{\alpha\beta}(x)$$

where $A_{\alpha\beta}(x)$ are the components in the x coordinate system, and $B_{ij}(y)$ are the corresponding components in the y coordinate system. It may be mentioned in passing that moment of inertia, which is a second-order tensor, has a transformation law of this form.

It appears, therefore, that the operation of covariant differentiation of a vector or tensor increases the covariance by one index; that is, the x^j covariant derivative of the contravariant vector A^i is $A^i_{,j}$, which is a mixed tensor, with one index of contravariance and one index of covariance. The x^j covariant derivative of the covariant vector A_i is $A_{i,j}$. This is a doubly covariant tensor or a tensor with two indices of covariance. The intrinsic derivative of the covariant form of the vector $\bar{A}$ is obtained from equation (1.12.32) in the following form:

$$\frac{d\bar{A}}{dt} = \left(\frac{dA_i}{dt} - \left\{ \begin{matrix} j \\ ik \end{matrix} \right\} A_j \frac{dx^k}{dt} \right) \bar{a}^i = A_{i,k} \frac{dx^k}{dt} \bar{a}^i \tag{1.12.33}$$

For a general space of three dimensions, equation (1.12.33) assumes the form

$$\frac{d\bar{A}}{dt} = \left(\frac{dA_1}{dt} - f_1 \right) \bar{a}^1 + \left(\frac{dA_2}{dt} - f_2 \right) \bar{a}^2 + \left(\frac{dA_3}{dt} - f_3 \right) \bar{a}^3 \tag{1.12.34}$$

where

$$f_1 = A_1 \left(\begin{Bmatrix} 1 \\ 11 \end{Bmatrix} \frac{dx^1}{dt} + \begin{Bmatrix} 1 \\ 12 \end{Bmatrix} \frac{dx^2}{dt} + \begin{Bmatrix} 1 \\ 13 \end{Bmatrix} \frac{dx^3}{dt} \right)$$

$$+ A_2 \left(\begin{Bmatrix} 2 \\ 11 \end{Bmatrix} \frac{dx^1}{dt} + \begin{Bmatrix} 2 \\ 12 \end{Bmatrix} \frac{dx^2}{dt} + \begin{Bmatrix} 2 \\ 13 \end{Bmatrix} \frac{dx^3}{dt} \right)$$

$$+ A_3 \left(\begin{Bmatrix} 3 \\ 11 \end{Bmatrix} \frac{dx^1}{dt} + \begin{Bmatrix} 3 \\ 12 \end{Bmatrix} \frac{dx^2}{dt} + \begin{Bmatrix} 3 \\ 13 \end{Bmatrix} \frac{dx^3}{dt} \right) \qquad (1.12.35)$$

$$f_2 = A_1 \left(\begin{Bmatrix} 1 \\ 21 \end{Bmatrix} \frac{dx^1}{dt} + \begin{Bmatrix} 1 \\ 22 \end{Bmatrix} \frac{dx^2}{dt} + \begin{Bmatrix} 1 \\ 23 \end{Bmatrix} \frac{dx^3}{dt} \right)$$

$$+ A_2 \left(\begin{Bmatrix} 2 \\ 21 \end{Bmatrix} \frac{dx^1}{dt} + \begin{Bmatrix} 2 \\ 22 \end{Bmatrix} \frac{dx^2}{dt} + \begin{Bmatrix} 2 \\ 23 \end{Bmatrix} \frac{dx^3}{dt} \right)$$

$$+ A_3 \left(\begin{Bmatrix} 3 \\ 21 \end{Bmatrix} \frac{dx^1}{dt} + \begin{Bmatrix} 3 \\ 22 \end{Bmatrix} \frac{dx^2}{dt} + \begin{Bmatrix} 3 \\ 23 \end{Bmatrix} \frac{dx^3}{dt} \right) \qquad (1.12.36)$$

$$f_3 = A_1 \left(\begin{Bmatrix} 1 \\ 31 \end{Bmatrix} \frac{dx^1}{dt} + \begin{Bmatrix} 1 \\ 32 \end{Bmatrix} \frac{dx^2}{dt} + \begin{Bmatrix} 1 \\ 33 \end{Bmatrix} \frac{dx^3}{dt} \right)$$

$$+ A_2 \left(\begin{Bmatrix} 2 \\ 31 \end{Bmatrix} \frac{dx^1}{dt} + \begin{Bmatrix} 2 \\ 32 \end{Bmatrix} \frac{dx^2}{dt} + \begin{Bmatrix} 2 \\ 33 \end{Bmatrix} \frac{dx^3}{dt} \right)$$

$$+ A_3 \left(\begin{Bmatrix} 3 \\ 31 \end{Bmatrix} \frac{dx^1}{dt} + \begin{Bmatrix} 3 \\ 32 \end{Bmatrix} \frac{dx^2}{dt} + \begin{Bmatrix} 3 \\ 33 \end{Bmatrix} \frac{dx^3}{dt} \right) \qquad (1.12.37)$$

As in the case of the intrinsic derivative of the contravariant vector, the intrinsic derivative of the covariant form of the vector $\overline{A}$ is seen to contain 27 Christoffel symbols. However, because of the symmetry implied by equation (1.12.26), the number of independent Christoffel symbols is again reduced to 18.

1.13 SPECIAL COORDINATE SYSTEMS

The large number of terms appearing in equations (1.12.19) and (1.12.33) is due to the generality of these equations which are applicable to any space of three dimensions. Fortunately, for the three-dimensional spaces most commonly used, both of these equations reduce to a more manageable form.

For example, if base vectors of unit length are denoted by $\hat{a}_i$ or $\hat{a}^i$, then in a cylindrical polar coordinate system (fig. 1.13.1):

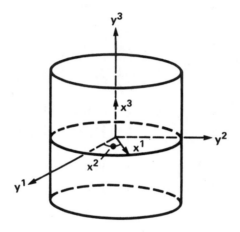

Figure 1.13.1.— Cylindrical coordinates.

$$\left.\begin{array}{ll} \bar{a}_1 = \hat{a}_1 & g_{11} = 1 \\[2mm] \bar{a}_2 = x^1 \hat{a}_2 & g_{22} = (x^1)^2 \\[2mm] \bar{a}_3 = \hat{a}_3 & g_{33} = 1 \end{array}\right\} \qquad (1.13.1)$$

and

$$\left.\begin{array}{ll} \bar{a}^1 = \hat{a}^1 & g^{11} = 1 \\[2mm] \bar{a}^2 = \dfrac{1}{x^1}\hat{a}^2 & g^{22} = \dfrac{1}{(x^1)^2} \\[2mm] \bar{a}^3 = \hat{a}^3 & g^{33} = 1 \end{array}\right\} \qquad (1.13.2)$$

As a consequence of equations (1.13.1) and (1.13.2) there are only two independent, nonzero Christoffel symbols in a cylindrical polar coordinate system. These are

$$\left.\begin{aligned} \left\{ \begin{matrix} 1 \\ 22 \end{matrix} \right\} &= -x^1 \\[2mm] \left\{ \begin{matrix} 2 \\ 12 \end{matrix} \right\} = \left\{ \begin{matrix} 2 \\ 21 \end{matrix} \right\} &= \frac{1}{x^1} \end{aligned}\right\} \qquad (1.13.3)$$

Hence, a contravariant vector referred to this coordinate system has a time rate of
change as follows:

$$\frac{d\bar{A}}{dt} = \left(\frac{dA^1}{dt} + A^2 \left\{ \begin{matrix} 1 \\ 22 \end{matrix} \right\} \frac{dx^2}{dt} \right) \bar{a}_1 + \left[\frac{dA^2}{dt} + \left\{ \begin{matrix} 2 \\ 12 \end{matrix} \right\} \left(A^1 \frac{dx^2}{dt} + A^2 \frac{dx^1}{dt} \right) \right] \bar{a}_2 + \frac{dA^3}{dt} \bar{a}_3$$

$$(1.13.4)$$

Likewise, the time rate of change of a covariant vector referred to this coordinate
system is given by

$$\frac{d\bar{A}}{dt} = \left(\frac{dA_1}{dt} - A_2 \left\{ \begin{matrix} 2 \\ 12 \end{matrix} \right\} \frac{dx^2}{dt} \right) \bar{a}^1 + \left(\frac{dA_2}{dt} - A_1 \left\{ \begin{matrix} 1 \\ 22 \end{matrix} \right\} \frac{dx^2}{dt} - A_2 \left\{ \begin{matrix} 2 \\ 21 \end{matrix} \right\} \frac{dx^1}{dt} \right) \bar{a}^2 + \frac{dA_3}{dt} \bar{a}^3$$

$$(1.13.5)$$

In spherical polar coordinates (fig. 1.13.2):

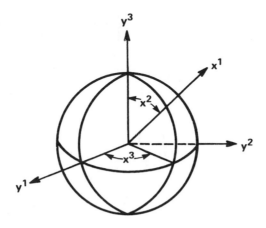

Figure 1.13.2.– Polar coordinates.

$$\left.\begin{array}{ll} \bar{a}_1 = \hat{a} & g_{11} = 1 \\[2mm] \bar{a}_2 = x^1 \hat{a}_2 & g_{22} = (x^1)^2 \\[2mm] \bar{a}_3 = x^1 \sin x^2 \hat{a}_3 & g_{33} = (x^1 \sin x^2)^2 \end{array}\right\} \tag{1.13.6}$$

and

$$\left.\begin{array}{ll} \bar{a}^1 = \hat{a}^1 & g^{11} = 1 \\[3mm] \bar{a}^2 = \dfrac{1}{x^1} \hat{a}^2 & g^{22} = \dfrac{1}{(x^1)^2} \\[4mm] \bar{a}^3 = \dfrac{1}{x^1 \sin x^2} \hat{a}^3 & g^{33} = \dfrac{1}{(x^1 \sin x^2)^2} \end{array}\right\} \tag{1.13.7}$$

In this case there are six independent, nonzero Christoffel symbols. These are

$$\left.\begin{array}{ll} \left\{\begin{matrix}1\\22\end{matrix}\right\} = -x^1 & \left\{\begin{matrix}2\\33\end{matrix}\right\} = -\sin x^2 \cos x^2 \\[4mm] \left\{\begin{matrix}2\\12\end{matrix}\right\} = \left\{\begin{matrix}1\\21\end{matrix}\right\} = \dfrac{1}{x^1} & \left\{\begin{matrix}3\\13\end{matrix}\right\} = \left\{\begin{matrix}3\\31\end{matrix}\right\} = \dfrac{1}{x^1} \\[4mm] \left\{\begin{matrix}1\\33\end{matrix}\right\} = -x^1 \sin^2 x^2 & \left\{\begin{matrix}3\\23\end{matrix}\right\} = \left\{\begin{matrix}3\\32\end{matrix}\right\} = \cot x^2 \end{array}\right\} \tag{1.13.8}$$

When the Christoffel symbols are substituted in equation (1.12.19), the time rate of change of a contravariant vector referred to a spherical coordinate system assumes the following form (ref. 13):

$$\frac{d\bar{A}}{dt} = \left(\frac{dA^1}{dt} + A^2 \left\{\begin{matrix}1\\22\end{matrix}\right\} \frac{dx^2}{dt} + A^3 \left\{\begin{matrix}1\\33\end{matrix}\right\} \frac{dx^3}{dt}\right) \bar{a}_1$$

$$+ \left[\frac{dA^2}{dt} + \left\{\begin{matrix}2\\12\end{matrix}\right\} \left(A^1 \frac{dx^2}{dt} + A^2 \frac{dx^1}{dt}\right) + A^3 \left\{\begin{matrix}2\\33\end{matrix}\right\} \frac{dx^3}{dt}\right] \bar{a}_2$$

$$+ \left[\frac{dA^3}{dt} + \left\{\begin{matrix}3\\13\end{matrix}\right\} \left(A^1 \frac{dx^3}{dt} + A^3 \frac{dx^1}{dt}\right) + \left\{\begin{matrix}3\\23\end{matrix}\right\} \left(A^2 \frac{dx^3}{dt} + A^3 \frac{dx^2}{dt}\right)\right] \bar{a}_3 \tag{1.13.9}$$

The corresponding rate of change of a covariant vector is obtained by substitution from equation (1.13.8) in equation (1.12.33). In this case,

$$\frac{d\bar{A}}{dt} = \left(\frac{dA_1}{dt} - A_2 \left\{\begin{matrix}2\\12\end{matrix}\right\} \frac{dx^2}{dt} - A_3 \left\{\begin{matrix}3\\13\end{matrix}\right\} \frac{dx^3}{dt}\right) \bar{a}^1$$

$$+ \left(\frac{dA_2}{dt} - A_1 \left\{\begin{matrix}1\\22\end{matrix}\right\} \frac{dx^2}{dt} - A_2 \left\{\begin{matrix}2\\21\end{matrix}\right\} \frac{dx^1}{dt} - A_3 \left\{\begin{matrix}3\\23\end{matrix}\right\} \frac{dx^3}{dt}\right) \bar{a}^2$$

$$+ \left[\frac{dA_3}{dt} - A_1 \left\{\begin{matrix}1\\33\end{matrix}\right\} \frac{dx^3}{dt} - A_2 \left\{\begin{matrix}2\\33\end{matrix}\right\} \frac{dx^3}{dt} - A_3 \left(\left\{\begin{matrix}3\\31\end{matrix}\right\} \frac{dx^1}{dt} + \left\{\begin{matrix}3\\32\end{matrix}\right\} \frac{dx^2}{dt}\right)\right] \bar{a}^3$$

$$(1.13.10)$$

1.13.1 Alternative Derivation of the Christoffel Symbols

In equations (1.12.3) and (1.12.4), the Christoffel symbols have been defined in terms of the scalar product of two of the base vectors. These symbols can also be derived from the equations of coordinate transformation by the following method, which is suitable for some applications.

In a rectangular Cartesian coordinate reference frame, with coordinates denoted by y^i, an element of arc of length $d\bar{s}$ may be expressed in the following form:

$$d\bar{s} = \hat{a}_\alpha \, dy^\alpha = \hat{a}_\beta \, dy^\beta$$

Therefore

$$ds^2 = \hat{a}_\alpha \cdot \hat{a}_\beta \, dy^\alpha \, dy^\beta = \delta_\beta{}^\alpha \, dy^\alpha \, dy^\beta = dy^\alpha \, dy^\alpha$$

Consider a curvilinear coordinate system with coordinates denoted by x^i, and assume that the x and y coordinates are related by a set of transformation equations as follows

$$y^i = y^i(x^1, x^2, x^3) \qquad (1.13.11)$$

The element of arc $d\bar{s}$ in the x coordinate system assumes the form

$$d\bar{s} = \bar{a}_i \, dx^i = \bar{a}_j \, dx^j$$

Therefore

$$ds^2 = (\bar{a}_i \, dx^i) \cdot (\bar{a}_j \, dx^j) = g_{ij} \, dx^i \, dx^j = dy^\alpha \, dy^\alpha$$

$$dy^\alpha \, dy^\alpha = \frac{\partial y^\alpha}{\partial x^i} \frac{\partial y^\alpha}{\partial x^j} \, dx^i \, dx^j = g_{ij} \, dx^i \, dx^j$$

and

$$g_{ij} = \frac{\partial y^\alpha}{\partial x^i} \frac{\partial y^\alpha}{\partial x^j} \tag{1.13.12}$$

If the transformation equation (1.13.11) is reversible and one-to-one, then

$$x^i = x^i(y^1, y^2, y^3) \tag{1.13.13}$$

By substitution from equation (1.13.12) in equation (1.12.3), the Christoffel symbol of the first kind assumes the following form:

$$[ij,k] = \frac{\partial^2 y^\alpha}{\partial x^i \partial x^j} \frac{\partial y^\alpha}{\partial x^k} \tag{1.13.14}$$

Likewise, substituting equation (1.13.12) in equation (1.12.4) gives for the Christoffel symbol of the second kind

$$\left\{ {i \atop jk} \right\} = \frac{\partial^2 y^\alpha}{\partial x^j \partial x^k} \frac{\partial x^i}{\partial y^\alpha} \tag{1.13.15}$$

1.14 THE DIFFERENTIAL OPERATOR ∇

As they stand, equations (1.12.18) and (1.12.32) do not appear to satisfy the definition of a tensor given by equations (1.8.2) and (1.8.4), since they are not bilinear in the base vectors. In order to remove this inconsistency, it is necessary to define covariant differentiation in terms of the differential operator ∇. This operator is defined as follows

$$\nabla = \bar{b}^j \frac{\partial}{\partial y^j} \qquad (1.14.1)$$

The proof that this operator is a differential invariant is straightforward.

The transformation law for a contravariant base vector is

$$\bar{b}^j(y) = \frac{\partial y^j}{\partial x^i} \, \bar{a}^i(x) \qquad (1.14.2)$$

Moreover

$$\frac{\partial}{\partial y^j} = \frac{\partial x^\beta}{\partial y^j} \frac{\partial}{\partial x^\beta} \qquad (1.14.3)$$

Therefore

$$\bar{b}^j \frac{\partial}{\partial y^j} = \bar{a}^i \frac{\partial y^j}{\partial x^i} \frac{\partial x^\beta}{\partial y^j} \frac{\partial}{\partial x^\beta}$$

$$= \bar{a}^i \delta_i{}^\beta \frac{\partial}{\partial x^\beta} \qquad (1.14.4)$$

where $\delta_i{}^\beta$ is the Kronecker delta. Hence

$$\bar{b}^j \frac{\partial}{\partial y^j} = \bar{a}^i \frac{\partial}{\partial x^i} \qquad (1.14.5)$$

As has been noted previously, the operation of covariant differentiation increases the covariance by one index. This can be seen more clearly if covariant differentiation is defined in terms of the differential operator ∇. Consider a scalar point function ϕ, which is a tensor of rank zero. When this function is operated upon with the differential operator ∇, a covariant vector of rank one is obtained as follows:

$$\nabla \phi = \bar{a}^i \frac{\partial \phi}{\partial x^i} = A_i \bar{a}^i \qquad (1.14.6)$$

where

$$A_i = \frac{\partial \phi}{\partial x^i} \qquad (1.14.7)$$

This is a covariant tensor of rank one. Hence, operation upon the scalar point function ϕ with the differential operator ∇ has resulted in a covariant tensor of rank one.

Next consider the result of operating upon the vector $\bar{A}$, which is a tensor of rank one, with the operator ∇. In this case the vector $\bar{A}$ can assume the following alternative forms:

$$\bar{A} = A^i a_i = A_j \bar{a}^j \qquad (1.14.8)$$

The result of operating upon the contravariant form gives

$$\nabla \bar{A} = \bar{a}^j \frac{\partial}{\partial x^j} (A^i \bar{a}_i) \qquad (1.14.9)$$

$$\nabla \bar{A} = \bar{a}^j \left(\frac{\partial A^i}{\partial x^j} \bar{a}_i + A^i \frac{\partial \bar{a}_i}{\partial x^j} \right) \qquad (1.14.10)$$

Substitution from equation (1.12.16) in equation (1.14.10) gives

$$\nabla \bar{A} = \bar{a}^j \left(\frac{\partial A^i}{\partial x^j} \bar{a}_i + A^i \left\{ \begin{matrix} \alpha \\ ij \end{matrix} \right\} \bar{a}_\alpha \right)$$

and since α and i are dummy indices, this equation can be rewritten as

$$\nabla \bar{A} = \bar{a}^j \left(\frac{\partial A^i}{\partial x^j} + \left\{ \begin{matrix} i \\ \alpha j \end{matrix} \right\} A^\alpha \right) \bar{a}_i \qquad (1.14.11)$$

Hence

$$\nabla \bar{A} = \left(\frac{\partial A^i}{\partial x^j} + \left\{ \begin{matrix} i \\ \alpha j \end{matrix} \right\} A^\alpha \right) \bar{a}^j \bar{a}_i \qquad (1.14.12)$$

This form shows that $\nabla \bar{A}$ in a mixed tensor of rank two, which justifies the following notation

$$\nabla \bar{A} = A^i{}_{,j} \bar{a}^j \bar{a}_i \tag{1.14.13}$$

where

$$A^i{}_{,j} = \frac{\partial A^i}{\partial x^j} + \left\{ \begin{matrix} i \\ \alpha j \end{matrix} \right\} A^\alpha \tag{1.14.14}$$

defines the covariant derivative of A^i with respect to x^j. Likewise, the result of operating upon the covariant form of the vector $\bar{A}$ gives

$$\nabla \bar{A} = \bar{a}^j \frac{\partial}{\partial x^j} (A_i \bar{a}^i) \tag{1.14.15}$$

$$\nabla \bar{A} = \bar{a}^j \left(\frac{\partial A_i}{\partial x^j} \bar{a}^i + A_i \frac{\partial \bar{a}^i}{\partial x^j} \right) \tag{1.14.16}$$

Substitution from equation (1.12.30) in equation (1.14.16) gives

$$\nabla \bar{A} = \bar{a}^j \left(\frac{\partial A_i}{\partial x^j} \bar{a}^i - A_i \left\{ \begin{matrix} i \\ \alpha j \end{matrix} \right\} \bar{a}^\alpha \right)$$

Again, interchanging the dummy indices to obtain a common vector coefficient

$$\nabla \bar{A} = \left(\frac{\partial A_i}{\partial x^j} - \left\{ \begin{matrix} \alpha \\ ij \end{matrix} \right\} A_\alpha \right) \bar{a}^j \bar{a}^i \tag{1.14.17}$$

Therefore

$$\nabla \bar{A} = A_{i,j} \bar{a}^j \bar{a}^i \tag{1.14.18}$$

where

$$A_{i,j} = \frac{\partial A_i}{\partial x^j} - \left\{\begin{matrix} \alpha \\ ij \end{matrix}\right\} A_\alpha \qquad\qquad (1.14.19)$$

defines the covariant derivative of A_i with respect to x^j.

Hence, operation upon

$$\bar{A} = A_i \bar{a}^i$$

with ∇ has again increased the covariance by one index.

1.15 THE RIEMANN-CHRISTOFFEL TENSOR AND THE RICCI TENSOR

The tensor known as the Riemann-Christoffel tensor, plays a basic role in many investigations of differential geometry, dynamics of rigid and deformable bodies, electrodynamics, and relativity. Those who are not interested in such applications may omit this section.

Since the covariant derivative of a tensor is a tensor, it can be differentiated covariantly again. However, in all cases, covariant differentiation of a tensor gives rise to a tensor having one more unit of covariant character than the given tensor.

If $\bar{A} = A_i \bar{a}^i$ is a covariant tensor of rank one, its covariant derivative with respect to x^j can be obtained from equation (1.14.17) where it is shown that

$$\nabla \bar{A} = \left(\frac{\partial A_i}{\partial x^j} - \left\{\begin{matrix} \alpha \\ ij \end{matrix}\right\} A_\alpha\right) \bar{a}^j \bar{a}^i$$

or

$$\nabla \bar{A} = A_{i,j} \bar{a}^j \bar{a}^i$$

where

$$A_{i,j} = \frac{\partial A_i}{\partial x^j} - \left\{\begin{matrix} \alpha \\ ij \end{matrix}\right\} A_\alpha$$

defines the covariant derivative of the covariant tensor component A_i with respect to x^j.

Consider next the tensor of rank two

$$A_{i,j}\bar{a}^j\bar{a}^i$$

The result of operating upon this tensor with the differential operator ∇ is a new tensor of rank three as follows:

$$\nabla(A_{i,j}\bar{a}^j\bar{a}^i) = \bar{a}^k \frac{\partial}{\partial x^k}(A_{i,j}\bar{a}^j\bar{a}^i) = \bar{a}^k\left(\frac{\partial A_{i,j}}{\partial x^k}\bar{a}^j\bar{a}^i + A_{i,j}\frac{\partial \bar{a}^j}{\partial x^k}\bar{a}^i + A_{i,j}\bar{a}^j\frac{\partial \bar{a}^i}{\partial x^k}\right)$$

$$(1.15.1)$$

Moreover, it has been shown that

$$\frac{\partial \bar{a}^j}{\partial x^k} = -\left\{\begin{matrix}j\\ \alpha k\end{matrix}\right\}\bar{a}^\alpha$$

$$\frac{\partial \bar{a}^i}{\partial x^k} = -\left\{\begin{matrix}i\\ \alpha k\end{matrix}\right\}\bar{a}^\alpha$$

Substitution of these values in equation (1.15.1) yields the following:

$$\nabla(A_{i,j}\bar{a}^j\bar{a}^i) = \bar{a}^k\left(\frac{\partial A_{i,j}}{\partial x^k}\bar{a}^j\bar{a}^i - A_{i,j}\left\{\begin{matrix}j\\ \alpha k\end{matrix}\right\}\bar{a}^\alpha\bar{a}^i - A_{i,j}\left\{\begin{matrix}i\\ \alpha k\end{matrix}\right\}\bar{a}^j\bar{a}^\alpha\right) \quad (1.15.2)$$

By interchanging the dummy indices to obtain a common factor, the following result is obtained:

$$\nabla(A_{i,j}\bar{a}^j\bar{a}^i) = \left(\frac{\partial A_{i,j}}{\partial x^k} - \left\{\begin{matrix}\alpha\\ jk\end{matrix}\right\}A_{i,\alpha} - \left\{\begin{matrix}\alpha\\ ik\end{matrix}\right\}A_{\alpha,j}\right)\bar{a}^k\bar{a}^j\bar{a}^i \quad (1.15.3)$$

$$= A_{i,jk}\bar{a}^k\bar{a}^j\bar{a}^i \quad (1.15.4)$$

where

$$A_{i,jk} = \frac{\partial A_{i,j}}{\partial x^k} - \left\{\begin{matrix}\alpha\\ jk\end{matrix}\right\}A_{i,\alpha} - \left\{\begin{matrix}\alpha\\ ik\end{matrix}\right\}A_{\alpha,j} \quad (1.15.5)$$

is the covariant derivative of the tensor component $A_{i,j}$, and is seen to be a covariant tensor of rank three.

By substitution of

$$A_{i,j} = \frac{\partial A_i}{\partial x^j} - \left\{ \begin{matrix} \alpha \\ ij \end{matrix} \right\} A_\alpha$$

in equation (1.15.5), $A_{i,jk}$ is obtained in the following form:

$$A_{i,jk} = \frac{\partial}{\partial x^k}\left(\frac{\partial A_i}{\partial x^j} - \left\{ \begin{matrix} \alpha \\ ij \end{matrix} \right\} A_\alpha \right) - \left\{ \begin{matrix} \alpha \\ jk \end{matrix} \right\}\left(\frac{\partial A_i}{\partial x^\alpha} - \left\{ \begin{matrix} \lambda \\ i\alpha \end{matrix} \right\} A_\lambda \right) - \left\{ \begin{matrix} \alpha \\ ik \end{matrix} \right\}\left(\frac{\partial A_\alpha}{\partial x^j} - \left\{ \begin{matrix} \beta \\ \alpha j \end{matrix} \right\} A_\beta \right)$$

Therefore

$$A_{i,jk} = \frac{\partial^2 A_i}{\partial x^k \partial x^j} - \left\{ \begin{matrix} \alpha \\ ij \end{matrix} \right\}\frac{\partial A_\alpha}{\partial x^k} - A_\alpha \frac{\partial}{\partial x^k}\left\{ \begin{matrix} \alpha \\ ij \end{matrix} \right\} - \left\{ \begin{matrix} \alpha \\ jk \end{matrix} \right\}\frac{\partial A_i}{\partial x^\alpha}$$

$$+ \left\{ \begin{matrix} \alpha \\ jk \end{matrix} \right\}\left\{ \begin{matrix} \lambda \\ i\alpha \end{matrix} \right\} A_\lambda - \left\{ \begin{matrix} \alpha \\ ik \end{matrix} \right\}\frac{\partial A_\alpha}{\partial x^j} + \left\{ \begin{matrix} \alpha \\ ik \end{matrix} \right\}\left\{ \begin{matrix} \beta \\ \alpha j \end{matrix} \right\} A_\beta \qquad (1.15.6)$$

It is interesting to examine the result of performing the operation of covariant differentiation of A_i, first with respect to k, and then with respect to j. This operation gives

$$A_{i,kj} = \frac{\partial}{\partial x^j}\left(\frac{\partial A_i}{\partial x^k} - \left\{ \begin{matrix} \alpha \\ ik \end{matrix} \right\} A_\alpha \right) - \left\{ \begin{matrix} \alpha \\ jk \end{matrix} \right\}\left(\frac{\partial A_i}{\partial x^\alpha} - \left\{ \begin{matrix} \lambda \\ i\alpha \end{matrix} \right\} A_\lambda \right) - \left\{ \begin{matrix} \alpha \\ ij \end{matrix} \right\}\left(\frac{\partial A_\alpha}{\partial x^k} - \left\{ \begin{matrix} \beta \\ \alpha k \end{matrix} \right\} A_\beta \right)$$

Therefore

$$A_{i,kj} = \frac{\partial^2 A_i}{\partial x^j \partial x^k} - \left\{ \begin{matrix} \alpha \\ ik \end{matrix} \right\}\frac{\partial A_\alpha}{\partial x^j} - A_\alpha \frac{\partial}{\partial x^j}\left\{ \begin{matrix} \alpha \\ ik \end{matrix} \right\} - \left\{ \begin{matrix} \alpha \\ jk \end{matrix} \right\}\frac{\partial A_i}{\partial x^\alpha}$$

$$+ \left\{ \begin{matrix} \alpha \\ jk \end{matrix} \right\}\left\{ \begin{matrix} \lambda \\ i\alpha \end{matrix} \right\} A_\lambda - \left\{ \begin{matrix} \alpha \\ ij \end{matrix} \right\}\frac{\partial A_\alpha}{\partial x^k} + \left\{ \begin{matrix} \alpha \\ ij \end{matrix} \right\}\left\{ \begin{matrix} \beta \\ \alpha k \end{matrix} \right\} A_\beta \qquad (1.15.7)$$

A sufficient condition for the equality of the mixed partial derivatives

$$\frac{\partial^2 f(x,y)}{\partial x \partial y} \quad \text{and} \quad \frac{\partial^2 f(x,y)}{\partial y \partial x} \tag{1.15.8}$$

is that the function $f(x,y)$ and its first two partial derivatives be continuous. However, this condition alone is not sufficient to ensure the equality of mixed covariant derivatives. This can be seen by subtracting equation (1.15.7) from equation (1.15.6) to obtain

$$A_{i,jk} - A_{i,kj} = \begin{Bmatrix} \alpha \\ ik \end{Bmatrix} \begin{Bmatrix} \beta \\ \alpha j \end{Bmatrix} A_\beta - A_\alpha \frac{\partial}{\partial x^k} \begin{Bmatrix} \alpha \\ ij \end{Bmatrix} - \begin{Bmatrix} \alpha \\ ij \end{Bmatrix} \begin{Bmatrix} \beta \\ \alpha k \end{Bmatrix} A_\beta + A_\alpha \frac{\partial}{\partial x^j} \begin{Bmatrix} \alpha \\ ik \end{Bmatrix} \tag{1.15.9}$$

The dummy indices in this equation may be interchanged to permit factorization as follows

$$A_{i,jk} - A_{i,kj} = \left[\begin{Bmatrix} \beta \\ ik \end{Bmatrix} \begin{Bmatrix} \alpha \\ \beta j \end{Bmatrix} - \begin{Bmatrix} \beta \\ ij \end{Bmatrix} \begin{Bmatrix} \alpha \\ \beta k \end{Bmatrix} + \frac{\partial}{\partial x^j} \begin{Bmatrix} \alpha \\ ik \end{Bmatrix} - \frac{\partial}{\partial x^k} \begin{Bmatrix} \alpha \\ ij \end{Bmatrix} \right] A_\alpha \tag{1.15.10}$$

As previously indicated, covariant differentiation of a tensor produces a tensor, and the sum or difference of two tensors is also a tensor. Hence

$$A_{i,jk} - A_{i,kj}$$

is a tensor. Moreover, this tensor is a covariant tensor of rank three. And since A_α is an arbitrary covariant tensor of rank one, its coefficient, namely, the quantity in square brackets, must also be a tensor. Indeed, this quantity must be a mixed tensor of rank four, since on inner multiplication by A_α a covariant tensor of rank three is obtained as follows:

$$R_{ijk}^\alpha A_\alpha = R_{ijk} \tag{1.15.11}$$

where

$$R_{ijk}^\alpha = \begin{Bmatrix} \beta \\ ik \end{Bmatrix} \begin{Bmatrix} \alpha \\ \beta j \end{Bmatrix} - \begin{Bmatrix} \beta \\ ij \end{Bmatrix} \begin{Bmatrix} \alpha \\ \beta k \end{Bmatrix} + \frac{\partial}{\partial x^j} \begin{Bmatrix} \alpha \\ ik \end{Bmatrix} - \frac{\partial}{\partial x^k} \begin{Bmatrix} \alpha \\ ij \end{Bmatrix} \tag{1.15.12}$$

This is the Riemann Christoffel tensor. It may be expressed in determinant form as follows (ref. 14):

$$R^{\alpha}_{ijk} = \begin{vmatrix} \begin{Bmatrix} \alpha \\ \beta j \end{Bmatrix} & \begin{Bmatrix} \alpha \\ \beta k \end{Bmatrix} \\ \begin{Bmatrix} \beta \\ ij \end{Bmatrix} & \begin{Bmatrix} \beta \\ ik \end{Bmatrix} \end{vmatrix} + \begin{vmatrix} \dfrac{\partial}{\partial x^j} & \dfrac{\partial}{\partial x^k} \\ \begin{Bmatrix} \alpha \\ ij \end{Bmatrix} & \begin{Bmatrix} \alpha \\ ik \end{Bmatrix} \end{vmatrix} \qquad (1.15.13)$$

Hence, if the order of covariant differentiation is to be immaterial this tensor must vanish. Therefore, a necessary and sufficient condition for the validity of inversion of the order of covariant differentiation is that (ref. 1)

$$R^{\alpha}_{ijk} \equiv 0 \qquad (1.15.14)$$

The number of components in this tensor of rank four depends on the dimensionality of the space. In a space of two dimensions the number of components will be 16; similarly, in a space of three dimensions it would have 81 components. In relativistic studies where a four-dimensional space time continuum is required, the number of components would be 256, and so on.

The definition of the Christoffel symbol of the second kind is

$$\begin{Bmatrix} k \\ ij \end{Bmatrix} = g^{k\alpha}[ij,\alpha]$$

where

$$[ij,\alpha] = \frac{1}{2}\left(\frac{\partial g_{i\alpha}}{\partial x^j} + \frac{\partial g_{j\alpha}}{\partial x^i} - \frac{\partial g_{ij}}{\partial x^\alpha} \right)$$

If the Christoffel symbols appearing in the Riemann-Christoffel tensor are replaced by these quantities, the result is an expression containing first and second partial derivatives of the g_{ij}. Moreover, the g_{ij} are themselves the coefficients in the fundamental quadratic form

$$ds^2 = g_{ij}dx^i dx^j \qquad (1.15.15)$$

where ds is the distance between two neighboring points with coordinates x^i and $x^i + dx^i$, and the g_{ij} are functions of x^i.

1.16 THE RICCI TENSOR

The Ricci tensor, which will be required in a subsequent application, can be obtained by contracting the Riemann-Christoffel tensor. And since the process of contraction reduces the rank of the contracted tensor by two, the result will be a tensor of rank two. In other words, by replacing k by α in the Riemann-Christoffel tensor, the Ricci tensor is obtained as follows:

$$R_{ij} \equiv R_{ij\alpha}^{\alpha} \qquad (1.16.1)$$

Since α appears twice in the terms on the right, it must be summed in accordance with the summation convention. In a four-dimensional space, equation (1.16.1) represents only 16 equations. Thus

$$R_{ij} = R_{ij1}^{1} + R_{ij2}^{2} + R_{ij3}^{3} + R_{ij4}^{4}$$

Substitution of $k = \alpha$ in equation (1.15.12) gives

$$R_{ij} = R_{ij\alpha}^{\alpha} = \begin{Bmatrix} \beta \\ i\alpha \end{Bmatrix} \begin{Bmatrix} \alpha \\ \beta j \end{Bmatrix} - \begin{Bmatrix} \beta \\ ij \end{Bmatrix} \begin{Bmatrix} \alpha \\ \beta\alpha \end{Bmatrix} + \frac{\partial}{\partial x^{j}} \begin{Bmatrix} \alpha \\ i\alpha \end{Bmatrix} - \frac{\partial}{\partial x^{\alpha}} \begin{Bmatrix} \alpha \\ ij \end{Bmatrix} \qquad (1.16.2)$$

Therefore

$$R_{ij} = \begin{vmatrix} \begin{Bmatrix} \alpha \\ \beta j \end{Bmatrix} & \begin{Bmatrix} \alpha \\ \beta\alpha \end{Bmatrix} \\ \\ \begin{Bmatrix} \beta \\ ij \end{Bmatrix} & \begin{Bmatrix} \beta \\ i\alpha \end{Bmatrix} \end{vmatrix} + \begin{vmatrix} \dfrac{\partial}{\partial x^{j}} & \dfrac{\partial}{\partial x^{\alpha}} \\ \\ \begin{Bmatrix} \alpha \\ ij \end{Bmatrix} & \begin{Bmatrix} \alpha \\ i\alpha \end{Bmatrix} \end{vmatrix} \qquad (1.16.3)$$

It can be shown that

$$\frac{\partial}{\partial x^{i}} \log \sqrt{g} = \begin{Bmatrix} \alpha \\ i\alpha \end{Bmatrix}$$

Therefore, R_{ij} can be rewritten in the following form:

$$R_{ij} = \begin{Bmatrix} \beta \\ i\alpha \end{Bmatrix} \begin{Bmatrix} \alpha \\ \beta j \end{Bmatrix} - \begin{Bmatrix} \beta \\ ij \end{Bmatrix} \frac{\partial}{\partial x^{\beta}} \log \sqrt{g} + \frac{\partial^{2}}{\partial x^{j} \partial x^{i}} \log \sqrt{g} - \frac{\partial}{\partial x^{\alpha}} \begin{Bmatrix} \alpha \\ ij \end{Bmatrix} \qquad (1.16.4)$$

69

which, in view of the definition of the Christoffel symbols, represents an expression containing first and second partial derivatives of the g_{ij}. From inspection of this result it can be seen that R_{ij} is symmetric. Hence, the number of distinct components of R_{ij} is $(1/2)n(n+1)$. In a four-dimensional manifold $n = 4$, and in this case R_{ij} has 10 components. It may be noted that in space devoid of matter, the equation

$$R_{ij} = 0$$

is the relativistic analog of Laplace's equation

$$\nabla^2 \phi = 0$$

where ϕ is a gravitational potential function in the Newtonian theory of gravitation. This will be discussed in more detail in subsequent sections.

1.17 CURVATURE TENSOR

The curvature tensor R_{ijkl} is defined by

$$R_{ijkl} = g_{i\alpha} R^{\alpha}_{jkl} \tag{1.17.1}$$

Therefore

$$R_{ijkl} = g_{i\alpha} \left[\frac{\partial}{\partial x^k} \left\{ {\alpha \atop jl} \right\} - \frac{\partial}{\partial x^l} \left\{ {\alpha \atop jk} \right\} + \left\{ {\beta \atop jl} \right\} \left\{ {\alpha \atop \beta k} \right\} - \left\{ {\beta \atop jk} \right\} \left\{ {\alpha \atop \beta l} \right\} \right] \tag{1.17.2}$$

This may be rewritten as follows:

$$R_{ijkl} = \frac{\partial}{\partial x^k} g_{i\alpha} \left\{ {\alpha \atop jl} \right\} - \frac{\partial g_{i\alpha}}{\partial x^k} \left\{ {\alpha \atop jl} \right\} - \frac{\partial}{\partial x^l} g_{i\alpha} \left\{ {\alpha \atop jk} \right\} + \frac{\partial g_{i\alpha}}{\partial x^l} \left\{ {\alpha \atop jk} \right\} + \left\{ {\beta \atop jl} \right\} [\beta k, i]$$

$$- \left\{ {\beta \atop jk} \right\} [\beta l, i] \tag{1.17.3}$$

It is easy to show that

$$\frac{\partial g_{ij}}{\partial x^k} = [ik,j] + [jk,i] \tag{1.17.4}$$

The Christoffel symbols of the first kind satisfy the following equations:

$$[ik,j] = \frac{1}{2}\left(\frac{\partial g_{ij}}{\partial x^k} + \frac{\partial g_{kj}}{\partial x^i} - \frac{\partial g_{ik}}{\partial x^j}\right)$$

$$[jk,i] = \frac{1}{2}\left(\frac{\partial g_{ji}}{\partial x^k} + \frac{\partial g_{ki}}{\partial x^j} - \frac{\partial g_{jk}}{\partial x^i}\right)$$

Therefore

$$[ik,j] + [jk,i] = \frac{\partial g_{ij}}{\partial x^k}$$

as required. Substitution of this result in equation (1.17.3) yields the following:

$$R_{ijkl} = \frac{\partial}{\partial x^k}[jl,i] - \frac{\partial}{\partial x^l}[jk,i] + \left\{{\alpha \atop jk}\right\}[il,\alpha] - \left\{{\alpha \atop jl}\right\}[ik,\alpha] \qquad (1.17.5)$$

This tensor can be written in determinantal form as follows:

$$R_{ijkl} = \begin{vmatrix} \dfrac{\partial}{\partial x^k} & \dfrac{\partial}{\partial x^l} \\[2ex] [jk,i] & [jl,i] \end{vmatrix} + \begin{vmatrix} \left\{{\alpha \atop jk}\right\} & \left\{{\alpha \atop jl}\right\} \\[2ex] [ik,\alpha] & [il,\alpha] \end{vmatrix} \qquad (1.17.6)$$

In order to determine the properties of the set of Riemann-Christoffel tensors R_{ijkl} the defined values of the Christoffel symbols are substituted in equation (1.17.5). When this is done, the following result is obtained:

$$R_{ijkl} = \frac{1}{2}\left(\frac{\partial^2 g_{il}}{\partial x^j \partial x^k} - \frac{\partial^2 g_{jl}}{\partial x^i \partial x^k} - \frac{\partial^2 g_{ik}}{\partial x^j \partial x^l} + \frac{\partial^2 g_{jk}}{\partial x^i \partial x^l}\right) + g^{\alpha\beta}([jk,\beta][il,\alpha] - [jl,\beta][ik,\alpha])$$

$$(1.17.7)$$

From this formula it follows that

$$R_{ijkl} = -R_{ijlk} \qquad (1.17.8)$$

$$R_{ijkl} = -R_{jikl} \qquad (1.17.9)$$

$$R_{ijkl} = R_{klij} \qquad\qquad (1.17.10)$$

and

$$R_{ijkl} + R_{iklj} + R_{iljk} = 0 \qquad\qquad (1.17.11)$$

Moreover, from the defining formula for R^i_{jkl}, it is seen that this mixed tensor is skew-symmetric with respect to the indices k and l. Therefore

$$R^i_{jkl} = -R^i_{jlk} \qquad\qquad (1.17.12)$$

1.18 EINSTEIN TENSOR

The following identity is due to Bianchi (ref. 15):

$$R_{ijkl,m} + R_{ijlm,k} + R_{ijmk,l} = 0 \qquad\qquad (1.18.1)$$

If (1.18.1) is multiplied by $g^{il}g^{jk}$ and the skew-symmetric properties of the curvature tensor are used, one obtains

$$g^{il}g^{jk}(R_{ijkl,m} + R_{ijlm,k} + R_{ijmk,l}) = 0$$

or

$$g^{jk}R^l_{jkl,m} - g^{jk}R^l_{jml,k} - g^{il}R^k_{imk,l} = 0 \qquad\qquad (1.18.2)$$

It has been shown that the contracted form of the Riemann-Christoffel tensor defines the Ricci tensor, that is

$$R^\alpha_{ij\alpha} = R_{ij} \qquad\qquad (1.18.3)$$

Hence

$$g^{jk}R_{jk,m} - g^{jk}R_{jm,k} - g^{il}R_{im,l} = 0$$

and therefore

$$R_{,m} - R^k_{m,k} - R^l_{m,l} = 0 \tag{1.18.4}$$

where

$$R = g^{jk}R_{jk} \tag{1.18.5}$$

Since k and l are dummy indices, equation (1.18.4) may be rewritten in the following form:

$$R_{,m} - 2R^k_{m,k} = 0 \tag{1.18.6}$$

or in the alternative form

$$\left(R^k_m - \frac{1}{2} \, \delta^k_m R \right)_{,k} = 0 \tag{1.18.7}$$

where

$$R^k_m = g^{jk}R_{jm} \tag{1.18.8}$$

The tensor

$$R_j{}^i - \frac{1}{2} \delta_j{}^i R = G_j{}^i \tag{1.18.9}$$

is the Einstein tensor, which plays an important role in the theory of relativity, and will be required in chapter 5.

1.19 REFERENCES

1. Sokolnikoff, Ivan S.: Tensor Analysis; Theory and Applications. John Wiley & Sons, Inc., 1960.

2. Wills, Albert P.: Vector Analysis, With an Introduction to Tensor Analysis. Dover Publications, Inc., 1958.

3. McConnell, Albert J.: Application of Tensor Analysis. Dover Publications, Inc., 1957.

4. Mathlab Group, Project MAC, M.I.T.: MACSYMA Reference Manual. Massachusetts Institute of Technology, 1957.

5. Spain, Barry: Tensor Calculus. Third ed., revised. Oliver and Boyd, Edinburgh, 1960.

6. Rutherford, Daniel E.: Vector Methods, Applied to Differential Geometry, Mechanics and Potential Theory. Sixth ed. Oliver and Boyd, Edinburgh, 1949.

7. Leipholz, Horst: Theory of Elasticity. Noordhoff International Publishing, Leyden, 1974.

8. Little, Robert W.: Elasticity. Prentice-Hall, 1973.

9. Margenau, Henry; and Murphy, George M.: The Mathematics of Physics and Chemistry. D. Van Nostrand Company, Inc., 1964.

10. Brillouin, Leon: Tensors in Mechanics and Elasticity. Academic Press, 1964.

11. Lieber, L. R.; and Lieber, H. G.: The Einstein Theory of Relativity. Rinehart and Company, Inc., 1945.

12. Lichnerowicz, A.; Leech, J. W.; and Newman, D. J.: Elements of Tensor Calculus. Methuen and Company, Ltd., 1962.

13. Howard, James C.: Application of Computers to the Formulation of Problems in Curvilinear Coordinate Systems. NASA TN D-3939, 1967.

14. Abram, J.: Tensor Calculus Through Differential Geometry. Butterworth and Company, 1965.

15. Rainich, G. Y.: Mathematics of Relativity. John Wiley & Sons, Inc., 1958.

AERONAUTICAL APPLICATIONS

2.1 TRANSFORMATIONS AND TRANSFORMATION LAWS

Many problems in engineering and physics involve the formulation of complex models of physical systems or processes, and the manipulation of sets of differential equations. In this and later chapters it will be seen that the formulation of mathematical models can be reduced to a series of routine operations, which can be performed without reference to the physics of the problem. Moreover, it will be demonstrated that the operations are purely mechanical and consist only of differentiation and summation.

The feasibility of applying this technique to the problem of deriving the equations of motion of a particle in any curvilinear coordinate system is demonstrated in chapter 3. In chapter 4 it is shown that the same method may be used to derive the Navier-Stokes equations of fluid motion and the corresponding continuity equation.

In order to reduce the equations of mathematical physics to a form which is amenable to routine operations of this type, it is desirable that the form chosen be invariant with respect to coordinate transformations. It has already been noted that a tensor formulation meets this requirement. When conventional methods are employed, the form which the equations assume depends on the coordinate system used to describe the problem. This dependence, which is due to the practice of expressing vectors in terms of their physical components, can be removed by the simple expedient of expressing all vectors in terms of their tensor components. These are related to the physical components by a simple transformation.

This chapter describes how the technique may be used to assist in the formulation of mathematical models of aeronautical systems. These models, which are frequently required for simulation and other purposes involve at least 12 equations: 3 force equations; 3 moment equations; 3 Euler angle equations to determine the spatial orientation of the body, and 3 equations to determine the location of the body in inertial space. An important aspect of the formulation of mathematical models of aeronautical systems is the specification of the system of forces and moments. In aeronautical situations the thrust and gravity forces can be formulated without difficulty, but the aerodynamic forces and moments require more detailed consideration. These are represented by the static forces and moments and the aerodynamic stability derivatives. For reasons which will become apparent as we proceed, these forces, moments, and derivatives have to be transformed from wind or wind-tunnel stability axes to aircraft body axes before the formulation can proceed.

For the benefit of those who have access to a digital computer equipped with a formula manipulation compiler, simple programs will be described which will facilitate the mechanization of these operations. This kind of computer operation is usually referred to as symbolic mathematical computation or symbolic and algebraic manipulation. The advantages of symbolic mathematical computation are most evident in the formulation of models analogous to those described in chapter 5; that is, the use of the method to derive cosmological models and associated trajectories. The field equations that govern the trajectories of bodies in cosmological space consist of 10 nonlinear partial differential equations for the 10 unknown potential functions. Each of these equations has a large number of terms, with each term a complicated mathematical expression. The formulation of these terms, and the derivation of the equations of the geodesics that describe the trajectories of bodies in the space defined by the postulated metric, require a substantial amount of algebraic manipulation and symbolic differentiation. Because of the compact nature of the tensor expressions, and the facility with which symbolic differentiation can be exploited by a simple computational algorithm, computation time is reduced and the errors to which human operators are prone can be avoided. Moreover, the computerized method enables the researcher to examine a greater number of possibilities and to explore cosmological situations that might otherwise be avoided if the time and labor involved were excessive. This is not to say that noncosmological models cannot be formulated with equal facility by a human operator. Bearing in mind that the only operations involved are differentiation and summation, it may, in fact, be more economical to formulate the majority of models manually.

2.2 AERONAUTICAL REFERENCE SYSTEMS

There are many coordinate systems in use in aeronautical research (ref. 1). Aerodynamic data obtained from wind-tunnel experiments may be referred to wind axes or to wind-tunnel stability axes. When the wind axes are used, the x-axis is aligned with the relative wind at all times. The wind-tunnel stability axes are the system about which most wind-tunnel data are obtained. For this system, the x-axis is in the same horizontal plane as the relative wind at all times. In addition to the wind axes and the wind-tunnel stability axes, there are other systems of axes fixed in the body and moving with the body. These are referred to as body axes. In aerospace applications, a body axis system has the x-axis fixed along the longitudinal center line of the body, the y-axis normal to the plane of symmetry, and the z-axis in the plane of symmetry. It should also be noted that when an aircraft is in horizontal flight, the z-axis points downward in the direction of the gravity vector, the x-axis points forward, and the y-axis points to the right (fig. 2.2.1).

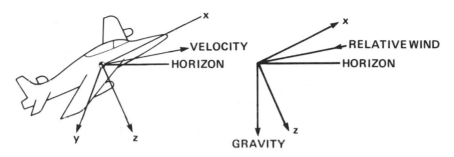

Figure 2.2.1.– Aeronautical reference systems.

The equations of motion of aerospace vehicles are formulated with respect to body axes. The main advantage of these axes in motion calculations is that vehicle moments and products of inertia about the axes are constants. When the body axes are chosen so that the products of inertia vanish, they are known as principal axes. A system of axes which is frequently used to study the stability of aircraft in the presence of disturbing forces that produce small perturbations is the flight stability axes. This is an orthogonal system fixed to the vehicle, the x-axis of which is aligned with the relative wind vector, when the vehicle is in a steady-state condition, but then rotates with the vehicle after a disturbance as the vehicle changes angle of attack and sideslip.

Although the equations of motion of aerospace vehicles are referred to body axes, the aerodynamic forces, moments, and stability derivatives are usually referred to wind axes or to wind-tunnel stability axes. Hence, before the equations of motion can be formulated with respect to body axes, the aerodynamic forces, moments, and derivatives must be transformed from the wind or wind-tunnel stability axes to the appropriate body axes (ref. 2).

2.3 TRANSFORMATION LAW FOR STATIC FORCES AND MOMENTS

The static aerodynamic forces and moments transform like the components of a contravariant vector; that is, if X_i denotes the aerodynamic force in the x frame of reference and Y^i denotes the corresponding transformed force in the y system of coordinates, then

$$Y^i = \frac{\partial y^i}{\partial x^n} X^n \tag{2.3.1}$$

where

$$y = y(x)$$

In this equation, X^j may denote either a force or a moment. For example

$$\left.\begin{array}{l} X^1 \quad \text{may be either } X \text{ or } l \\[2mm] X^2 \quad \text{may be either } Y \text{ or } m \\[2mm] X^3 \quad \text{may be either } Z \text{ or } n \end{array}\right\} \tag{2.3.2}$$

where X, Y, Z and l, m, n are aerodynamic forces and moments, respectively; that is, each of these pairs obeys the same transformation law in going from the x-coordinate frame to the y-coordinate frame.

In aeronautical language, the aerodynamic forces acting on a body that is moving through the atmosphere are defined in terms of the force coefficients as follows:

$$X = \bar{q} S C_x \quad \text{in the } x \text{ direction}$$

$$Y = \bar{q} S C_y \quad \text{in the } y \text{ direction}$$

$$Z = \bar{q} S C_z \quad \text{in the } z \text{ direction}$$

where $\bar{q} = 1/2(\rho V^2)$ is the dynamic pressure, when ρ and v are the density and velocity, respectively; and S is a reference area.

Likewise, the aerodynamic moments about the axes are defined in terms of the moment coefficients as follows:

$$l = \bar{q}SbC_l \quad \text{about the } x \text{ axis}$$

$$m = \bar{q}ScC_m \quad \text{about the } y \text{ axis}$$

$$n = \bar{q}SbC_n \quad \text{about the } z \text{ axis}$$

where b denotes wing span and c denotes wing chord length.

The body axes are related to the wind-tunnel axes as shown in figure 2.3.1. To bring a reference frame from the wind axes into coincidence with the body axes involves a negative rotation (B) about the z axis followed by a positive rotation (A) about the y axis.

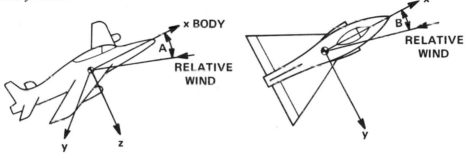

Figure 2.3.1.— Angle of attack A and angle of sideslip B.

Instead of using (xyz), let $(y^1y^2y^3)$ denote the body axes, and $(x^1x^2x^3)$ the wind-tunnel axes. With this notation, the coordinate transformation is given by

$$\begin{pmatrix} y^1 \\ y^2 \\ y^3 \end{pmatrix} = \begin{pmatrix} \cos A & 0 & -\sin A \\ 0 & 1 & 0 \\ \sin A & 0 & \cos A \end{pmatrix} \begin{pmatrix} \cos B & -\sin B & 0 \\ \sin B & \cos B & 0 \\ 0 & 0 & 1 \end{pmatrix} \begin{pmatrix} x^1 \\ x^2 \\ x^3 \end{pmatrix}$$

$$\left.\begin{array}{l} y^1 = x^1 \cos A \cos B - x^2 \cos A \sin B - x^3 \sin A \\[6pt] y^2 = x^1 \sin B + x^2 \cos B \\[6pt] y^3 = x^1 \sin A \cos B - x^2 \sin A \sin B + x^3 \cos A \end{array}\right\} \quad (2.3.3)$$

Since this transformation is orthogonal, the inverse transformation is simply the transpose of the preceding matrix, and can be written by inspection as follows:

$$\left.\begin{array}{l} x^1 = y^1 \cos A \cos B + y^2 \sin B + y^3 \sin A \cos B \\[2mm] x^2 = -y^1 \cos A \sin B + y^2 \cos B - y^3 \sin A \sin B \\[2mm] x^3 = -y^1 \sin A + y^3 \cos A \end{array}\right\} \qquad (2.3.4)$$

Given the transformation (2.3.3), equation (2.3.1) may be expanded as follows:

$$\left.\begin{array}{l} Y^1 = X^1 \cos A \cos B - X^2 \cos A \sin B - X^3 \sin A \\[2mm] Y^2 = X^1 \sin B + X^2 \cos B \\[2mm] Y^3 = X^1 \sin A \cos B - X^2 \sin A \sin B + X^3 \cos A \end{array}\right\} \qquad (2.3.5)$$

where

$$\begin{array}{ll} Y^1 = X' & X^1 = X \\[2mm] Y^2 = Y' & X^2 = Y \\[2mm] Y^3 = Z' & X^3 = Z \end{array}$$

Therefore,

$$\left.\begin{array}{l} X' = X \cos A \cos B - Y \cos A \sin B - Z \sin A \\[2mm] Y' = X \sin B + Y \cos B \\[2mm] Z' = X \sin A \cos B - Y \sin A \sin B + Z \cos A \end{array}\right\} \qquad (2.3.6)$$

In view of the relationships indicated in equation (2.3.2), the static moment coefficients obey the same transformation law. Hence

$$\left.\begin{array}{l} l' = l \cos A \cos B - m \cos A \sin B - n \sin A \\[2mm] m' = l \sin B + m \cos B \\[2mm] n' = l \sin A \cos B - m \sin A \sin B + n \cos A \end{array}\right\} \qquad (2.3.7)$$

In these transformations the primed quantities refer to body axes components, and unprimed quantities to the wind-tunnel axes components.

2.4 TRANSFORMATION LAW FOR THE STATIC-STABILITY DERIVATIVES

The static stability derivatives obey the same transformation law as the force and moment coefficients. From equation (2.3.1)

$$Y_A{}^i = \frac{\partial Y^i}{\partial A} = \frac{\partial}{\partial A}\left(\frac{\partial y^i}{\partial x^n}\right) X^n$$

Therefore,

$$Y_A{}^i = \frac{\partial y^i}{\partial x^n} X_A{}^n + X^n \frac{\partial}{\partial A}\left(\frac{\partial y^i}{\partial x^n}\right)$$

However, in a static situation

$$\frac{\partial y^i}{\partial x^n} - \text{constant}$$

Therefore,

$$Y_A{}^i = \frac{\partial y^i}{\partial x^n} X_A{}^n \qquad (2.4.1)$$

Likewise

$$Y_B{}^i = \frac{\partial y^i}{\partial x^n} X_B{}^n \qquad (2.4.2)$$

where

$$Y_A{}^1 = X_A{}', l_A{}' \quad ; \qquad X_A{}^1 = X_A, l_A$$

$$Y_A{}^2 = Y_A{}', m_A{}' \quad ; \qquad X_A{}^2 = Y_A, m_A$$

$$Y_A{}^3 = Z_A{}', n_A{}' \quad ; \qquad X_A{}^3 = Z_A, n_A$$

Therefore,

$$\left.\begin{array}{l} X_A{}' = X_A \cos A \cos B - Y_A \cos A \sin B - Z_A \sin A \\[2mm] Y_A{}' = X_A \sin B + Y_A \cos B \\[2mm] Z_A{}' = X_A \sin A \cos B - Y_A \sin A \sin B + Z_A \cos A \end{array}\right\} \qquad (2.4.3)$$

and

$$\left.\begin{array}{l} l_A{}' = l_A \cos A \cos B - m_A \cos A \sin B - n_A \sin A \\[2mm] m_A{}' = l_A \sin B + m_A \cos B \\[2mm] n_A{}' = l_A \sin A \cos B - m_A \sin A \sin B + n_A \cos A \end{array}\right\} \qquad (2.4.4)$$

Similarly

$$\left.\begin{array}{l} l_B{}' = l_B \cos A \cos B - m_B \cos A \sin B - n_B \sin A \\[2mm] m_B{}' = l_B \sin B + m_B \cos B \\[2mm] n_B{}' = l_B \sin A \cos B - m_B \sin A \sin B + n_B \cos A \end{array}\right\} \qquad (2.4.5)$$

Likewise

$$\left.\begin{array}{l} X_B{}' = X_B \cos A \cos B - Y_B \cos A \sin B - Z_B \sin A \\[2mm] Y_B{}' = X_B \sin B + Y_B \cos B \\[2mm] Z_B{}' = X_B \sin A \cos B - Y_B \sin A \sin B + Z_B \cos A \end{array}\right\} \qquad (2.4.6)$$

2.5 INVERSE TRANSFORMATION LAW FOR STATIC FORCES AND MOMENTS

The equations in the preceding section transform the forces and moments and the static stability derivatives from wind-tunnel axes to body axes. As already indicated, it is frequently necessary to convert these measured quantities from body axes to wind-tunnel axes. In this case the inverse transformation is required.

A useful property of tensor transformations may be stated as follows: Let the components of a mixed tensor in the x-coordinate system be denoted by

$$X^{j_1, \ldots, j_s}_{i_1, \ldots, i_r}$$

and its components in the y system by

$$Y^{j_1, \ldots, j_s}_{i_1, \ldots, i_r}$$

Then from the law of transformation of mixed tensors

$$Y^{j_1, \ldots, j_s}_{i_1, \ldots, i_r} = \frac{\partial y^{j_1}}{\partial x^{\beta_1}} \cdots \frac{\partial y^{j_s}}{\partial x^{\beta_s}} \frac{\partial x^{\alpha_1}}{\partial y^{i_1}} \cdots \frac{\partial x^{\alpha_r}}{\partial y^{i_r}} X^{\beta_1, \ldots, \beta_s}_{\alpha_1, \ldots, \alpha_r} \qquad (2.5.1)$$

On the other hand, the components

$$X^{\beta_1, \ldots, \beta_s}_{\alpha_1, \ldots, \alpha_r}$$

of the same tensor in the x frame of reference are given by the formula

$$X^{\beta_1, \ldots, \beta_s}_{\alpha_1, \ldots, \alpha_r} = \frac{\partial y^{i_1}}{\partial x^{\alpha_1}} \cdots \frac{\partial y^{i_r}}{\partial x^{\alpha_r}} \frac{\partial x^{\beta_1}}{\partial y^{j_1}} \cdots \frac{\partial x^{\beta_s}}{\partial y^{j_s}} Y^{j_1, \ldots, j_s}_{i_1, \ldots, i_r} \qquad (2.5.2)$$

It should be noted that (2.5.2) is obtained from (2.5.1) by treating the partial derivatives in (2.5.1) as though they were fractions and products appearing in simple algebraic expressions. Hence, by using this very useful property of tensor equations, the inverse transformation can be obtained from equation (2.3.1) as follows:

$$X^n = \frac{\partial x^n}{\partial y^i} Y^i \qquad (2.5.3)$$

On using the equations (2.3.4), the transformation (2.5.3) assumes the following form for the forces

$$
\left.
\begin{aligned}
X^1 &= X = X' \cos A \cos B + Y' \sin B + Z' \sin A \cos B \\
X^2 &= Y = -X' \cos A \sin B + Y' \cos B - Z' \sin A \sin B \\
X^3 &= Z = -X' \sin A + Z' \cos A
\end{aligned}
\right\}
\qquad (2.5.4)
$$

Similarly, by using the relationships given in (2.3.2), the inverse transformation for the moment coefficients is

$$
\left.
\begin{aligned}
X^1 &= l = l' \cos A \cos B + m' \sin B + n' \sin A \cos B \\
X^2 &= m = -l' \cos A \sin B + m' \cos B - n' \sin A \sin B \\
X^3 &= n = -l' \sin A + n' \cos A
\end{aligned}
\right\}
\qquad (2.5.5)
$$

Likewise, the static stability derivatives are

$$
\left.
\begin{aligned}
X^1 &= X_A = X_A' \cos A \cos B + Y_A' \sin B + Z_A' \sin A \cos B \\
X^2 &= Y_A = -X_A' \cos A \sin B + Y_A' \cos B - Z_A' \sin A \sin B \\
X^3 &= Z_A = -X_A' \sin A + Z_A' \cos A
\end{aligned}
\right\}
\qquad (2.5.6)
$$

and

$$
\left.
\begin{aligned}
X^1 &= l_A = l_A' \cos A \cos B + m_A' \sin B + n_A' \sin A \cos B \\
X^2 &= m_A = -l_A' \cos A \sin B + m_A' \cos B - n_A' \sin A \sin B \\
X^3 &= n_A = -l_A' \sin A + n_A' \cos A
\end{aligned}
\right\}
\qquad (2.5.7)
$$

The static stability derivatives with respect to the angle B have exactly the same form. These are

$$
\left.
\begin{aligned}
X^1 &= X_B = X_B' \cos A \cos B + Y_B' \sin B + Z_B' \sin A \cos B \\
X^2 &= Y_B = -X_B' \cos A \sin B + Y_B' \cos B - Z_B' \sin A \sin B \\
X^3 &= Z_B = -X_B' \sin A + Z_B' \cos A
\end{aligned}
\right\}
\qquad (2.5.8)
$$

and

$$X^1 = l_B = l_B{'}\cos A \cos B + m_B{'}\sin B + n_B{'}\sin A \cos B$$

$$X^2 = m_B = -l_B{'}\cos A \sin B + m_B{'}\cos B - n_B{'}\sin A \sin B \qquad (2.5.9)$$

$$X^3 = n_B = -l_B{'}\sin A + n_B{'}\cos A$$

Although a computer would not be necessary to formulate the simple relationships described so far, a computer program will be described in the next section to prepare potential computer users for more complicated formulations.

2.6 TRANSFORMATION BY COMPUTER

For the benefit of those who are not familiar with computers or computer programming, it should be emphasized that a computer program is simply a list of instructions that a computer can accept and execute. There are a variety of computer languages that can be used to express the user's wishes in a form that is acceptable to a given computer. When it is required to use a computer to solve a particular problem, a program must be written in one of the languages that the computer will accept, instructing the computer what to do. When the instructions are written in the prescribed form, they are transferred to punched cards, before being presented to the computer, or typed at a terminal which is connected to the computer. If the instructions are coded correctly, the problem solution will be printed out on paper tape, or presented in some other form specified by the user.

A simple program that can be used to transform the static force and moment coefficients and the corresponding derivatives from wind-tunnel axes to body axes will be described. The program uses the coordinate transformation equations (2.3.3) as input to permit expansion of equation (2.3.1).

If a coefficient or derivative in wind-tunnel axes be denoted by $C(I)$, and the corresponding transformed coefficients be denoted by $TC(I)$, where $I = 1,2,3$, then a suitable program will consist of only a few instructions.

Just as a human operator cannot expand equation (2.3.1) unless he knows the special form $y = y(x)$ given by equation (2.3.3), the computer must likewise be told what this relationship is. Therefore, the first statement in the program gives the computer this information. However, the information cannot be given in the form in which it is written in equation (2.3.3); it must be presented in a modified form that the computer will accept. If the information is not in the precise language that is acceptable to the computer, it will be rejected. In the case of a FORMAC program (ref. 3), the computer will accept the following input statement, where asterisks denote multiplication.

```
LET(Y(1)=X(1)*COS(A)*COS(B)-X(2)*COS(A)*SIN(B)-X(3)*SIN(A));

LET(Y(2)=X(1)*SIN(B)+X(2)*COS(B));

LET(Y(3)=X(1)*SIN(A)*COS(B)-X(2)*SIN(A)*SIN(B)+X(3)*COS(A));
```

In order to make certain that the transformation equations have been coded correctly, it is advisable to instruct the computer to print out the transformation equations before printing out the problem solution, that is, the expanded form of equation (2.3.1). In this way the user can make certain that equation (2.3.1) has been coded without error. With this objective in mind, the next program statement or instruction to the computer is

```
PRINT_OUT(Y(1);Y(2);Y(3));
```

It should be noted that these program statements must always be reproduced exactly. Semicolons cannot be replaced by commas. The position of semicolons must be strictly adhered to. If a program statement ends with a semicolon, it cannot be replaced by a period. Likewise, the number and positions of parentheses is invariable. Computers are usually quite inflexible in their insistence on precisely worded statements.

Having received the transformation equations and printed them out, the computer will behave in a very human fashion and stop working if it is not told what to do next. So the next program statement is an instruction to expand equation (2.3.1) in a manner that exploits the advantages of the summation convention. In this connection, it is perhaps worth repeating that the summation convention, which is a characteristic property of all tensor expressions, will be used repeatedly in the examples that follow.

An instruction to a computer to perform a series of operations in a repetitive manner takes the form of a "DO" statement; and the group of instructions involved in carrying out the repetitions constitute what programmers call a "DO" loop. The "DO" statement prescribes the range of the operation. For example, in expanding equation (2.3.1), the indices "i" and "n" each assume the values 1,2,3 in turn, and two "DO" statements appear in the program as follows:

```
DO I=I TO 3 BY 1;

DO N=I TO 3 BY 1;
```

The following statement is for initializing the program

```
LET(TC(I)=0);
```

A key statement, which is the target of the "DO" statements, instructs the computer how to expand equation (2.3.1). It is a statement that will appear frequently in subsequent applications; its implementation permits the computer to differentiate mathematical expressions symbolically. When expressed in computer language, equation (2.3.1) assumes the following form:

```
LET(TCC(I)=(DERIV(Y(I),X(N)))*(C(N)));
```

This statement is accompanied by the following supplementary statement which tells the computer to add the results of each step in the operation prior to incrementing the indices.

```
LET(TC(I)=TC(I)+TCC(I));
```

When the index "N" reaches the value 3, the computer is instructed to terminate that phase of the operation by the following statement:

```
END;
```

It is then told to print out the result by the statement,

```
PRINT_OUT(TC(I));
```

It then proceeds to increment the index "I" until all three equations have been formulated. At this point the computer encounters the final statement in the program. It is

```
END;
```

The reader may wonder why two "END" statements are necessary. The first "END" statement encountered terminates what programmers call the inner loop:

this part of the program manages the formulation of individual equations, and terminates the operation when the index "N" equals 3. The second "END" statement terminates the program when the index "I" equals 3, that is, when all equations have been formulated. When this occurs, the problem solution is printed out. The following list of input equations and formulated transformation equations are reproduced from the actual computer printout:

Y(1) = COS (B) COS(A) X(1) - COS (A) SIN (B) X(2) - SIN (A) X(3)

Y(2) = SIN (B) X(1) + COS (B) X(2)

Y(3) = COS (B) SIN (A) X(1) - SIN (B) SIN (A) X(2) + COS (A) X(3)

TC(1) = COS (B) COS (A) C(1) - COS (A) SIN (B) C(2) - SIN (A) C(3)

TC(2) = SIN (B) C(1) + COS (B) C(2)

TC(3) = COS (B) SIN(A) C(1) - SIN (B) SIN (A) C(2) + COS (A) C(3)

On substitution of the appropriate symbols in these output expressions, equations (2.3.6) and (2.3.7) and (2.4.3) through (2.4.6) are reproduced. For example, if the transformation equations for the static force coefficients are required, the following symbol substitutions should be made

$$TC(1) = X' \quad ; \quad C(1) = X$$

$$TC(2) = Y' \quad ; \quad C(2) = Y$$

$$TC(3) = Z' \quad ; \quad C(3) = Z$$

The appropriate symbol substitutions for the moment transformations are

$$TC(1) = l' \quad ; \quad C(1) = l$$

$$TC(2) = m' \quad ; \quad C(2) = m$$

$$TC(3) = n' \quad ; \quad C(3) = n$$

Static stability derivatives with respect to "A" are reproduced when the output symbols are assigned as follows:

$$TC(1) = l_A' \quad ; \quad C(1) = l_A$$

$$TC(2) = m_A' \quad ; \quad C(2) = m_A$$

$$TC(3) = n_A' \quad ; \quad C(3) = n_A$$

and similarly for the static stability derivatives with respect to "B."

The same program can be used to carry out the inverse transformations, provided the coordinate transformation equations (2.3.4) are used instead of equations (2.3.3), and the transformed coefficients $TC(I)$ obtained from equation (2.5.3) instead of equation (2.3.1).

When inverse transformations are required, the input statement and the statement controlling differentiation are modified to correspond to equations (2.3.4) and (2.5.3), respectively.

In this case, the input equations are

```
LET(X(1)-Y(1)*COS(A)*COS(B)+Y(2)*SIN(B)+Y(3)*SIN(A)COS(B));

LET(X(2)=-Y(1)*COS(A)*SIN(B)+Y(2)*COS(B)-Y(3)*SIN(A)*SIN(B));

LET(X(3)=-Y(1)*SIN(A)+Y(3)*COS(A));
```

and the inverse transformation equation is

```
LET(TCC(M)=(DERIV(X(M),Y(I)))*(C(M)));
```

When these modified statements are substituted in the preceding program, the following output is obtained:

```
X(1) = COS(B) COS(A) Y(1) + SIN(B) Y(2) + COS(B) SIN(A) Y(3)

X(2) = -COS(A) SIN(B) Y(1) + COS(B) Y(2) - SIN(B) SIN(A) Y(3)

X(3) = -SIN(A) Y(1) + COS(A) Y(3)
```

$$TC(1) = COS(B) \; COS(A) \; C(1) + SIN(B) \; C(2) + COS(B) \; SIN(A) \; C(3)$$

$$TC(2) = -COS(A) \; SIN(B) \; C(1) + COS(B) \; C(2) - SIN(B) \; SIN(A) \; C(3)$$

$$TC(3) = -SIN(A) \; C(1) + COS(A) \; C(3)$$

In these output expressions, a coefficient or stability derivative in the body axis system is denoted by $C(M)$, and the corresponding transformed coefficient in wind-tunnel axes is denoted by $TC(M)$, where $M = 1.2.3$. For example, the symbol substitutions for the force coefficients are

$$TC(1) = X \quad ; \quad C(1) = X'$$

$$TC(2) = Y \quad ; \quad C(2) = Y'$$

$$TC(3) = Z \quad ; \quad C(3) = Z'$$

Likewise, the symbol substitutions for the moment coefficients are

$$TC(1) = l \quad ; \quad C(1) = l'$$

$$TC(2) = m \quad ; \quad C(2) = m'$$

$$TC(3) = n \quad ; \quad C(3) = n'$$

Static stability derivatives with respect to A and B are obtained in a similar fashion. When interpreted in this manner, the program output gives the inverse transformations for static force and moment coefficients and the static stability derivatives by reproducing equations (2.5.4) through (2.5.9).

2.7 TRANSFORMATION LAW FOR AERODYNAMIC STABILITY DERIVATIVES

A necessary preliminary to the formulation of mathematical models of aeronautical systems is the transformation of aerodynamic stability derivatives from wind or wind-tunnel stability axes to body axes. It will be seen that the aerodynamic stability derivatives transform like the components of a mixed tensor, having one index of covariance and one index of contravariance (ref. 4). Moreover, due to the equivalence of covariant and contravariant transformations in orthogonal Cartesian systems of coordinates, it will be seen that the transformations can be treated as doubly covariant or doubly contravariant if this simplifies the formulation.

The aerodynamic stability derivatives measure the rates of change of aerodynamic forces and moments with respect to motion vector components. In keeping with the usual practice in aerodynamic formulations, motion vector components will refer specifically to components of the linear velocity vector, components of the angular velocity vector, and components of the corresponding linear and angular acceleration vectors. The transformation law for these derivatives may be obtained as follows: let Y^i be a force or moment in the y system of axes, and let $U^j(y)$ be a motion vector component in this system of axes. Similarly, let X^α be a force or moment in the x system of axes, and let $U^\beta(x)$ be a motion vector component in this system of axes. Then the stability derivatives with respect to motion components, as measured in the y system of axes, are related to the corresponding derivatives in the x system of axes by the following equation:

$$\frac{\partial Y^i}{\partial U^j(y)} = \frac{\partial Y^i}{\partial X^\alpha} \frac{\partial X^\alpha}{\partial U^\beta(x)} \frac{\partial U^\beta(x)}{\partial U^j(y)} \tag{2.7.1}$$

It should be noted that force, moment, and motion components obey the same transformation law as the system coordinates, that is

$$y^i = \frac{\partial y^i}{\partial x^\alpha} x^\alpha$$

Hence

$$Y^i = \frac{\partial y^i}{\partial x^\alpha} X^\alpha \tag{2.7.2}$$

$$U^j(y) = \frac{\partial y^j}{\partial x^\beta} U^\beta(x) \tag{2.7.3}$$

Therefore, from equations (2.5.2) and (2.7.3)

$$U^\beta(x) = \frac{\partial x^\beta}{\partial y^j} U^j(y) \tag{2.7.4}$$

Substitution from (2.7.2) and (2.7.4) in (2.7.1) gives

$$\frac{\partial Y^i}{\partial U^j(y)} = \left(\frac{\partial y^i}{\partial x^\alpha} \frac{\partial x^\beta}{\partial y^j} \right) \frac{\partial X^\alpha}{\partial U^\beta(x)} \tag{2.7.5}$$

Equation (2.7.5) may be rewritten as follows: Let

$$\frac{\partial Y^i}{\partial U^j(y)} = Y_j{}^i$$

and let

$$\frac{\partial X^\alpha}{\partial U^\beta(x)} = X_\beta{}^\alpha$$

Therefore

$$Y_j{}^i = \frac{\partial y^i}{\partial x^\alpha} \frac{\partial x^\beta}{\partial y^j} X_\beta{}^\alpha \tag{2.7.6}$$

where the superscript denotes the component of the aerodynamic force or moment, and the subscript denotes the motion vector component with respect to which the derivative is obtained.

Equation (2.7.6) shows that the aerodynamic stability derivatives transform like the components of a mixed tensor, having one index of contravariance and one index of covariance. Being a tensor of rank two, equation (2.7.6) represents nine equations, with each equation having, in general, nine terms.

Note that once the tensor law (2.7.6) is established, the specialized form of the transformation equations can be obtained without further reference to the physics of the problem. Moreover, the derivations involved are purely mechanical operations and can be performed by anyone who can differentiate.

Since α and β can each assume the values 1,2,3, we have, by summing first on α

$$Y_j{}^i = \frac{\partial y^i}{\partial x^1} \frac{\partial x^\beta}{\partial y^j} X_\beta{}^1 + \frac{\partial y^i}{\partial x^2} \frac{\partial x^\beta}{\partial y^j} X_\beta{}^2 + \frac{\partial y^i}{\partial x^3} \frac{\partial x^\beta}{\partial y^j} X_\beta{}^3$$

Summing next on β yields the following nine terms:

$$Y_j{}^i = \frac{\partial y^i}{\partial x^1} \frac{\partial x^1}{\partial y^j} X_1{}^1 + \frac{\partial y^i}{\partial x^1} \frac{\partial x^2}{\partial y^j} X_2{}^1 + \frac{\partial y^i}{\partial x^1} \frac{\partial x^3}{\partial y^j} X_3{}^1$$

$$+ \frac{\partial y^i}{\partial x^2} \frac{\partial x^1}{\partial y^j} X_1{}^2 + \frac{\partial y^i}{\partial x^2} \frac{\partial x^2}{\partial y^j} X_2{}^2 + \frac{\partial y^i}{\partial x^2} \frac{\partial x^3}{\partial y^j} X_3{}^2$$

$$+ \frac{\partial y^i}{\partial x^3} \frac{\partial x^1}{\partial y^j} X_1{}^3 + \frac{\partial y^i}{\partial x^3} \frac{\partial x^2}{\partial y^j} X_2{}^3 + \frac{\partial y^i}{\partial x^3} \frac{\partial x^3}{\partial y^j} X_3{}^3 \qquad (2.7.7)$$

Note that while the indices i and j retain the same values throughout, α and β each assume the values 1,2,3 in turn. Since i,j = 1,2,3, there are nine tensor components. These are

$$Y_1{}^1 = \frac{\partial y^1}{\partial x^\alpha} \frac{\partial x^\beta}{\partial y^1} X_\beta{}^\alpha \quad ; \qquad Y_2{}^1 = \frac{\partial y^1}{\partial x^\alpha} \frac{\partial x^\beta}{\partial y^2} X_\beta{}^\alpha$$

$$Y_3{}^1 = \frac{\partial y^1}{\partial x^\alpha} \frac{\partial x^\beta}{\partial y^3} X_\beta{}^\alpha \quad ; \qquad Y_1{}^2 = \frac{\partial y^2}{\partial x^\alpha} \frac{\partial x^\beta}{\partial y^1} X_\beta{}^\alpha$$

$$Y_2{}^2 = \frac{\partial y^2}{\partial x^\alpha} \frac{\partial x^\beta}{\partial y^2} X_\beta{}^\alpha \quad ; \qquad Y_3{}^2 = \frac{\partial y^2}{\partial x^\alpha} \frac{\partial x^\beta}{\partial y^3} X_\beta{}^\alpha$$

$$Y_1{}^3 = \frac{\partial y^3}{\partial x^\alpha} \frac{\partial x^\beta}{\partial y^1} X_\beta{}^\alpha \quad ; \qquad Y_2{}^3 = \frac{\partial y^3}{\partial x^\alpha} \frac{\partial x^\beta}{\partial y^2} X_\beta{}^\alpha$$

and

$$Y_3{}^3 = \frac{\partial y^3}{\partial x^\alpha} \frac{\partial x^\beta}{\partial y^3} X_\beta{}^\alpha$$

Moreover, for α,β = 1,2,3, it has been shown that each of these transformation equations has, in general, nine terms. Hence, there are, in general, a total of 81 terms to be formulated, using the transformation equations (2.3.3) and (2.3.4). The coefficient of each wind axes component consists of the product of two partial differential coefficients. From equation (2.3.3) the partial differential coefficients of y with respect to x are

$$\frac{\partial y^1}{\partial x^1} = \cos A \, \cos B \quad ; \qquad \frac{\partial y^1}{\partial x^2} = -\cos A \, \sin B$$

$$\frac{\partial y^1}{\partial x^3} = -\sin A \qquad ; \qquad \frac{\partial y^2}{\partial x^1} = \sin B$$

$$\frac{\partial y^2}{\partial x^2} = \cos B \qquad ; \qquad \frac{\partial y^2}{\partial x^3} = 0 \qquad\qquad (2.7.8)$$

$$\frac{\partial y^3}{\partial x^1} = \sin A \, \cos B \quad ; \qquad \frac{\partial y^3}{\partial x^2} = -\sin A \, \sin B$$

$$\frac{\partial y^3}{\partial x^3} = \cos A$$

From equations (2.3.4), the partial differential coefficients of x with respect to y are

$$\frac{\partial x^1}{\partial y^1} = \cos A \, \cos B \quad ; \qquad \frac{\partial x^1}{\partial y^2} = \sin B$$

$$\frac{\partial x^1}{\partial y^3} = \sin A \, \cos B \quad ; \qquad \frac{\partial x^2}{\partial y^1} = -\cos A \, \sin B$$

$$\frac{\partial x^2}{\partial y^2} = \cos B \qquad ; \qquad \frac{\partial x^2}{\partial y^3} = -\sin A \, \sin B \qquad (2.7.9)$$

$$\frac{\partial x^3}{\partial y^1} = -\sin A \qquad ; \qquad \frac{\partial x^3}{\partial y^2} = 0$$

$$\frac{\partial x^3}{\partial y^3} = \cos A$$

In terms of these partial differential coefficients, each tensor component Y_j^i can be formulated in accordance with the transformation law (2.7.6). From equations (2.7.7) through (2.7.9), the individual components are

$$Y_1^1 = \frac{\partial y^1}{\partial x^1}\frac{\partial x^1}{\partial y^1}X_1^1 + \frac{\partial y^1}{\partial x^1}\frac{\partial x^2}{\partial y^1}X_2^1 + \frac{\partial y^1}{\partial x^1}\frac{\partial x^3}{\partial y^1}X_3^1$$

$$+ \frac{\partial y^1}{\partial x^2}\frac{\partial x^1}{\partial y^1}X_1^2 + \frac{\partial y^1}{\partial x^2}\frac{\partial x^2}{\partial y^1}X_2^2 + \frac{\partial y^1}{\partial x^2}\frac{\partial x^3}{\partial y^1}X_3^2$$

$$+ \frac{\partial y^1}{\partial x^3}\frac{\partial x^1}{\partial y^1}X_1^3 + \frac{\partial y^1}{\partial x^3}\frac{\partial x^2}{\partial y^1}X_2^3 + \frac{\partial y^1}{\partial x^3}\frac{\partial x^3}{\partial y^1}X_3^3$$

$$Y_1^1 = (\cos^2 A \cos^2 B)X_1^1 - (\cos^2 A \cos B \sin B)X_2^1$$

$$- (\sin A \cos A \cos B)X_3^1 - (\cos^2 A \cos B \sin B)X_1^2$$

$$+ (\cos^2 A \sin^2 B)X_2^2 + (\sin A \cos A \sin B)X_3^2$$

$$- (\sin A \cos A \cos B)X_1^3 + (\sin A \cos A \sin B)X_2^3 + (\sin^2 A)X_3^3$$

$$Y_2^1 = \frac{\partial y^1}{\partial x^1}\frac{\partial x^1}{\partial y^2}X_1^1 + \frac{\partial y^1}{\partial x^1}\frac{\partial x^2}{\partial y^2}X_2^1 + \frac{\partial y^1}{\partial x^1}\frac{\partial x^3}{\partial y^2}X_3^1$$

$$+ \frac{\partial y^1}{\partial x^2}\frac{\partial x^1}{\partial y^2}X_1^2 + \frac{\partial y^1}{\partial x^2}\frac{\partial x^2}{\partial y^2}X_2^2 + \frac{\partial y^1}{\partial x^2}\frac{\partial x^3}{\partial y^2}X_3^2$$

$$+ \frac{\partial y^1}{\partial x^3}\frac{\partial x^1}{\partial y^2}X_1^3 + \frac{\partial y^1}{\partial x^3}\frac{\partial x^2}{\partial y^2}X_2^3 + \frac{\partial y^1}{\partial x^3}\frac{\partial x^3}{\partial y^2}X_3^3$$

$$Y_2^1 = (\cos A \cos B \sin B)X_1^1 + (\cos A \cos^2 B)X_2^1 - (\cos A \sin^2 B)X_1^2$$

$$- (\cos A \sin B \cos B)X_2^2 - (\sin A \sin B)X_1^3 - (\sin A \cos B)X_2^3$$

$$Y_3{}^1 = \frac{\partial y^1}{\partial x^1} \frac{\partial x^1}{\partial y^3} X_1{}^1 + \frac{\partial y^1}{\partial x^1} \frac{\partial x^2}{\partial y^3} X_2{}^1 + \frac{\partial y^1}{\partial x^1} \frac{\partial x^3}{\partial y^3} X_3{}^1$$

$$+ \frac{\partial y^1}{\partial x^2} \frac{\partial x^1}{\partial y^3} X_1{}^2 + \frac{\partial y^1}{\partial x^2} \frac{\partial x^2}{\partial y^3} X_2{}^2 + \frac{\partial y^1}{\partial x^2} \frac{\partial x^3}{\partial y^3} X_3{}^2$$

$$+ \frac{\partial y^1}{\partial x^3} \frac{\partial x^1}{\partial y^3} X_1{}^3 + \frac{\partial y^1}{\partial x^3} \frac{\partial x^2}{\partial y^3} X_2{}^3 + \frac{\partial y^1}{\partial x^3} \frac{\partial x^3}{\partial y^3} X_3{}^3$$

$$Y_3{}^1 = (\sin A \cos A \cos^2 B)X_1{}^1 - (\sin A \cos A \cos B \sin B)X_2{}^1$$

$$+ (\cos^2 A \cos B)X_3{}^1 - (\sin A \cos A \sin B \cos B)X_1{}^2$$

$$+ (\sin A \cos A \sin^2 B)X_2{}^2 - (\cos^2 A \sin B)X_3{}^2 - (\sin^2 A \cos B)X_1{}^3$$

$$+ (\sin^2 A \sin B)X_2{}^3 - (\sin A \cos A)X_3{}^3$$

$$Y_1{}^2 = \frac{\partial y^2}{\partial x^1} \frac{\partial x^1}{\partial y^1} X_1{}^1 + \frac{\partial y^2}{\partial x^1} \frac{\partial x^2}{\partial y^1} X_2{}^1 + \frac{\partial y^2}{\partial x^1} \frac{\partial x^3}{\partial y^1} X_3{}^1$$

$$+ \frac{\partial y^2}{\partial x^2} \frac{\partial x^1}{\partial y^1} X_1{}^2 + \frac{\partial y^2}{\partial x^2} \frac{\partial x^2}{\partial y^1} X_2{}^2 + \frac{\partial y^2}{\partial x^2} \frac{\partial x^3}{\partial y^1} X_3{}^2$$

$$+ \frac{\partial y^2}{\partial x^3} \frac{\partial x^1}{\partial y^1} X_1{}^3 + \frac{\partial y^2}{\partial x^3} \frac{\partial x^2}{\partial y^1} X_2{}^3 + \frac{\partial y^2}{\partial x^3} \frac{\partial x^3}{\partial y^1} X_3{}^3$$

$$Y_1{}^2 = (\cos A \cos B \sin B)X_1{}^1 - (\cos A \sin^2 B)X_2{}^1 - (\sin A \sin B)X_3{}^1$$

$$+ (\cos A \cos^2 B)X_1{}^2 - (\cos A \cos B \sin B)X_2{}^2 - (\sin A \cos B)X_3{}^2$$

$$Y_2{}^2 = \frac{\partial y^2}{\partial x^1}\frac{\partial x^1}{\partial y^2}X_1{}^1 + \frac{\partial y^2}{\partial x^1}\frac{\partial x^2}{\partial y^2}X_2{}^1 + \frac{\partial y^2}{\partial x^1}\frac{\partial x^3}{\partial y^2}X_3{}^1$$

$$+ \frac{\partial y^2}{\partial x^2}\frac{\partial x^1}{\partial y^2}X_1{}^2 + \frac{\partial y^2}{\partial x^2}\frac{\partial x^2}{\partial y^2}X_2{}^2 + \frac{\partial y^2}{\partial x^2}\frac{\partial x^3}{\partial y^2}X_3{}^2$$

$$+ \frac{\partial y^2}{\partial x^3}\frac{\partial x^1}{\partial y^2}X_1{}^3 + \frac{\partial y^2}{\partial x^3}\frac{\partial x^2}{\partial y^2}X_2{}^3 + \frac{\partial y^2}{\partial x^3}\frac{\partial x^3}{\partial y^2}X_3{}^3$$

$$Y_2{}^2 = (\sin^2 B)X_1{}^1 + (\sin B \cos B)X_2{}^1 + (\sin B \cos B)X_1{}^2 + (\cos^2 B)X_2{}^2$$

$$Y_3{}^2 = \frac{\partial y^2}{\partial x^1}\frac{\partial x^1}{\partial y^3}X_1{}^1 + \frac{\partial y^2}{\partial x^1}\frac{\partial x^2}{\partial y^3}X_2{}^1 + \frac{\partial y^2}{\partial x^1}\frac{\partial x^3}{\partial y^3}X_3{}^1$$

$$+ \frac{\partial y^2}{\partial x^2}\frac{\partial x^1}{\partial y^3}X_1{}^2 + \frac{\partial y^2}{\partial x^2}\frac{\partial x^2}{\partial y^3}X_2{}^2 + \frac{\partial y^2}{\partial x^2}\frac{\partial x^3}{\partial y^3}X_3{}^2$$

$$+ \frac{\partial y^2}{\partial x^3}\frac{\partial x^1}{\partial y^3}X_1{}^3 + \frac{\partial y^2}{\partial x^3}\frac{\partial x^2}{\partial y^3}X_2{}^3 + \frac{\partial y^2}{\partial x^3}\frac{\partial x^3}{\partial y^3}X_3{}^3$$

$$Y_3{}^2 = (\sin A \sin B \cos B)X_1{}^1 - (\sin A \sin^2 B)X_2{}^1 + (\sin B \cos A)X_3{}^1$$

$$+ (\sin A \cos^2 B)X_1{}^2 - (\sin A \sin B \cos B)X_2{}^2 + (\cos A \cos B)X_3{}^2$$

$$Y_1{}^3 = \frac{\partial y^3}{\partial x^1}\frac{\partial x^1}{\partial y^1}X_1{}^1 + \frac{\partial y^3}{\partial x^1}\frac{\partial x^2}{\partial y^1}X_2{}^1 + \frac{\partial y^3}{\partial x^1}\frac{\partial x^3}{\partial y^1}X_3{}^1$$

$$+ \frac{\partial y^3}{\partial x^2}\frac{\partial x^1}{\partial y^1}X_1{}^2 + \frac{\partial y^3}{\partial x^2}\frac{\partial x^2}{\partial y^1}X_2{}^2 + \frac{\partial y^3}{\partial x^2}\frac{\partial x^3}{\partial y^1}X_3{}^2$$

$$+ \frac{\partial y^3}{\partial x^3}\frac{\partial x^1}{\partial y^1}X_1{}^3 + \frac{\partial y^3}{\partial x^3}\frac{\partial x^2}{\partial y^1}X_2{}^3 + \frac{\partial y^3}{\partial x^3}\frac{\partial x^3}{\partial y^1}X_3{}^3$$

$$Y_1{}^3 = (\sin A \cos A \cos^2 B)X_1{}^1 - (\sin A \cos A \sin B \cos B)X_2{}^1$$

$$- (\sin^2 A \cos B)X_3{}^1 - (\sin A \cos A \sin B \cos B)X_1{}^2$$

$$+ (\sin A \sin^2 B \cos A)X_2{}^2 + (\sin^2 A \sin B)X_3{}^2 + (\cos^2 A \cos B)X_1{}^3$$

$$- (\cos^2 A \sin B)X_2{}^3 - (\sin A \cos A)X_3{}^3$$

$$Y_2{}^3 = \frac{\partial y^3}{\partial x^1}\frac{\partial x^1}{\partial y^2} X_1{}^1 + \frac{\partial y^3}{\partial x^1}\frac{\partial x^2}{\partial y^2} X_2{}^1 + \frac{\partial y^3}{\partial x^1}\frac{\partial x^3}{\partial y^2} X_3{}^1$$

$$+ \frac{\partial y^3}{\partial x^2}\frac{\partial x^1}{\partial y^2} X_1{}^2 + \frac{\partial y^3}{\partial x^2}\frac{\partial x^2}{\partial y^2} X_2{}^2 + \frac{\partial y^3}{\partial x^2}\frac{\partial x^3}{\partial y^2} X_3{}^2$$

$$+ \frac{\partial y^3}{\partial x^3}\frac{\partial x^1}{\partial y^2} X_1{}^3 + \frac{\partial y^3}{\partial x^3}\frac{\partial x^2}{\partial y^2} X_2{}^3 + \frac{\partial y^3}{\partial x^3}\frac{\partial x^3}{\partial y^2} X_3{}^3$$

$$Y_2{}^3 = (\sin A \sin B \cos B)X_1{}^1 + (\sin A \cos^2 B)X_2{}^1 - (\sin A \sin^2 B)X_1{}^2$$

$$- (\sin A \sin B \cos B)X_2{}^2 + (\cos A \sin B)X_1{}^3 + (\cos A \cos B)X_2{}^3$$

$$Y_3{}^3 = \frac{\partial y^3}{\partial x^1}\frac{\partial x^1}{\partial y^3} X_1{}^1 + \frac{\partial y^3}{\partial x^1}\frac{\partial x^2}{\partial y^3} X_2{}^1 + \frac{\partial y^3}{\partial x^1}\frac{\partial x^3}{\partial y^3} X_3{}^1$$

$$+ \frac{\partial y^3}{\partial x^2}\frac{\partial x^1}{\partial y^3} X_1{}^2 + \frac{\partial y^3}{\partial x^2}\frac{\partial x^2}{\partial y^3} X_2{}^2 + \frac{\partial y^3}{\partial x^2}\frac{\partial x^3}{\partial y^3} X_3{}^2$$

$$+ \frac{\partial y^3}{\partial x^3}\frac{\partial x^1}{\partial y^3} X_1{}^3 + \frac{\partial y^3}{\partial x^3}\frac{\partial x^2}{\partial y^3} X_2{}^3 + \frac{\partial y^3}{\partial x^3}\frac{\partial x^3}{\partial y^3} X_3{}^3$$

$$Y_3{}^3 = (\sin^2 A \cos^2 B)X_1{}^1 - (\sin^2 A \sin B \cos B)X_2{}^1 + (\sin A \cos A \cos B)X_3{}^1$$

$$- (\sin^2 A \sin B \cos B)X_1{}^2 + (\sin^2 A \sin^2 B)X_2{}^2 - (\sin A \cos A \sin B)X_3{}^2$$

$$+ (\sin A \cos A \cos B)X_1{}^3 - (\sin A \cos A \sin B)X_2{}^3 + (\cos^2 A)X_3{}^3$$

When interpreted in terms of conventional aeronautical symbolism, each of these tensor components represents eight aerodynamic derivatives: four velocity derivatives and four acceleration derivatives. Hence, the transformation law given by equation (2.7.6) represents a total of 72 transformation equations for the velocity and acceleration derivatives. The tensor components representing the velocity derivatives may be interpreted as follows:

$$X_1{}^1 = X_u, X_p, l_u, l_p \qquad X_1{}^2 = Y_u, Y_p, m_u, m_p \qquad X_1{}^3 = Z_u, Z_p, n_u, n_p$$

$$X_2{}^1 = X_v, X_q, l_v, l_q \qquad X_2{}^2 = Y_v, Y_q, m_v, m_q \qquad X_2{}^3 = Z_v, Z_q, n_v, n_q$$

$$X_3{}^1 = X_w, X_r, l_w, l_r \qquad X_3{}^2 = Y_w, Y_r, m_w, m_r \qquad X_3{}^3 = Z_w, Z_r, n_w, n_r$$

The derivatives with respect to the acceleration components are obtained by replacing the velocity component subscripts with the acceleration component subscripts. Meaning may be assigned to the $Y_j{}^i$ components in the same way with the understanding that these represent the transformed derivatives. For example

$$Y_1{}^1 = X'_u, X'_p, l'_u, l'_p$$

$$Y_1{}^2 = Y'_u, Y'_p, m'_u, m'_p$$

$$Y_1{}^3 = Z'_u, Z'_p, n'_u, n'_p$$

2.8 COMPUTER TRANSFORMATIONS OF AERODYNAMIC STABILITY DERIVATIVES

2.8.1 Direct Transformations

A program to expand equation (2.7.6) requires that both the direct (2.3.3) and reverse (2.3.4) coordinate transformation equations be used as input. When both coordinate transformation equations are used as input, it is expedient to redefine the functional relationship

$$x = x(y) \qquad (2.8.1)$$

as follows:

$$w = w(z) \qquad (2.8.2)$$

where w takes the place of x and z takes the place of y.

The static force and moment coefficients and their derivatives could be specified with one index, but the aerodynamic stability derivatives require, for their complete specification, two indices. Hence, an aerodynamic stability derivative in wind-tunnel axes will be denoted by $C(I,J)$, and the corresponding transformed derivative by $TC(I,J)$. The first index in the derivative symbol denotes the force or moment component being considered, and the second one specifies the motion vector component with respect to which the derivative is obtained.

Apart from a few auxiliary statements and DO loops, which are conventional programming steps, the key program statement for the present application, as in the preceding one, is the statement that causes the computer to differentiate symbolically. The present application requires that equation (2.7.6) be programmed to facilitate the derivation of the transformation equations for the aerodynamic stability derivatives. Apart from the fact that the program for the transformation of the static aerodynamic coefficients required only two DO loops, and the present application requires four such loops, the programs are very similar. After substitution of the functional relationship, equation (2.8.2), for equation (2.8.1), the statement controlling differentiation takes the form

```
LET(TCC(I,J)=(DERIV(Y(I),X(M)))*(DERIV(W(N),Z(J)))*(C(M,N)));
```

The entire program and the corresponding output follow.

```
LET(Y(1)=X(1)*COS(A)*COS(B)-X(2)*COS(A)*SIN(B)-X(3)*SIN(A));

LET(Y(2)=X(1)*SIN(B)+X(2)*COS(B));

LET(Y(3)=X(1)*SIN(A)*COS(B)-X(2)*SIN(A)*SIN(B)+X(3)*COS(A));

PRINT_OUT(Y(1);Y(2);Y(3));
```

```
PUT SKIP(5);

LET(W(1)=Z(1)*COS(A)*COS(B)+Z(2)*SIN(B)+Z(3)*SIN(A)*COS(B));

LET(W(2)=-Z(1)*COS(A)*SIN(B)+Z(2)*COS(B)-Z(3)*SIN(A)*SIN(B));

LET(W(3)=-Z(1)*SIN(A)+Z(3)*COS(A));


DO I=1 TO 3 BY 1;

DO J=1 TO 3 BY 1;

LET(I="I");

LET(J="J");

LET(TC(I,J)=0);

DO M=1 TO 3 BY 1;

DO N=1 TO 3 BY 1;

LET(M="M");

LET(N="N");

LET(TCC(I,J)=(DERIV(Y(I),X(M)))*(DERIV(W(N),Z(J)))*(C(M,N)));

LET(TC(I,J)=TC(I,J)+TCC(I,J));

END;

END;

PRINT_OUT(TC(I,J));

PUT SKIP(5);

END;

END;
```

This program may be said to consist of the single statement

```
LET(TCC(I,J)=(DERIV(Y(I),X(M)))*(DERIV(W(N),Z(J)))*(C(M,N)));
```

which enables the computer to differentiate symbolically. With one exception, the remaining statements are conventional programming steps that instruct the computer how to manage each stage of the differentiation process and carry out the necessary summations. The exception referred to above is the group of statements: LET(I = "I"); etc. These statements are required to facilitate operations involving both numerical processing and symbolic manipulation. The statement: PUT SKIP(5); is for editing purposes and instructs the printer to skip five lines between each batch of output.

Activation of the preceding program resulted in the following output:

$$Y(1) = COS(B) \; COS(A) \; X(1) - COS(A) \; SIN(B) \; X(2) - SIN(A) \; X(3)$$

$$Y(2) = SIN(B) \; X(1) + COS(B) \; X(2)$$

$$Y(3) = COS(B) \; SIN(A) \; X(1) - SIN(B) \; SIN(A) \; X(2) + COS(A) \; X(3)$$

$$
\begin{aligned}
TC(1,1) = \; & -COS(B)COS^2(A)SIN(B)C(1,2)-COS(B)COS(A)SIN(A)C(1,3) \\
& -COS(B)COS^2(A)SIN(B)C(2,1)+COS^2(A)SIN^2(B)C(2,2) \\
& +COS(A)SIN(B)SIN(A)C(2,3)-COS(B)COS(A)SIN(A)C(3,1) \\
& +COS(A)SIN(B)SIN(A)C(3,2)+SIN^2(A)C(3,3) \\
& +COS^2(B)COS^2(A)C(1,1)
\end{aligned}
$$

$$
\begin{aligned}
TC(1,2) = \; & COS^2(B)COS(A)C(1,2)-COS(A)SIN^2(B)C(2,1) \\
& -COS(B)COS(A)SIN(B)C(2,2)-SIN(B)SIN(A)C(3,1) \\
& -COS(B)SIN(A)C(3,2)+COS(B)COS(A)SIN(B)C(1,1)
\end{aligned}
$$

$$TC(1,3) = -COS(B)COS(A)SIN(B)SIN(A)C(1,2)+COS(B)COS^2(A)C(1,3)$$
$$-COS(B)COS(A)SIN(B)SIN(A)C(2,1)+COS(A)SIN^2(B)SIN(A)C(2,2)$$
$$-COS^2(A)SIN(B)C(2,3)-COS(B)SIN^2(A)C(3,1)$$
$$+SIN(B)SIN^2(A)C(3,2)-COS(A)SIN(A)C(3,3)$$
$$+COS^2(B)COS(A)SIN(A)C(1,1)$$

$$TC(2,1) = -COS(A)SIN^2(B)C(1,2)-SIN(B)SIN(A)C(1,3)$$
$$+COS^2(B)COS(A)C(2,1)-COS(B)COS(A)SIN(B)C(2,2)$$
$$-COS(B)SIN(A)C(2,3)+COS(B)COS(A)SIN(B)C(1,1)$$

$$TC(2,2) = COS(B)SIN(B)C(1,2)+COS(B)SIN(B)C(2,1)+COS^2(B)C(2,2)$$
$$+SIN^2(B)C(1,1)$$

$$TC(2,3) = -SIN^2(B)SIN(A)C(1,2)+COS(A)SIN(B)C(1,3)$$
$$+COS^2(B)SIN(A)C(2,1)-COS(B)SIN(B)SIN(A)C(2,2)$$
$$+COS(B)COS(A)C(2,3)+COS(B)SIN(B)SIN(A)C(1,1)$$

$$TC(3,1) = -COS(B)COS(A)SIN(B)SIN(A)C(1,2)-COS(B)SIN^2(A)C(1,3)$$
$$-COS(B)COS(A)SIN(B)SIN(A)C(2,1)+COS(A)SIN^2(B)SIN(A)C(2,2)$$
$$+SIN(B)SIN^2(A)C(2,3)+COS(B)COS^2(A)C(3,1)$$
$$-COS^2(A)SIN(B)C(3,2)-COS(A)SIN(A)C(3,3)$$
$$+COS^2(B)COS(A)SIN(A)C(1,1)$$

$$TC(3,2) = COS^2(B)SIN(A)C(1,2)-SIN^2(B)SIN(A)C(2,1)$$
$$-COS(B)SIN(B)SIN(A)C(2,2)+COS(A)SIN(B)C(3,1)$$
$$+COS(B)COS(A)C(3,2)+COS(B)SIN(B)SIN(A)C(1,1)$$

$$TC(3,3) = -COS(B)SIN(B)SIN^2(A)C(1,2)+COS(B)COS(A)SIN(A)C(1,3)$$
$$-COS(B)SIN(B)SIN^2(A)C(2,1)+SIN^2(B)SIN^2(A)C(2,2)$$
$$-COS(A)SIN(B)SIN(A)C(2,3)+COS(B)COS(A)SIN(A)C(3,1)$$
$$-COS(A)SIN(B)SIN(A)C(3,2)+COS^2(A)C(3,3)$$
$$+COS^2(B)SIN^2(A)C(1,1)$$

Readers are reminded that, in computer notation, an aerodynamic stability derivative in wind-tunnel axes is denoted by $C(I,J)$ and the corresponding transformed derivative by $TC(I,J)$.

The manual derivation of the preceding section uses the notation X_j^i to denote the aerodynamic stability derivative in wind-tunnel axes and Y_j^i to denote the transformed derivative. These two methods are seen to produce identical results.

In this particular case, manual derivation proved to be the quicker method. Although the actual computing time was quite small, less than 1 min, the programming and debugging time exceeded the time required to formulate the equations manually. It should be pointed out, however, that this disadvantage is due to the batch processing mode that FORMAC users are required to use. Those who have access to a computer system with an interactive mode language, such as that described in the final chapter, would find that the computerized formulation is quicker.

The preceding output gives the computerized version of the transformation equations for the velocity and acceleration derivatives. For example, when the transformed derivative given by $TC(1,1)$ is transcribed from the computer output and interpreted in accordance with the definitions assigned to the identifying indices for angular velocity derivatives, it represents the following equation:

$$X_p' = [X_p \cos^2 B - (X_q + Y_p)\sin B \cos B + Y_q \sin^2 B] \cos^2 A$$
$$+ Z_r \sin^2 A - [(X_r + Z_p)\cos B - (Y_r + Z_q)\sin B] \sin A \cos A$$

It is instructive to dwell on this rather complicated equation for a moment and examine its meaning and the meanings of the individual terms and coefficients. This equation gives the value of the derivatives in body coordinates (primed quantities) in terms of the corresponding derivatives measured in the wind tunnel (the unprimed quantities).

It will be recalled that the aerodynamic forces acting on a body which is moving through the atmosphere are defined in terms of the force coefficients C_x, C_y, and C_z. The magnitude of the force in the x direction being

$$\bar{q}SC_x = \frac{1}{2}\rho V^2 SC_x = X$$

When the x direction is in the direction of the velocity vector, which is the direction of motion of the body, aeronautical engineers usually refer to these forces as drag forces and define a drag coefficient as follows:

$$C_D = -C_x$$

In terms of this coefficient the aerodynamic drag force is

$$\bar{q}SC_D = \frac{1}{2}\rho V^2 SC_D$$

If we think of X as a measure of the drag force, the symbol $C(1,1)$ corresponds to X_p and is a measure of the rate of change of the drag force with respect to p, the angular velocity of the body about the x-axis, or as it is called, the rolling angular velocity. Likewise, the symbol $C(1,2)$ corresponds to X_q and is a measure of the rate of change of the drag force with respect to the pitching velocity q. Similarly, the symbol $C(1,3)$ corresponds to X_r and is a measure of the rate of change of the drag force with respect to the yawing velocity r.

As in the case of the static forces and moments, the transformation of the aerodynamic stability derivatives from wind axes to body axes involves the angles A and B (see sketch (f)). The angle A is the angle of attack, which is the angle between the component of the wind vector in the plane of symmetry and the longitudinal axis of the aircraft. The angle B, on the other hand, is the angle of sideslip, which is the angle between the wind vector and the plane of symmetry.

The remaining transformation equations for the drag force derivatives with respect to the angular velocity components follow the same pattern. These are

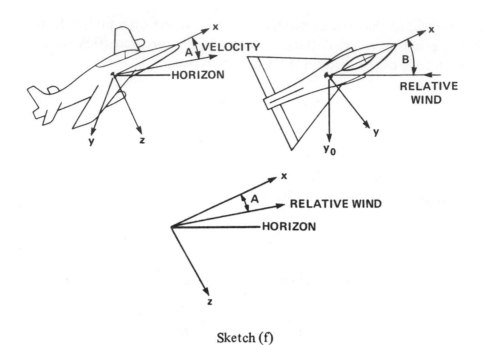

Sketch (f)

$$X_q' = [X_q \cos^2 B - Y_p \sin^2 B + (X_p - Y_q)\sin B \cos B]\cos A$$

$$- (Z_q \cos B + Z_p \sin B)\sin A$$

$$X_r' = (X_r \cos B - Y_r \sin B)\cos^2 A - (Z_p \cos B - Z_q \sin B)\sin^2 A$$

$$+ [X_p \cos^2 B + Y_q \sin^2 B - (X_q + Y_p)\sin B \cos B - Z_r]\sin A \cos A$$

In the computed transformation equations the symbols $C(2,1)$, $C(2,2)$, and $C(2,3)$ correspond, respectively, to Y_p, Y_q, and Y_r. The quantity Y is a measure of the side force acting on the aircraft, and the derivatives Y_p, Y_q, Y_r determine the rates of change of this force with respect to the rolling, pitching, and yawing velocities of the vehicle. The computed transformations are

$$Y_p' = [Y_p \cos^2 B - X_q \sin^2 B + (X_p - Y_q)\sin B \cos B]\cos A$$

$$- (Y_r \cos B + X_r \sin B)\sin A$$

$$Y_q' = Y_q \cos^2 B + X_p \sin^2 B + (X_q + Y_p)\sin B \cos B$$

$$Y_r' = (Y_r \cos B + X_r \sin B)\cos A$$

$$+ [Y_p \cos^2 B - X_q \sin^2 B + (X_p - Y_q)\sin B \cos B]\sin A$$

Instead of the quantity Z which determines the lift force acting on an aircraft, aeronautical engineers use a lift coefficient which is defined by the relation

$$\frac{1}{2}\rho V^2 S C_L = -Z$$

In terms of this notation, the derivatives Z_p, Z_q, Z_r determine the rates of change of the lift force with respect to the rolling, pitching, and yawing velocities of the aircraft. In the computed transformation equations, the following substitutions are required: the symbols $C(3,1)$, $C(3,2)$, and $C(3,3)$ correspond, respectively, to Z_p, Z_q, and Z_r. When these substitutions are made in the computer printout, the transformation equations assume the following form:

$$Z_p' = (Z_p \cos B - Z_q \sin B)\cos^2 A - (X_r \cos B - Y_r \sin B)\sin^2 A$$

$$+ [X_p \cos^2 B + Y_q \sin^2 B - (X_q + Y_p)\sin B \cos B - Z_r]\sin A \cos A$$

$$Z_q' = (Z_q \cos B + Z_p \sin B)\cos A$$

$$+ [X_q \cos^2 B - Y_p \sin^2 B + (X_p - Y_q)\sin B \cos B]\sin A$$

$$Z_r' = Z_r \cos^2 A + [X_p \cos^2 B + Y_q \sin^2 B - (X_q + Y_p)\sin B \cos B]\sin^2 A$$

$$+ [(X_r + Z_p)\cos B - (Y_r + Z_q)\sin B]\sin A \cos A$$

The same computer output gives the transformation equations for the moment derivatives also. As already indicated, the first index in the symbol $C(I,J)$ or $TC(I,J)$ denotes the force or moment component being considered, and the second one specifies the motion vector component with respect to which the derivative is obtained. Hence, if the first index is used to denote moment components instead of force components, the moment derivatives are obtained. For example, the symbol

$C(1,1)$, which was previously used to denote the force derivative X_p, may also be used to denote the moment derivative l_p, where the quantity l is a measure of the aerodynamic rolling moment. Likewise, $C(1,2)$, which was previously used to denote X_q, may now be used to denote l_q, the rolling moment derivative with respect to the pitching velocity. Finally, $C(1,3)$, which has been used to denote X_r, gives the derivative l_r, that is, the rolling moment derivative with respect to the yawing velocity. These are

$$l_p' = [l_p \cos^2 B - (l_q + m_p)\sin B \cos B + m_q \sin^2 B]\cos^2 A + n_r \sin^2 A$$

$$+ [-(l_r + n_p)\cos B + (m_r + n_q)\sin B]\sin A \cos A$$

$$l_q' = [l_q \cos^2 B - m_p \sin^2 B + (l_p - m_q)\sin B \cos B]\cos A$$

$$- (n_q \cos B + n_p \sin B)\sin A$$

$$l_r' = (l_r \cos B - m_r \sin B)\cos^2 A - (n_p \cos B - n_q \sin B)\sin^2 A$$

$$+ [l_p \cos^2 B + m_q \sin^2 B - (l_q + m_p)\sin B \cos B - n_r]\sin A \cos A$$

Again, the symbol $C(2,1)$, which was previously used to denote the force derivative Y_p, may also be used to denote the moment derivative m_p, where the quantity m is a measure of the aerodynamic pitching moment. Moreover, the symbol $C(2,2)$, which was used to represent the force derivative Y_q, is used in the present context to represent the pitching moment derivative with respect to the pitching velocity, that is, m_q, and the symbol $C(2,3)$ gives the pitching moment derivative with respect to the yawing velocity, that is, m_r. The transformation equations for the aerodynamic pitching moment derivatives with respect to the angular velocity components are

$$m_p' = [m_p \cos^2 B - l_q \sin^2 B + (l_p - m_q)\sin B \cos B]\cos A$$

$$- (m_r \cos B + l_r \sin B)\sin A$$

$$m_q' = m_q \cos^2 B + l_p \sin^2 B + (l_q + m_p)\sin B \cos B$$

$$m_r' = (m_r \cos B + l_r \sin B)\cos A$$

$$+ [m_p \cos^2 B - l_q \sin^2 B + (l_p - m_q)\sin B \cos B]\sin A$$

The remaining moment derivatives are n_p, the yawing moment derivative with respect to the rolling velocity; n_q, the yawing moment derivative with respect to the pitching velocity; and n_r, the yawing moment derivative with respect to the yawing velocity. The symbols $C(3,1)$, $C(3,2)$, and $C(3,3)$ correspond respectively to n_p, n_q, and n_r. The transformation equations for the aerodynamic yawing moment derivatives with respect to the angular velocity components are

$$n_p' = (n_p \cos B - n_q \sin B)\cos^2 A - (l_r \cos B - m_r \sin B)\sin^2 A$$

$$+ [l_p \cos^2 B + m_q \sin^2 B - (l_q + m_p)\sin B \cos B - n_r]\sin A \cos A$$

$$n_q' = (n_q \cos B + n_p \sin B)\cos A$$

$$+ [l_q \cos^2 B - m_p \sin^2 B + (l_p - m_q)\sin B \cos B]\sin A$$

$$n_r' = n_r \cos^2 A + [l_p \cos^2 B + m_q \sin^2 B - (l_q + m_p)\sin B \cos B]\sin^2 A$$

$$+ [(l_r + n_p)\cos B - (m_r + n_q)\sin B]\sin A \cos A$$

It is hoped that the reader will be sufficiently impressed with the compactness of the tensor notation and the simplicity of the computer programs for symbolic manipulation, that he will be encouraged to write some programs of his own. If he does, he will discover that there are many formulations that are amenable to the technique of symbolic manipulation.

The information contained in the tensor transformation equation (2.7.6) and its expanded form as given by the computer output, illustrates again the advantages of the tensor method and the facility with which the summation convention can be exploited by a simple computational algorithm.

Another set of derivatives which plays an important role in the study of the response of an aircraft to aerodynamic forces is the set of aerodynamic stability derivatives with respect to the linear velocity components u,v,w. These derivatives are obtained in the same manner as the aerodynamic stability derivatives with respect to the angular velocity components. Referring again to the computerized version of equation (2.7.6), it will be recalled that the first index of the symbol $C(I,J)$ denotes the force or moment component being considered, while the second index specifies the motion vector component with respect to which the derivative is obtained. In the preceding formulation, all derivatives were obtained with respect to the angular velocity components, but the present application requires that all derivatives be obtained with respect to the linear velocity components. In order to convert from computer output to conventional aeronautical symbolism, the following substitutions are required: In the present context, the symbol $C(1,1)$ denotes X_u where X_u is a measure of the rate of change of the aerodynamic drag force with respect to the velocity component along the x reference axis (see sketch (g)).

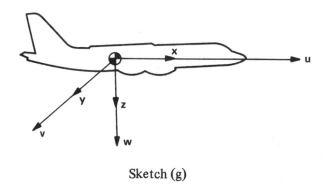

Sketch (g)

The rate of change of the drag force with respect to the velocity component along the y reference axis, that is, the lateral velocity, is X_v, which in the present context will be denoted by the symbol $C(1,2)$. The symbol $C(1,3)$ corresponds to X_w, which is the rate of change of the drag force with respect to the velocity component along the z-axis of the aircraft.

The transformation equations for these three components are

$$X_u' = [X_u \cos^2 B - (X_v + Y_u)\sin B \cos B + Y_v \sin^2 B] \cos^2 A + Z_w \sin^2 A$$

$$+ [-(X_w + Z_u)\cos B + (Y_w + Z_v)\sin B] \sin A \cos A$$

$$X_v' = [X_v \cos^2 B - Y_u \sin^2 B + (X_u - Y_v)\sin B \cos B] \cos A$$

$$- (Z_v \cos B + Z_u \sin B)\sin A$$

$$X_w' = (X_w \cos B - Y_w \sin B)\cos^2 A - (Z_u \cos B - Z_v \sin B)\sin^2 A$$

$$+ [X_u \cos^2 B + Y_v \sin^2 B - (X_v + Y_u)\sin B \cos B - Z_w]\sin A \cos A$$

The symbol $C(2,1)$ denotes Y_u, which determines the rate of change of the side force with respect to the u component of velocity. Likewise, $C(2,2)$ denotes Y_v, which determines the rate of change of the side force with respect to the v component of velocity; and $C(2,3)$ represents Y_w, the rate of change of the side force with respect to the w component of velocity.

These three components transform as follows:

$$Y_u' = [Y_u \cos^2 B - X_v \sin^2 B + (X_u - Y_v)\sin B \cos B]\cos A$$

$$- (Y_w \cos B + X_w \sin B)\sin A$$

$$Y_v' = Y_v \cos^2 B + X_u \sin^2 B + (X_v + Y_u)\sin B \cos B$$

$$Y_w' = (Y_w \cos B + X_w \sin B)\cos A$$

$$+ [Y_u \cos^2 B - X_v \sin^2 B + (X_u - Y_v)\sin B \cos B]\sin A$$

Proceeding in the same manner, the symbols $C(3,1)$, $C(3,2)$, and $C(3,3)$ correspond, respectively, to Z_u, Z_v, and Z_w, the rates of change of the aerodynamic lift force with respect to the velocity components u,v,w.

In this case the transformation equations are

$$Z_u' = (Z_u \cos B - Z_v \sin B)\cos^2 A - (X_w \cos B - Y_w \sin B)\sin^2 A$$

$$+ [X_u \cos^2 B + Y_v \sin^2 B - (X_v + Y_u)\sin B \cos B - Z_w]\sin A \cos A$$

$$Z_v' = (Z_v \cos B + Z_u \sin B)\cos A$$

$$+ [X_v \cos^2 B - Y_u \sin^2 B + (X_u - Y_v)\sin B \cos B]\sin A$$

$$Z_w' = Z_w \cos^2 A + [X_u \cos^2 B + Y_v \sin^2 B - (X_v + Y_u)\sin B \cos B]\sin^2 A$$

$$+ [(X_w + Z_u)\cos B - (Y_w + Z_v)\sin B]\sin A \cos A$$

The influence of the linear velocity components on the aerodynamic moments is determined by a set of stability derivatives analogous to the preceding force derivatives. Again, we can avail ourselves of the expanded form of equation (2.7.6) as given by the computer output. In the present application, the first index of the symbol $C(I,J)$ refers to the aerodynamic moment being considered, while the second one specifies the linear velocity component with respect to which the derivative is obtained; that is, $C(1,1)$ corresponds to l_u, which is the rate of change of the rolling moment with respect to the u component of velocity.

The rate of change of the rolling moment with respect to the v component of velocity is l_v and is given in the present context by the symbol $C(1,2)$. The symbol $C(1,3)$ denotes l_w, the rate of change of the rolling moment with respect to the w component of velocity. The transformation equations for these three components are

$$l_u' = [l_u \cos^2 B - (l_v + m_u)\sin B \cos B + m_v \sin^2 B] \cos^2 A$$

$$+ n_w \sin^2 A + [-(l_w + n_u)\cos B + (m_w + n_v)\sin B] \sin A \cos A$$

$$l_v' = [l_v \cos^2 B - m_u \sin^2 B + (l_u - m_v)\sin B \cos B] \cos A$$

$$- (n_v \cos B + n_u \sin B)\sin A$$

$$l_w' = (l_w \cos B - m_v \sin B)\cos^2 A - (n_u \cos B - n_v \sin B)\sin^2 A$$

$$+ [l_u \cos^2 B + m_v \sin^2 B - (l_v + m_u)\sin B \cos B - n_w] \sin A \cos A$$

It is seen that the amount of information contained in equation (2.7.6) and its expanded form as given by the computer output is quite large. At each step in the formulation a reinterpretation of the significance of the indices in the symbol $C(I,J)$ yields additional transformations. For example, if we wish to transform the pitching moment derivatives from wind-tunnel axes to body axes, the symbols $C(2,J)$ and $TC(2,J)$ would be used. In this case, the symbol $C(2,1)$ would correspond to m_u, where m_u is the rate of change of the pitching moment with respect to the u component of velocity. The symbol $C(2,2)$ can be replaced by m_v, the pitching moment derivative with respect to the v component of velocity, and m_w can be substituted for the symbol $C(2,3)$ in the computed transformation equations. The transformation equations for the pitching moment derivatives are

$$m_u' = [m_u \cos^2 B - l_v \sin^2 B + (l_u - m_v)\sin B \cos B]\cos A$$

$$- (m_w \cos B + l_w \sin B)\sin A$$

$$m_v' = m_v \cos^2 B + l_u \sin^2 B + (l_v + m_u)\sin B \cos B$$

$$m_w' = (m_w \cos B + l_w \sin B)\cos A$$

$$+ [m_u \cos^2 B - l_v \sin^2 B + (l_u - m_v)\sin B \cos B]\sin A$$

The yawing moment derivatives with respect to the linear velocity components complete the list of aerodynamic stability derivatives required to study the response of an aircraft to aerodynamic forces of this type. In this case, the symbols $C(3,J)$ and $TC(3,J)$ are required. The symbol $C(3,1)$ represents n_u, which determines the rate of change of the yawing moment with respect to the u component of velocity. The symbol $C(3,2)$ can be replaced by n_v, the rate of change of the yawing moment with respect to the v component of velocity. Lastly, n_w can be substituted for $C(3,3)$ in the computed transformation equations, and the yawing moment derivatives transform as follows:

$$n_u' = (n_u \cos B - n_v \sin B)\cos^2 A - (l_w \cos B - m_w \sin B)\sin^2 A$$

$$+ [l_u \cos^2 B + m_v \sin^2 B - (l_v + m_u)\sin B \cos B - n_w]\sin A \cos A$$

$$n_v' = (n_v \cos B + n_u \sin B)\cos A$$

$$\cdot \quad {}_{v} \cos^2 B - m_u \sin^2 B + (l_u - m_v)\sin B \cos B]\sin A$$

$$n_w' = n_w \cos^2 A + [l_u \cos^2 B + m_v \sin^2 B - (l_v + m_u)\sin B \cos B]\sin^2 A$$

$$+ [(l_w + n_u)\cos B - (m_w + n_v)\sin B]\sin A \cos A$$

The aerodynamic stability derivatives with respect to the components of linear and angular acceleration follow exactly the same pattern. These are obtained by a reinterpretation of the second index in the derivative symbol. Although the first index of the derivative symbol $C(I,J)$ still denotes the force or moment component being considered, the second index now specifies the linear or angular acceleration component with respect to which the derivative is obtained. When substitutions are made in accordance with these interpretations, the computer output yields an additional 36 equations for the acceleration derivatives. Hence, the compact tensor equation (2.7.6) and the corresponding computer output represent a total of 72 transformation equations for the aerodynamic stability derivatives.

2.8.2 *Aerodynamic Stability Derivatives as Second Order Contravariant Tensors*

As indicated in section 1.4, it is possible to avoid the use of the inverse transformation $x = x(y)$, if the coordinate transformations are orthogonal Cartesian. Equation (1.4.12) shows that for orthogonal Cartesian transformations

$$\frac{\partial y^i}{\partial x^j} = \frac{\partial x^j}{\partial y^i} \tag{2.8.3}$$

Substitution of this relationship in equation (2.7.6) gives

$$Y^{ij} = \frac{\partial y^i}{\partial x^\alpha} \frac{\partial y^j}{\partial x^\beta} X^{\alpha\beta} \tag{2.8.4}$$

where the first superscript again denotes the component of the aerodynamic force or moment, and the second superscript denotes the motion vector component with respect to which the derivatives are obtained. The form of equation (2.8.4) shows that it is only necessary to use the direct coordinate transformation $y = y(x)$.

In case some readers are not quite convinced by the arguments of section 1.4, the computer program will be modified to process equation (2.8.4) for comparison with the result of processing equation (2.7.6).

The modified program (which only requires the direct coordinate transformation equations as input) and the resulting output are as follows:

```
LET(Y(1)=X(1)*COS(A)*COS(B)-X(2)*COS(A)*SIN(B)-X(3)*SIN(A));

LET(Y(2)=X(1)*SIN(B)+X(2)*COS(B));

LET(Y(3)=X(1)*SIN(A)*COS(B)-X(2)*SIN(A)*SIN(B)+X(3)*COS(A));

PRINT OUT(Y(1);Y(2);Y(3));
```

```
PUT SKIP(5);
DO I=1 TO 3 BY 1;
DO J=1 TO 3 BY 1;
LET(I="I");
LET(J="J");
LET(TC(I,J)=0);
DO M=1 TO 3 BY 1;
DO N=1 TO 3 BY 1;
LET(M="M");
LET(N="N");
LET(TCC(I,J)=(DERIV(Y(I),X(M)))*(DERIV(Y(J),X(N)))*(C(M,N)));
LET(TC(I,J)=TC(I,J)+TCC(I,J));
END;
END;
PRINT_OUT(TC(I,J));
PUT SKIP(5);
END;
END;
```

$$Y(1) = COS(B)COS(A)X(1)-COS(A)SIN(B)X(2)-SIN(A)X(3)$$

$$Y(2) = SIN(B)X(1)+COS(B)X(2)$$

$$Y(3) = COS(B)SIN(A)X(1)-SIN(B)SIN(A)X(2)+COS(A)X(3)$$

$$TC(1,1) = -COS(B)COS^2(A)SIN(B)C(1,2)-COS(B)COS(A)SIN(A)C(1,3)$$
$$-COS(B)COS^2(A)SIN(B)C(2,1)+COS^2(A)SIN^2(B)C(2,2)$$
$$+COS(A)SIN(B)SIN(A)C(2,3)-COS(B)COS(A)SIN(A)C(3,1)$$
$$+COS(A)SIN(B)SIN(A)C(3,2)+SIN^2(A)C(3,3)$$
$$+COS^2(B)COS^2(A)C(1,1)$$

$$TC(1,2) = COS^2(B)COS(A)C(1,2)-COS(A)SIN^2(B)C(2,1)$$
$$-COS(B)COS(A)SIN(B)C(2,2)-SIN(B)SIN(A)C(3,1)$$
$$-COS(B)SIN(A)C(3,2)+COS(B)COS(A)SIN(B)C(1,1)$$

$$TC(1,3) = -COS(B)COS(A)SIN(B)SIN(A)C(1,2)+COS(B)COS^2(A)C(1,3)$$
$$-COS(B)COS(A)SIN(B)SIN(A)C(2,1)+COS(A)SIN^2(B)SIN(A)C(2,2)$$
$$-COS^2(A)SIN(B)C(2,3)-COS(B)SIN^2(A)C(3,1)$$
$$+SIN(B)SIN^2(A)C(3,2)-COS(A)SIN(A)C(3,3)$$
$$+COS^2(B)COS(A)SIN(A)C(1,1)$$

$$TC(2,1) = -COS(A)SIN^2(B)C(1,2)-SIN(B)SIN(A)C(1,3)$$
$$+COS^2(B)COS(A)C(2,1)-COS(B)COS(A)SIN(B)C(2,2)$$
$$-COS(B)SIN(A)C(2,3)+COS(B)COS(A)SIN(B)C(1,1)$$

$$TC(2,2) = COS(B)SIN(B)C(1,2)+COS(B)SIN(B)C(2,1)$$
$$+COS^2(B)C(2,2)+SIN^2(B)C(1,1)$$

$$TC(2,3) = -SIN^2(B)SIN(A)C(1,2)+COS(A)SIN(B)C(1,3)$$
$$+COS^2(B)SIN(A)C(2,1)-COS(B)SIN(B)SIN(A)C(2,2)$$
$$+COS(B)COS(A)C(2,3)+COS(B)SIN(B)SIN(A)C(1,1)$$

```
TC(3,1) = -COS(B)COS(A)SIN(B)SIN(A)C(1,2)-COS(B)SIN²(A)C(1,3)

           -COS(B)COS(A)SIN(B)SIN(A)C(2,1)+COS(A)SIN²(B)SIN(A)C(2,2)

           +SIN(B)SIN²(A)C(2,3)+COS(B)COS²(A)C(3,1)

           -COS²(A)SIN(B)C(3,2)-COS(A)SIN(A)C(3,3)

           +COS²(B)COS(A)SIN(A)C(1,1)

TC(3,2) = COS²(B)SIN(A)C(1,2)-SIN²(B)SIN(A)C(2,1)

           -COS(B)SIN(B)SIN(A)C(2,2)+COS(A)SIN(B)C(3,1)

           +COS(B)COS(A)C(3,2)+COS(B)SIN(B)SIN(A)C(1,1)

TC(3,3) = -COS(B)SIN(B)SIN²(A)C(1,2)+COS(B)COS(A)SIN(A)C(1,3)

           -COS(B)SIN(B)SIN²(A)C(2,1)+SIN²(B)SIN²(A)C(2,2)

           -COS(A)SIN(B)SIN(A)C(2,3)+COS(B)COS(A)SIN(A)C(3,1)

           -COS(A)SIN(B)SIN(A)C(3,2)+COS²(A)C(3,3)

           +COS²(B)SIN²(A)C(1,1)
```

Hence, for a given coordinate transformation $y = y(x)$, which is orthogonal Cartesian, the second-order contravariant transformation gives the same result as the mixed tensor transformation. By using this property of orthogonal Cartesian transformation, the need for the reverse transformation $x = x(y)$ is eliminated.

2.8.3 Inverse Transformations

Equations (2.7.6) and (2.8.4) give the direct transformation of aerodynamic stability derivatives from wind-tunnel axes to body axes. It has been demonstrated that these two transformation laws give identical results if coordinate transformations are confined to orthogonal Cartesian systems.

Use of the simpler form (2.8.4) avoids the need for the reverse coordinate transformation $x = x(y)$. Hence, if

$$Y^{ij} = \frac{\partial y^i}{\partial x^\alpha} \frac{\partial y^j}{\partial x^\beta} X^{\alpha\beta} \tag{2.8.5}$$

equation (2.5.2) permits the inverse transformation to be written in the form

$$X^{\alpha\beta} = \frac{\partial x^\alpha}{\partial y^i} \frac{\partial x^\beta}{\partial y^j} Y^{ij} \tag{2.8.6}$$

Substitution from equation (2.8.3) in equation (2.8.6) gives

$$X^{\alpha\beta} = \frac{\partial y^i}{\partial x^\alpha} \frac{\partial y^j}{\partial x^\beta} Y^{ij} \tag{2.8.7}$$

When rewritten in covariant form, to conform to the notation already established for covariant and contravariant tensors, this equation becomes

$$X_{\alpha\beta} = \frac{\partial y^i}{\partial x^\alpha} \frac{\partial y^j}{\partial x^\beta} Y_{ij} \tag{2.8.8}$$

or

$$X_{ij} = \frac{\partial y^m}{\partial x^i} \frac{\partial y^n}{\partial x^j} Y_{mn} \tag{2.8.9}$$

As a consequence of equation (2.8.3) the distinction between contravariant and covariant tensors disappears when coordinate transformations are confined to orthogonal Cartesian systems.

2.8.4 Inverse Computer Transformations

A computer program to process equation (2.8.9) requires only the direct coordinate transformation $y = y(x)$ as input.

The symbol $C(I,J)$ will again be used to denote an aerodynamic stability derivative in computer notation. However, in this case $C(I,J)$ will denote a body axis component and $TC(I,J)$ will refer to a component relative to wind-tunnel axes.

When I,J are permitted to assume the values 1,2,3, the computer program and the corresponding output assume the following form:

```
LET(Y(1)=X(1)*COS(A)*COS(B)-X(2)*COS(A)*SIN(B)-X(3)SIN(A));

LET(Y(2)=X(1)*SIN(B)+X(2)*COS(B));

LET(Y(3)=X(1)*SIN(A)*COS(B)-X(2)*SIN(A)*SIN(B)+X(3)*COS(A));

PRINT_OUT(Y(1),Y(2);Y(3));

PUT SKIP(5);

DO I=1 TO 3 BY 1;

LET(I="I");

DO J=1 TO 3 BY 1;

LET(J="J");

LET(TC(I,J)=0);

DO M-1 TO 3 BY 1;

LET(M="M");

DO N=1 TO 3 BY 1;

LET(N="N");

LET(TCC(I,J)=(DERIV(Y(M),X(I)))*(DERIV(Y(N),X(J)))*(C(M,N)));

LET(TC(I,J)=TC(I,J)+TCC(I,J));

END;

END;

PRINT_OUT(TC(I,J));

PUT SKIP(5);

END;

END;
```

$Y(1) = COS(B)COS(A)X(1)-COS(A)SIN(B)X(2)-SIN(A)X(3)$

$Y(2) = SIN(B)X(1)+COS(B)X(2)$

$Y(3) = COS(B)SIN(A)X(1)-SIN(B)SIN(A)X(2)+COS(A)X(3)$

$TC(1,1) = COS(B)COS(A)SIN(B)C(1,2)+COS^2(B)COS(A)SIN(A)(C(1,3)$

$\qquad +COS(B)COS(A)SIN(B)C(2,1)+SIN^2(B)C(2,2)$

$\qquad +COS(B)SIN(B)SIN(A)C(2,3)+COS^2(B)COS(A)SIN(A)C(3,1)$

$\qquad +COS(B)SIN(B)SIN(A)C(3,2)+COS^2(B)SIN^2(A)C(3,3)$

$\qquad +COS^2(B)COS^2(A)C(1,1)$

$TC(1,2) = COS^2(B)COS(A)C(1,2)-COS(B)COS(A)SIN(B)SIN(A)C(1,3)$

$\qquad -COS(A)SIN^2(B)C(2,1)+COS(B)SIN(B)C(2,2)$

$\qquad -SIN^2(B)SIN(A)C(2,3)-COS(B)COS(A)SIN(B)SIN(A)C(3,1)$

$\qquad +COS^2(B)SIN(A)C(3,2)-COS(B)SIN(B)SIN^2(A)C(3,3)$

$\qquad -COS(B)COS^2(A)SIN(B)C(1,1)$

$TC(1,3) = COS(B)COS^2(A)C(1,3)-SIN(B)SIN(A)C(2,1)$

$\qquad +COS(A)SIN(B)C(2,3)-COS(B)SIN^2(A)C(3,1)$

$\qquad +COS(B)COS(A)SIN(A)C(3,3)-COS(B)COS(A)SIN(A)C(1,1)$

$TC(2,1) = -COS(A)SIN^2(B)C(1,2)-COS(B)COS(A)SIN(B)SIN(A)C(1,3)$

$\qquad +COS^2(B)COS(A)C(2,1)+COS(B)SIN(B)C(2,2)$

$\qquad +COS^2(B)SIN(A)C(2,3)-COS(B)COS(A)SIN(B)SIN(A)C(3,1)$

$\qquad -SIN^2(B)SIN(A)C(3,2)-COS(B)SIN(B)SIN^2(A)C(3,3)$

$\qquad -COS(B)COS^2(A)SIN(B)C(1,1)$

$$TC(2,2) = -\cos(B)\cos(A)\sin(B)C(1,2)+\cos(A)\sin^2(B)\sin(A)C(1,3)$$
$$-\cos(B)\cos(A)\sin(B)C(2,1)+\cos^2(B)C(2,2)$$
$$-\cos(B)\sin(B)\sin(A)C(2,3)+\cos(A)\sin^2(B)\sin(A)C(3,1)$$
$$-\cos(B)\sin(B)\sin(A)C(3,2)+\sin^2(B)\sin^2(A)C(3,3)$$
$$+\cos^2(A)\sin^2(B)C(1,1)$$

$$TC(2,3) = -\cos^2(A)\sin(B)C(1,3)-\cos(B)\sin(A)C(2,1)$$
$$+\cos(B)\cos(A)C(2,3)+\sin(B)\sin^2(A)C(3,1)$$
$$-\cos(A)\sin(B)\sin(A)C(3,3)+\cos(A)\sin(B)\sin(A)C(1,1)$$

$$TC(3,1) = -\sin(B)\sin(A)C(1,2)-\cos(B)\sin^2(A)C(1,3)$$
$$+\cos(B)\cos^2(A)C(3,1)+\cos(A)\sin(B)C(3,2)$$
$$+\cos(B)\cos(A)\sin(A)C(3,3)-\cos(B)\cos(A)\sin(A)C(1,1)$$

$$TC(3,2) = -\cos(B)\sin(A)C(1,2)+\sin(B)\sin^2(A)C(1,3)$$
$$-\cos^2(A)\sin(B)C(3,1)+\cos(B)\cos(A)C(3,2)$$
$$-\cos(A)\sin(B)\sin(A)C(3,3)+\cos(A)\sin(B)\sin(A)C(1,1)$$

$$TC(3,3) = -\cos(A)\sin(A)C(1,3)-\cos(A)\sin(A)C(3,1)$$
$$+\cos^2(A)C(3,3)+\sin^2(A)C(1,1)$$

Interpretation of these results requires that the body axes derivatives $C(I,J)$ be treated as primed quantities and wind-tunnel derivatives $TC(I,J)$ as unprimed quantities. As indicated previously, the first index denotes the component of the aerodynamic force or moment, and the second index the component of the motion vector with respect to which the derivative is obtained.

When interpreted in terms of conventional aeronautical symbolism, the inverse aerodynamic derivatives with respect to p, q, r are

$$X_p = [X_p{}' \cos^2 A + Z_r{}' \sin^2 A + (X_r{}' + Z_p{}')\sin A \cos A] \cos^2 B$$
$$+ Y_q{}' \sin^2 B + [(X_q{}' + Y_p{}')\cos A + (Y_r{}' + Z_q{}')\sin A] \sin B \cos B$$

$$X_q = (X_q{}' \cos A + Z_q{}' \sin A)\cos^2 B - (Y_p{}' \cos A + Y_r{}' \sin A)\sin^2 B$$
$$- [X_p{}' \cos^2 A + Z_r{}' \sin^2 A + (X_r{}' + Z_p{}')\sin A \cos A - Y_q{}'] \sin B \cos B$$

$$X_r = [X_r{}' \cos^2 A - Z_p{}' \sin^2 A - (X_p{}' - Z_r{}')\sin A \cos A] \cos B$$
$$+ (Y_r{}' \cos A - Y_p{}' \sin A)\sin B$$

$$Y_p = (Y_p{}' \cos A + Y_r{}' \sin A)\cos^2 B - (X_q{}' \cos A + Z_q{}' \sin A)\sin^2 B$$
$$- [X_p{}' \cos^2 A + Z_r{}' \sin^2 A + (X_r{}' + Z_p{}')\sin A \cos A - Y_q{}'] \sin B \cos B$$

$$Y_q = Y_q{}' \cos^2 B + [X_p{}' \cos^2 A + Z_r{}' \sin^2 A + (X_r{}' + Z_p{}')\sin A \cos A] \sin^2 B$$
$$- [(X_q{}' + Y_p{}')\cos A + (Y_r{}' + Z_q{}')\sin A] \sin B \cos B$$

$$Y_r = (Y_r{}' \cos A - Y_p{}' \sin A)\cos B$$
$$+ [-X_r{}' \cos^2 A + Z_p{}' \sin^2 A + (X_p{}' - Z_r{}')\sin A \cos A] \sin B$$

$$Z_p = [Z_p{}' \cos^2 A - X_r{}' \sin^2 A - (X_p{}' - Z_r{}')\sin A \cos A] \cos B$$
$$+ (Z_q{}' \cos A - X_q{}' \sin A)\sin B$$

$$Z_q = (Z_q{}' \cos A - X_q{}' \sin A)\cos B$$
$$+ [-Z_p{}' \cos^2 A + X_r{}' \sin^2 A + (X_p{}' - Z_r{}')\sin A \cos A] \sin B$$

$$Z_r = Z_r{}' \cos^2 A + X_p{}' \sin^2 A - (X_r{}' + Z_p{}')\sin A \cos A$$

$$l_p = [l_p{}' \cos^2 A + n_r{}' \sin^2 A + (l_r{}' + n_p{}')\sin A \cos A] \cos^2 B$$
$$+ m_q \sin^2 B + [(l_q{}' + m_p{}')\cos A + (m_r{}' + n_q{}')\sin A] \sin B \cos B$$

$$l_q = (l_q' \cos A + n_q' \sin A)\cos^2 B - (m_p' \cos A + m_r' \sin A)\sin^2 B$$

$$- [l_p' \cos^2 A + n_r' \sin^2 A + (l_r' + n_p')\sin A \cos A - m_q']\sin B \cos B$$

$$l_r = [l_r' \cos^2 A - n_p' \sin^2 A - (l_p' - n_r')\sin A \cos A]\cos B$$

$$+ (m_r' \cos A - m_p' \sin A)\sin B$$

$$m_p = (m_p' \cos A + m_r' \sin A)\cos^2 B - (l_q' \cos A + n_q' \sin A)\sin^2 B$$

$$- [l_p' \cos^2 A + n_r' \sin^2 A + (l_r' + n_p')\sin A \cos A - m_q']\sin B \cos B$$

$$m_q = m_q' \cos^2 B + [l_p' \cos^2 A + n_r' \sin^2 A + (l_r' + n_p')\sin A \cos A]\sin^2 B$$

$$- [(l_q' + m_p')\cos A + (m_r' + n_q')\sin A]\sin B \cos B$$

$$m_r = (m_r' \cos A - m_p' \sin A)\cos B$$

$$+ [-l_r' \cos^2 A + n_p' \sin^2 A + (l_p' - n_r')\sin A \cos A]\sin B$$

$$n_p = [n_p' \cos^2 A - l_r' \sin^2 A - (l_p' - n_r')\sin A \cos A]\cos B$$

$$+ (n_q' \cos A - l_q' \sin A)\sin B$$

$$n_q = (n_q' \cos A - l_q' \sin A)\cos B$$

$$+ [-n_p' \cos^2 A + l_r' \sin^2 A + (l_p - n_r')\sin A \cos A]\sin B$$

$$n_r = n_r' \cos^2 A + l_p' \sin^2 A - (l_r' + n_p')\sin A \cos A$$

The inverse aerodynamic stability derivatives with respect to u, v, w are

$$X_u = [X_u' \cos^2 A + Z_w' \sin^2 A + (X_w' + Z_u')\sin A \cos A]\cos^2 B$$

$$+ Y_v' \sin^2 B + [(X_v' + Y_u')\cos A + (Y_w' + Z_v')\sin A]\sin B \cos B$$

$$X_v = (X_v{'} \cos A + Z_v{'} \sin A)\cos^2 B - (Y_u{'} \cos A + Y_w{'} \sin A)\sin^2 B$$
$$- [X_u{'} \cos^2 A + Z_w{'} \sin^2 A + (X_w{'} + Z_u{'})\sin A \cos A - Y_v{'}] \sin B \cos B$$

$$X_w = [X_w{'} \cos^2 A - Z_u{'} \sin^2 A - (X_u{'} - Z_w{'})\sin A \cos A] \cos B$$
$$+ (Y_w{'} \cos A - Y_u{'} \sin A)\sin B$$

$$Y_u = (Y_u{'} \cos A + Y_w{'} \sin A)\cos^2 B - (X_v{'} \cos A + Z_v{'} \sin A)\sin^2 B$$
$$- [X_u{'} \cos^2 A + Z_w{'} \sin^2 A + (X_w{'} + Z_u{'})\sin A \cos A - Y_v{'}] \sin B \cos B$$

$$Y_v = Y_v{'} \cos^2 B + [X_u{'} \cos^2 A + Z_w{'} \sin^2 A + (X_w{'} + Z_u{'})\sin A \cos A] \sin^2 B$$
$$+ [(X_v{'} + Y_u{'})\cos A + (Y_w{'} + Z_v{'})\sin A] \sin B \cos B$$

$$Y_w = (Y_w{'} \cos A - Y_u{'} \sin A)\cos B$$
$$+ [-X_w{'} \cos^2 A + Z_u{'} \sin^2 A + (X_u{'} - Z_w{'})\sin A \cos A] \sin B$$

$$Z_u = [Z_u{'} \cos^2 A - X_w{'} \sin^2 A - (X_u{'} - Z_w{'})\sin A \cos A] \cos B$$
$$+ (Z_v{'} \cos A - X_v{'} \sin A)\sin B$$

$$Z_v = (Z_v{'} \cos A - X_v{'} \sin A)\cos B$$
$$+ [-Z_u{'} \cos^2 A + X_w{'} \sin A + (X_u{'} - Z_w{'})\sin A \cos A] \sin B$$

$$Z_w = Z_w{'} \cos^2 A + X_u{'} \sin^2 A - (X_w{'} + Z_u{'})\sin A \cos A$$

$$l_u = [l_u{'} \cos^2 A + n_w{'} \sin^2 A + (l_w{'} + n_u{'})\sin A \cos A] \cos^2 B$$
$$+ m_v{'} \sin^2 B + [(l_v{'} + m_u{'})\cos A + (m_w{'} + n_v{'})\sin A] \sin B \cos B$$

$$l_v = (l_v{'} \cos A + n_v{'} \sin A)\cos^2 B - (m_u{'} \cos A + m_w{'} \sin A)\sin^2 B$$
$$+ [l_u{'} \cos^2 A + n_w{'} \sin^2 A + (l_w{'} + n_u{'})\sin A \cos A - m_v{'}] \sin B \cos B$$

$$l_w = [l_w{}' \cos^2 A - n_u{}' \sin^2 A - (l_u{}' - n_w{}')\sin A \cos A]\cos B$$

$$+ (m_w{}' \cos A - m_u{}' \sin A)\sin B$$

$$m_u = (m_u{}' \cos A + m_w{}' \sin A)\cos^2 B - (l_v{}' \cos A + n_v{}' \sin A)\sin^2 B$$

$$- [l_u{}' \cos^2 A + n_w{}' \sin^2 A + (l_w{}' + n_u{}')\sin A \cos A - m_v{}']\sin B \cos B$$

$$m_v = m_v{}' \cos^2 B + [l_u{}' \cos^2 A + n_w{}' \sin^2 A + (l_w{}' + n_u{}')\sin A \cos A]\sin^2 B$$

$$- [(l_v{}' + m_u{}')\cos A + (m_w{}' + n_v{}')\sin A]\sin B \cos B$$

$$m_w = (m_w{}' \cos A - m_u{}' \sin A)\cos B$$

$$+ [-l_w{}' \cos^2 A + n_u{}' \sin^2 A + (l_u{}' - n_w{}')\sin A \cos A]\sin B$$

$$n_u = [n_u{}' \cos^2 A - l_w{}' \sin^2 A - (l_u{}' - n_w{}')\sin A \cos A]\cos B$$

$$+ (n_v{}' \cos A - l_v{}' \sin A)\sin B$$

$$n_v = (n_v{}' \cos A - l_v{}' \sin A)\cos B$$

$$+ [-n_u{}' \cos^2 A + l_w{}' \sin^2 A + (l_u{}' - n_w{}')\sin A \cos A]\sin B$$

$$n_w = n_w{}' \cos^2 A + l_u{}' \sin^2 A - (l_w{}' + n_u{}')\sin A \cos A$$

2.9 TRANSFORMATION OF MOMENTS AND PRODUCTS OF INERTIA

The inertia tensor was discussed briefly in section 1.3. It was shown there that the inertia properties of a rigid-body were defined by the dyadic $\overline{\Phi}$, where

$$\overline{\Phi} = I_{XX}\hat{i}\hat{i} - I_{XY}\hat{i}\hat{j} - I_{XZ}\hat{i}\hat{k} - I_{YX}\hat{j}\hat{i} + I_{YY}\hat{j}\hat{j} - I_{YZ}\hat{j}\hat{k} - I_{ZX}\hat{k}\hat{i} - I_{ZY}\hat{k}\hat{j} + I_{ZZ}\hat{k}\hat{k}$$

$$(2.9.1)$$

In equation (2.9.1) the coefficients of the dyads are the moments and products of inertia, and $\hat{i}, \hat{j}, \hat{k}$ are a triad of mutually orthogonal unit vectors. This equation can be written in compact tensor notation as follows:

$$\overline{\Phi} = I_{\alpha\beta}\hat{a}^{\alpha}\hat{a}^{\beta} \tag{2.9.2}$$

where

$$\hat{a}^1 = \hat{i}$$

$$\hat{a}^2 = \hat{j}$$

$$\hat{a}^3 = \hat{k}$$

and

$$I_{11} = I_{XX} \;; \qquad I_{12} = -I_{XY} \;; \qquad I_{13} = -I_{XZ}$$

$$I_{21} = -I_{YX} \;; \qquad I_{22} = I_{YY} \;; \qquad I_{23} = -I_{YZ}$$

$$I_{31} = -I_{ZX} \;; \qquad I_{32} = -I_{ZY} \;; \qquad I_{33} = I_{ZZ}$$

Due to the equivalence of covariant and contravariant transformations in orthogonal Cartesian coordinate systems, equation (2.9.2) can be written in the alternative form

$$\overline{\Phi} = I^{\alpha\beta}\hat{a}_{\alpha}\hat{a}_{\beta} \tag{2.9.3}$$

where

$$\hat{a}_{\alpha} = \hat{a}^{\alpha}$$

and

$$I^{\alpha\beta} = I_{\alpha\beta}$$

The invariance of $\overline{\Phi}$ with respect to a coordinate transformation from the x coordinate system, to the y coordinate system, requires that

$$I^{ij}(y)\hat{b}_i(y)\hat{b}_j(y) = i^{\alpha\beta}(x)\hat{a}_{\alpha}(x)\hat{a}_{\beta}(x) \tag{2.9.4}$$

where

$\hat{b}_i(y)$ unit vectors in the y-coordinate system

$\hat{a}_\alpha(x)$ unit vectors in the x-coordinate system

Forming the scalar product of each side of equation (2.9.4) with $\hat{b}_j(y)$ gives

$$I^{ij}(y)\hat{b}_i(y) = I^{\alpha\beta}(x)\hat{a}_\alpha(x)\left\{\hat{b}_j(y) \cdot \hat{a}_\beta(x)\right\} \tag{2.9.5}$$

It is shown in section 1.5, equation (1.5.5), that

$$\hat{b}_j(y) \cdot \hat{a}_\beta(x) = \frac{\partial y^j}{\partial x^\beta} \tag{2.9.6}$$

Substitution from (2.9.6) in (2.9.5) gives

$$I^{ij}(y)\hat{b}_i(y) = \frac{\partial y^j}{\partial x^\beta} I^{\alpha\beta}(x)\hat{a}_\alpha(x) \tag{2.9.7}$$

Forming the scalar product of this equation with $\hat{b}_i(y)$ yields

$$I^{ij}(y) = \frac{\partial y^j}{\partial x^\beta} I^{\alpha\beta}(x)\left\{\hat{b}_i(y) \cdot \hat{a}_\alpha(x)\right\}$$

Using equation (1.5.5) again gives the transformation in the following form:

$$I^{ij}(y) = \frac{\partial y^i}{\partial x^\alpha} \frac{\partial y^j}{\partial x^\beta} I^{\alpha\beta}(x) \tag{2.9.8}$$

This is the transformation law for the components of a contravariant tensor of rank two (see eq. (1.8.3)). Equation (2.9.8) may be rewritten to conform to established terminology

$$I_{ij}(y) = \frac{\partial y^i}{\partial x^\alpha} \frac{\partial y^j}{\partial x^\beta} I_{\alpha\beta}(x) \tag{2.9.9}$$

Equation (2.9.8) is in complete agreement with equation (2.8.4). Therefore, the program used to process equation (2.8.4) can be used with equal facility to process equation (2.9.8) or (2.9.9). Hence, if a moment or product of inertia in wind-tunnel axes be denoted by $C(I,J)$, and the corresponding transformed inertia component be denoted by $TC(I,J)$, where $I,J = 1,2,3$, the output will be the transformed inertia components. There is one precaution to be observed, however. In interpreting the output, it should be noted that

$$C(1,1) = I_{XX} \; ; \qquad C(1,2) = -I_{XY} \; ; \qquad C(1,3) = -I_{XZ}$$

$$C(2,1) = -I_{YX} \; ; \qquad C(2,2) = I_{YY} \; ; \qquad C(2,3) = -I_{YZ}$$

$$C(3,1) = -I_{ZX} \; ; \qquad C(3,2) = -I_{ZY} \; ; \qquad C(3,3) = I_{ZZ}$$

Since these transformations are frequently used in aeronautical studies, they will be reproduced here for the convenience of readers.

2.9.1 Direct Transformations

When expressed in terms of conventional mathematical symbolism, the transformed inertia components assume the following form:

$$I'_{XX} = (I_{XX} \cos^2 B + 2I_{XY} \sin B \cos B + I_{YY} \sin^2 B)\cos A + I_{ZZ} \sin^2 A$$
$$+ (2I_{XZ} \cos B - 2I_{YZ} \sin B)\sin A \cos A$$

$$I'_{YY} = I_{YY} \cos^2 B + I_{XX} \sin^2 B - 2I_{XY} \sin B \cos B$$

$$I'_{ZZ} = I_{ZZ} \cos^2 A + (I_{XX} \cos^2 B + I_{YY} \sin^2 B + 2I_{XY} \sin B \cos B)\sin^2 A$$
$$- (2I_{XZ} \cos B - 2I_{YZ} \sin B)\sin A \cos A$$

$$I'_{XY} = [I_{XY}(\cos^2 B - \sin^2 B) - (I_{XX} - I_{YY})\sin B \cos B]\cos A$$
$$- (I_{YZ} \cos B + I_{XZ} \sin B)\sin A$$

$$I'_{XZ} = (I_{XZ} \cos B - I_{YZ} \sin B)\cos^2 A - (I_{XZ} \cos B - I_{YZ} \sin B)\sin^2 A$$

$$- (I_{XX} \cos^2 B + I_{YY} \sin^2 B + 2I_{XY} \sin B \cos B - I_{ZZ})\sin A \cos A$$

$$I'_{YZ} = (I_{YZ} \cos B + I_{XZ} \sin B)\cos A$$

$$+ [I_{XY}(\cos^2 B - \sin^2 B) - (I_{XX} - I_{YY})\sin B \cos B] \sin A$$

2.9.2 Inverse Transformation

The inverse transformation for inertia components is obtained by solving equation (2.9.8) for $I^{\alpha\beta}(x)$.

$$I^{\alpha\beta}(x) = \frac{\partial x^{\alpha}}{\partial y^{i}} \frac{\partial x^{\beta}}{\partial y^{j}} I^{ij}(y) \qquad (2.9.10)$$

Substitution from equation (2.8.3) in (2.9.10) gives

$$I^{\alpha\beta}(x) = \frac{\partial y^{i}}{\partial x^{\alpha}} \frac{\partial y^{j}}{\partial x^{\beta}} I^{ij}(y) \qquad (2.9.11)$$

Since i,j,β,α are dummy indices, equation (2.9.11) may be rewritten as

$$I^{ij}(x) = \frac{\partial y^{\alpha}}{\partial x^{i}} \frac{\partial y^{\beta}}{\partial x^{j}} I^{\alpha\beta}(y) \qquad (2.9.12)$$

or

$$I^{ij}(x) = \frac{\partial y^{m}}{\partial x^{i}} \frac{\partial y^{n}}{\partial x^{j}} I^{mn}(y) \qquad (2.9.13)$$

In order to be consistent with the notation previously established, the inertia components should be expressed in covariant form. Therefore

$$I_{ij}(x) = \frac{\partial y^{m}}{\partial x^{i}} \frac{\partial y^{n}}{\partial x^{j}} I_{mn}(y) \qquad (2.9.14)$$

This equation is in complete agreement with equation (2.8.9); therefore the program used to process that equation can be used to transform the inertia components from body axes to wind-tunnel axes. The results are given in section 2.8, and expressed in the following conventional mathematical notation:

$$I_{XX} = (I'_{XX} \cos^2 A + I'_{ZZ} \sin^2 A - 2I'_{XZ} \sin A \cos A)\cos^2 B + I'_{YY} \sin^2 B$$

$$- 2(I'_{XY} \cos A + I'_{YZ} \sin A)\sin B \cos B$$

$$I_{YY} = I'_{XX} \cos^2 B + (I'_{XX} \cos^2 A + I'_{ZZ} \sin^2 A - 2I'_{XZ} \sin A \cos A)\sin^2 B$$

$$+ 2(I'_{XY} \cos A + I'_{YZ} \sin A)\sin B \cos B$$

$$I_{ZZ} = I'_{ZZ} \cos^2 A + I'_{XX} \sin^2 A + 2I'_{XZ} \sin A \cos A$$

$$I_{XY} = (I'_{XY} \cos A + I'_{YZ} \sin A)\cos^2 B - (I'_{XY} \cos A + I'_{YZ} \sin A)\sin^2 B$$

$$+ (I'_{XX} \cos^2 A + I'_{ZZ} \sin^2 A - 2I'_{XZ} \sin A \cos A - I'_{YY})\sin B \cos B$$

$$I_{XZ} = [I'_{XZ}(\cos^2 A - \sin^2 A) + (I'_{XX} - I'_{ZZ})\sin A \cos A] \cos B$$

$$+ (I'_{YZ} \cos A - I'_{XY} \sin A)\sin B$$

$$I_{YZ} = (I'_{YZ} \cos A - I'_{XY} \sin A)\cos B + [I'_{XZ}(\sin^2 A - \cos^2 A)$$

$$- (I'_{XX} - I'_{ZZ})\sin A \cos A] \sin B$$

2.10 THE FORMULATION OF MATHEMATICAL MODELS OF AIRCRAFT

The response of an aircraft to the aerodynamic, thrust, gravity, and inertia forces acting on the vehicle in flight is determined by formulating a mathematical model of the system, solving the equations of the model, and using the solution to drive a simulator (ref. 5). If the mathematical model is a true representation of the aircraft and its environment, the response of the simulator will indicate to the aircraft engineer how the aircraft will behave in an actual flight environment.

The formulation of models of aeronautical systems for simulation and other purposes involves at least 12 equations — 3 force equations, 3 moment equations,

3 Euler angle equations to determine the spatial orientation of the body, and 3 equations to determine the location of the body in inertial space (ref. 1). Moreover, if the spatial orientation is determined by the use of direction cosines rather than Euler angles, the three Euler angle equations must be replaced by nine direction cosine equations. In view of this complexity, the formulation should be rendered amenable to mechanized procedures, routine manual derivations, or both.

An important aspect of the formulation of mathematical models of aeronautical systems is the specification of the system of forces and moments. In aeronautical applications, the thrust and gravity forces can be formulated without difficulty, but, as already indicated, the aerodynamic forces and moments require more detailed consideration. These are represented by the static forces and moments, the control derivatives, and the aerodynamic stability derivatives. As demonstrated in previous sections, these forces and moments have to be transformed from wind or wind-tunnel stability axes to aircraft body axes before the formulation can proceed. The equations of motion of aerospace vehicles are formulated with respect to body axes. The main advantage of these axes in motion calculations is that vehicle moments and products of inertia about the axes are constants. When the body axes are chosen so that the products of inertia vanish, they are known as principal axes. A system of axes which is frequently used to study the stability of aircraft in the presence of disturbing forces that produce small perturbations is the flight stability system. This is an orthogonal system fixed to the vehicle, the y^1 axis of which is aligned with the relative wind vector when the vehicle is in a steady-state condition, but then rotates with the vehicle after a disturbance as the vehicle changes angle of attack and sideslip. Some of these axes are shown in figure 2.10.1.

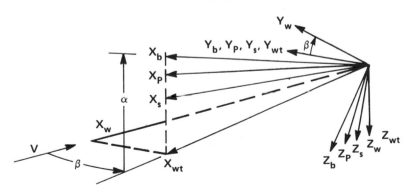

Figure 2.10.1.— Systems of reference axes, including body, principal, wind, flight stability, and wind-tunnel stability.

2.11 AERODYNAMIC FORCES

Two typical functional relations for the aerodynamic force and moment components $F_A{}^i$ acting on an aircraft in flight are

$$F_A{}^i = F_A{}^i(v^i, p^i, \dot{v}^i, \dot{p}^i, \delta_{c_i}, \dot{\delta}_{c_i})$$

and

$$F_A{}^i = F_A{}^i(A, B, V, \dot{A}, \dot{B}, \dot{V}, p^i, \dot{p}^i, \delta_{c_i}, \dot{\delta}_{c_i})$$

The independent variables in the first equation consist of the components of linear and angular velocity v^i, p^i; the components of linear and angular acceleration $\dot{v}^i$, $\dot{p}^i$; the control displacements δ_{c_i}; and the rate of change of these displacements $\dot{\delta}_{c_i}$. The corresponding variables in the second equation are the angle of attack A, the angle of sideslip B, and their rates of change $\dot{A}$ and $\dot{B}$; the linear velocity V of the aircraft and its linear acceleration $\dot{V}$; the components of angular velocity p^i and angular acceleration $\dot{p}^i$; and the control displacements δ_{c_i} and their rates of change $\dot{\delta}_{c_i}$.

To be of practical value, much simplification of the above functional relations is required. By assuming that the motion is limited to small perturbations, it is possible to simplify the mathematics and still obtain solutions of practical value. For small perturbations, the resulting forces and moments are given by the linear portions of a Taylor series expansion about the equilibrium state. These are

$$F^i_{Ao} + \left(\frac{\partial F_A{}^i}{\partial S^\tau} \right) \Delta S^\tau$$

where

F^i_{Ao} equilibrium values of the aerodynamic forces and moments

$\left(\dfrac{\partial F_A{}^i}{\partial S^\tau} \right)_o$ corresponding aerodynamic stability derivatives with respect to the state variables S^τ, measured at the equilibrium point

S^T state variables

ΔS^T small perturbations in the state variables

    Throughout this chapter the aerodynamic stability derivatives with respect to the variables v^i, p^i have been obtained. Although it has not been stated explicitly, only the dimensional forms of the aerodynamic forces, moments, and derivatives have been used. For many applications, the nondimensional forms of these parameters are preferred. However, it is less complicated to carry out the many transformations involved in the formulation of mathematical models if the dimensional forms are used. Consequently, the dimensional forms of the forces, moments, and derivatives will be used in this section, and only stability derivatives with respect to v^i and p^i will be considered. The same procedure may, of course, be used to transform the stability derivatives with respect to $A, B, V,$ and p^i.

    It has been demonstrated in section 2.3, that aerodynamic forces and moments which are measured in wind axes or wind-tunnel stability axes may be transformed to body axes when the corresponding coordinate transformation equations are known. The body axes coordinates y^i are related to the wind axes $x_w{}^i$ by equations of the form

$$y^i = y^i(x_w{}^i, A, B) \tag{2.11.1}$$

$$x_w{}^i = x_w{}^i(y^i, A, B) \tag{2.11.2}$$

where A and B are the angles of attack and sideslip, respectively. Since the transformation (2.11.1) represents a negative rotation B about the $x_w{}^3$ axis, followed by a positive rotation A about the resulting x^2 axis, it may be expressed in the following alternative forms

$$
\begin{pmatrix} y^1 \\ y^2 \\ y^3 \end{pmatrix} =
\begin{pmatrix} \cos A & 0 & -\sin A \\ 0 & 1 & 0 \\ \sin A & 0 & \cos A \end{pmatrix}
\begin{pmatrix} \cos B & -\sin B & 0 \\ \sin B & \cos B & 0 \\ 0 & 0 & 1 \end{pmatrix}
\begin{pmatrix} x_w{}^1 \\ x_w{}^2 \\ x_w{}^3 \end{pmatrix}
$$

$$y^1 = x_w{}^1 \cos B \cos A - x_w{}^2 \cos A \sin B - x_w{}^3 \sin A$$

$$y^2 = x_w{}^1 \sin B + x_w{}^2 \cos B$$

$$y^3 = x_w{}^1 \sin A \cos B - x_w{}^2 \sin A \sin B + x_w{}^3 \cos A$$

$$(2.11.3)$$

The transformation (2.11.2) is obtained by solving equation (2.11.3) for $x_w{}^i$.

$$
\begin{pmatrix} x_w{}^1 \\ x_w{}^2 \\ x_w{}^3 \end{pmatrix} = \begin{pmatrix} \cos B & \sin B & 0 \\ -\sin B & \cos B & 0 \\ 0 & 0 & 1 \end{pmatrix} \begin{pmatrix} \cos A & 0 & \sin A \\ 0 & 1 & 0 \\ -\sin A & 0 & \cos A \end{pmatrix} \begin{pmatrix} y^1 \\ y^2 \\ y^3 \end{pmatrix}
$$

$$x_w{}^1 = y^1 \cos A \cos B + y^2 \sin B + y^3 \sin A \cos B$$

$$x_w{}^2 = -y^1 \cos A \sin B + y^2 \cos B - y^3 \sin A \sin B$$

$$x_w{}^3 = -y^1 \sin A + y^3 \cos A$$

$$(2.11.4)$$

Moreover, if a static force or moment in wind axes be denoted by $C(\alpha)$, and the corresponding transformed coefficient be denoted by $TC(i)$, where $i = 1,2,3$, then as indicated in equation (2.3.1)

$$F_A{}^i = TC(i) = \frac{\partial y^i}{\partial x^\alpha} C(\alpha) \qquad (2.11.5)$$

These coefficients represent the aerodynamic forces and moments acting during a state of equilibrium. If small perturbations about the equilibrium condition occur, then, as previously indicated, the resulting forces and moments are given by the linear portions of the Taylor series expansion

$$F^i_{Ao} + \left(\frac{\partial F_A{}^i}{\partial S^\tau} \right)_o \Delta S^\tau$$

where

$$F^i_{Ao} = \frac{\partial y^i}{\partial x^\alpha} \, C_o(\alpha)$$

If an aerodynamic stability derivative in wind axes be denoted by $C(\alpha,\beta)$, and the corresponding transformed derivative be denoted by $TC(i,\tau)$, where $i,\tau,\alpha,\beta = 1,2,3$, equation (2.7.6) gives

$$\frac{\partial F_A^{\,i}}{\partial S^\tau} = TC(i,\tau) = \frac{\partial y^i}{\partial x^\alpha} \, \frac{\partial x^\beta}{\partial y^\tau} \, C(\alpha,\beta) \qquad (2.11.6)$$

Hence, if $C_o(\alpha)$ be equilibrium values of the static force and moment coefficients in wind axes, and small perturbations are assumed, the resulting aerodynamic forces and moments $F_A^{\,i}$ in body axes are

$$F_A^{\,i} = \frac{\partial y^i}{\partial x^\alpha} \, C_o(\alpha) + \frac{\partial y^i}{\partial y^\alpha} \, \frac{\partial x^\beta}{\partial y^\tau} \, v^\tau C_{\overline{V}}(\alpha,\beta) + \frac{\partial y^i}{\partial x^\alpha} \, \frac{\partial x^\beta}{\partial y^\tau} \, p^\tau C_{\overline{\omega}}(\alpha,\beta) \quad (2.11.7)$$

where v^τ and p^τ are perturbation components of the linear and angular velocity vectors, respectively, and the subscripts $\overline{V}$ and $\overline{\omega}$ denote differentiation with respect to linear and angular velocity components, respectively.

The aircraft's control surfaces give rise to additional aerodynamic forces and moments. These forces and moments are represented by control derivatives that obey the same transformation law as the static forces and moments, that is

$$TC_c(i) = \frac{\partial y^i}{\partial x^\alpha} \, C_c(\alpha) \qquad (2.11.8)$$

where the subscript c denotes a control force or moment derivative. The corresponding control forces $f_c^{\,i}$ are obtained by multiplying the control derivatives by the appropriate control increments δ_{c_i}. These are

$$f_c^{\,i} = \frac{\partial y^i}{\partial x^\alpha} \, C_c(\alpha)\delta_{c_i} \qquad (2.11.9)$$

When combined, equations (2.11.7) and (2.11.9) yield the total aerodynamic forces for the case considered.

$$F_A{}^i = \frac{\partial y^i}{\partial x^\alpha} C_0(\alpha) + \frac{\partial y^i}{\partial x^\alpha}\frac{\partial x^\beta}{\partial y^\tau} v^\tau C_{\overline{V}}(\alpha,\beta) + \frac{\partial y^i}{\partial x^\alpha}\frac{\partial x^\beta}{\partial y^\tau} p^\tau C_{\overline{\omega}}(\alpha,\beta) + \frac{\partial y^i}{\partial x^\alpha} C_c(\alpha)\delta_{c_i}$$

$$(2.11.10)$$

It should be noted, however, that the aerodynamic stability derivatives, with respect to components of the linear and angular acceleration vectors, have been omitted in this formulation.

Equation (2.11.3) may be used to evaluate the partial differential coefficients in equation (2.11.10). However, this operation may be simplified by taking advantage of the following relationship, which is valid in all orthogonal Cartesian coordinate systems.

$$\frac{\partial y^\tau}{\partial x^\beta} = \frac{\partial x^\beta}{\partial y^\tau}$$

Substitution of this result in equation (2.11.10) yields

$$F_A{}^i = \frac{\partial y^i}{\partial x^\alpha} C_0(\alpha) + \frac{\partial y^i}{\partial x^\alpha}\frac{\partial y^\tau}{\partial x^\beta} v^\tau C_{\overline{V}}(\alpha,\beta) + \frac{\partial y^i}{\partial x^\alpha}\frac{\partial y^\tau}{\partial x^\beta} p^\tau C_{\overline{\omega}}(\alpha,\beta) + \frac{\partial y^i}{\partial x^\alpha} C_c(\alpha)\delta_{c_i}$$

$$(2.11.11)$$

Note that this modification eliminates the need for the transformation equation (2.11.4). Only the direct transformation (2.11.3) is now required.

This equation represents six equations: three force equations and three moment equations. In the form given, the coefficients $C(\alpha)$, $C_c(\alpha)$, and $C(\alpha,\beta)$ are subject to a dual interpretation; that is, $C(\alpha)$ represents either a force or a moment and $C_c(\alpha)$ represents either a control force derivative or a control moment derivative. Moreover, $C_{\overline{V}}(\alpha,\beta)$ are aerodynamic stability derivatives of either force or moment components with respect to linear motion vector components. Likewise, $C_{\overline{\omega}}(\alpha,\beta)$ are aerodynamic stability derivatives of either force or moment components with respect to angular motion vector components. It is seen that use of the summation convention permits a compact formulation of the aerodynamic forces and moments.

The advantage of the notation used is evident when one considers that, in general, the aerodynamic stability derivatives with respect to linear and angular velocity components give rise to 324 terms. An additional 324 terms would be required, in general, if the acceleration derivatives were included.

Although the form shown is amenable to symbolic mathematical computation, it is more convenient from the point of view of manual formulation to avoid the dual interpretation of the coefficients, and separate the force and moment equations.

For $i = 1,2,3$, the aerodynamic forces $F_A{}^i$ are

$$F_A{}^i = \frac{\partial y^i}{\partial x^\alpha} f_o(\alpha) + \frac{\partial y^i}{\partial x^\alpha} \frac{\partial y^\tau}{\partial x^\beta} v^\tau f_{\overline{V}}(\alpha,\beta) + \frac{\partial y^i}{\partial x^\alpha} \frac{\partial y^\tau}{\partial x^\beta} p^\tau f_{\overline{\omega}}(\alpha,\beta) + \frac{\partial y^i}{\partial x^\alpha} f_c(\alpha)\delta_{c_i}$$

where

$f_o(\alpha)$ static forces

$f_{\overline{V}}(\alpha,\beta)$ stability derivatives of the aerodynamic forces with respect to linear velocity components

$f_{\overline{\omega}}(\alpha,\beta)$ stability derivatives of the aerodynamic forces with respect to angular velocity components

$f_c(\alpha)$ control force derivatives

The aerodynamic moments have exactly the same form. These are

$$M_A{}^i = \frac{\partial y^i}{\partial x^\alpha} m_o(\alpha) + \frac{\partial y^i}{\partial x^\alpha} \frac{\partial y^\tau}{\partial x^\beta} v^\tau m_{\overline{V}}(\alpha,\beta) + \frac{\partial y^i}{\partial x^\alpha} \frac{\partial y^\tau}{\partial x^\beta} p^\tau m_{\overline{\omega}}(\alpha,\beta) + \frac{\partial y^i}{\partial x^\alpha} m_c(\alpha)\delta_{c_i}$$

where

m_o static moments

$m_{\overline{V}}(\alpha,\beta)$ stability derivatives of the aerodynamic moments with respect to linear velocity components

$m_{\overline{\omega}}(\alpha,\beta)$ stability derivatives of the aerodynamic moments with respect to angular velocity components

$m_c(\alpha)$ control moment derivatives

2.12 THRUST FORCES

The same procedure may be employed to transform the components of the thrust vectors to body axes. It is assumed that there are n thrust generating systems T_n. Each thrust vector is referred to a thrust axes system, with origin at the point of application of the thrust vector. The axes are chosen such that each thrust vector coincides with the $x_n{}^1$ axis of the system. Moreover, each thrust vector is then transformed to a coordinate system $Y_n{}^i$ which has the same origin as the thrust axes, but is parallel to the body axes system. Finally, the components of thrust in the $Y_n{}^i$ system of axes are transformed to the body axes system, which has its origin at the center of gravity of the aircraft. Each thrust axis $x_n{}^1$ is related to the system $Y_n{}^i$ by transformation equations of the form

$$Y_n{}^i = Y_n{}^i(x_n{}^1, \theta_T{}^n, \psi_T{}^n) \tag{2.12.1}$$

The index n denotes which thrust generating system is being considered, and the subscript T denotes thrust, that is, $\theta_T{}^n$ and $\psi_T{}^n$ determine the orientation of the nth thrust vector.

As sketch (h) shows, the coordinates $Y_n{}^i$ are related to the thrust axes coordinates $x_n{}^1$ by the transformation equations

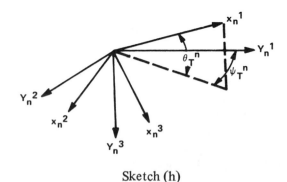

Sketch (h)

$$Y_n^1 = x_n^1 \cos \theta_{T^n} \cos \psi_{T^n}$$

$$Y_n^2 = x_n^1 \cos \theta_{T^n} \sin \psi_{T^n} \qquad (2.12.2)$$

$$Y_n^3 = -x_n^1 \sin \theta_{T^n}$$

Hence, the components of the thrust vector T_n in the Y^i system of axes are

$$\frac{\partial Y_n^1}{\partial x_n^1} T_n \; ; \qquad \frac{\partial Y_n^2}{\partial x_n^1} T_n \; ; \qquad \frac{\partial Y_n^3}{\partial x_n^1} T_n$$

These are also the components of thrust in the y^i system of coordinates, which has its origin at the center of gravity of the aircraft. The thrust components due to all thrust generating systems are obtained from the equation

$$F_T^i = \frac{\partial Y_n^i}{\partial x_n^1} T_n \qquad (2.12.3)$$

The individual components are obtained by summing on n. For $n = 4$, these are

$$F_T^1 = \frac{\partial Y_1^1}{\partial x_1^1} T_1 + \frac{\partial Y_2^1}{\partial x_2^1} T_2 + \frac{\partial Y_3^1}{\partial x_3^1} T_3 + \frac{\partial Y_4^1}{\partial x_4^1} T_4$$

$$F_T^2 = \frac{\partial Y_1^2}{\partial x_1^1} T_1 + \frac{\partial Y_2^2}{\partial x_2^1} T_2 + \frac{\partial Y_3^2}{\partial x_3^1} T_3 + \frac{\partial Y_4^2}{\partial x_4^1} T_4$$

$$F_T^3 = \frac{\partial Y_1^3}{\partial x_1^1} T_1 + \frac{\partial Y_2^3}{\partial x_2^1} T_2 + \frac{\partial Y_3^3}{\partial x_3^1} T_3 + \frac{\partial Y_4^3}{\partial x_4^1} T_4$$

When the coefficients are evaluated from equation (2.12.2), we obtain

$$F_T^1 = T_1 \cos \theta_{T^1} \cos \psi_{T^1} + T_2 \cos \theta_{T^2} \cos \psi_{T^2}$$

$$+ T_3 \cos \theta_{T^3} \cos \psi_{T^3} + T_4 \cos \theta_{T^4} \cos \psi_{T^4}$$

$$F_T^2 = T_1 \cos\theta_T{}^1 \sin\psi_T{}^1 + T_2 \cos\theta_T{}^2 \sin\psi_T{}^2$$

$$+ T_2 \cos\theta_T{}^3 \sin\psi_T{}^3 + T_4 \cos\theta_T{}^4 \sin\psi_T{}^4$$

$$F_T^3 = T_1 \sin\theta_T{}^1 + T_2 \sin\theta_T{}^2 + T_3 \sin\theta_T{}^3 + T_4 \sin\theta_T{}^4$$

2.13 THRUST MOMENTS

The moments produced by all thrust generating systems T_n are $M_T{}^i$, where $i = 1,2,3$, that is

$$M_T{}^i = T_n \left(y_n{}^j \frac{\partial Y_n{}^k}{\partial x_n{}^1} - y_n{}^k \frac{\partial Y_n{}^j}{\partial x_n{}^1} \right) \tag{2.13.1}$$

where $y_n{}^i$ are the coordinates of the point of application of the nth thrust vector and i,j,k are in cyclic order.

Assuming again that there are four thrust generating systems, and summing on n, we obtain

$$M_T{}^i = T_1 \left(y_1{}^j \frac{\partial Y_1{}^k}{\partial x_1{}^1} - y_1{}^k \frac{\partial Y_1{}^j}{\partial x_1{}^1} \right) + T_2 \left(y_2{}^j \frac{\partial Y_2{}^k}{\partial x_2{}^1} - y_2{}^k \frac{\partial Y_2{}^j}{\partial x_2{}^1} \right)$$

$$+ T_3 \left(y_3{}^j \frac{\partial Y_3{}^k}{\partial x_3{}^1} - y_3{}^k \frac{\partial Y_3{}^j}{\partial x_3{}^1} \right) + T_4 \left(y_4{}^j \frac{\partial Y_4{}^k}{\partial x_4{}^1} - y_4{}^k \frac{\partial Y_4{}^j}{\partial x_4{}^1} \right)$$

By assigning appropriate values to the superscripts i,j,k, and remembering that these have to be in cyclic order, we obtain the three thrust moments as follows

$$M_T{}^1 = T_1 \left(y_1{}^2 \frac{\partial Y_1{}^3}{\partial x_1{}^1} - y_1{}^3 \frac{\partial Y_1{}^2}{\partial x_1{}^1} \right) + T_2 \left(y_2{}^2 \frac{\partial Y_2{}^3}{\partial x_2{}^1} - y_2{}^3 \frac{\partial Y_2{}^2}{\partial x_2{}^1} \right)$$

$$+ T_3 \left(y_3{}^2 \frac{\partial Y_3{}^3}{\partial x_3{}^1} - y_3{}^3 \frac{\partial Y_3{}^2}{\partial x_3{}^1} \right) + T_4 \left(y_4{}^2 \frac{\partial Y_4{}^3}{\partial x_4{}^1} - y_4{}^3 \frac{\partial Y_4{}^2}{\partial x_4{}^1} \right)$$

$$M_T{}^2 = T_1\left(y_1{}^3 \frac{\partial Y_1{}^1}{\partial x_1{}^1} - y_1{}^1 \frac{\partial Y_1{}^3}{\partial x_1{}^1}\right) + T_2\left(y_2{}^3 \frac{\partial Y_2{}^1}{\partial x_2{}^1} - y_2{}^1 \frac{\partial Y_2{}^3}{\partial x_2{}^1}\right)$$

$$+ T_3\left(y_3{}^3 \frac{\partial Y_3{}^1}{\partial x_3{}^1} - y_3{}^1 \frac{\partial Y_3{}^3}{\partial x_3{}^1}\right) + T_4\left(y_4{}^3 \frac{\partial Y_4{}^1}{\partial x_4{}^1} - y_4{}^1 \frac{\partial Y_4{}^3}{\partial x_4{}^1}\right)$$

$$M_T{}^3 = T_1\left(y_1{}^1 \frac{\partial Y_1{}^2}{\partial x_1{}^1} - y_1{}^2 \frac{\partial Y_1{}^1}{\partial x_1{}^1}\right) + T_2\left(y_2{}^1 \frac{\partial Y_2{}^2}{\partial x_2{}^1} - y_2{}^2 \frac{\partial Y_2{}^1}{\partial x_2{}^1}\right)$$

$$+ T_3\left(y_3{}^1 \frac{\partial Y_3{}^2}{\partial x_3{}^1} - y_3{}^2 \frac{\partial Y_3{}^1}{\partial x_3{}^1}\right) + T_4\left(y_4{}^1 \frac{\partial Y_4{}^2}{\partial x_4{}^1} - y_4{}^2 \frac{\partial Y_4{}^1}{\partial x_4{}^1}\right)$$

Substitution from equation (2.12.2) in these equations gives the moment components in terms of the orientation of each thrust vector. These are

$$M_T{}^1 = -[T_1(y_1{}^2 \sin\theta_T{}^1 + y_1{}^3 \cos\theta_T{}^1 \sin\psi_T{}^1)$$

$$+ T_2(y_2{}^2 \sin\theta_T{}^2 + y_2{}^3 \cos\theta_T{}^2 \sin\psi_T{}^2)$$

$$+ T_3(y_3{}^2 \sin\theta_T{}^3 + y_3{}^3 \cos\theta_T{}^3 \sin\psi_T{}^3)$$

$$+ T_4(y_4{}^2 \sin\theta_T{}^4 + y_4{}^3 \cos\theta_T{}^4 \sin\psi_T{}^4)]$$

$$M_T{}^2 = [T_1(y_1{}^3 \cos\theta_T{}^1 \cos\psi_T{}^1 + y_1{}^1 \sin\theta_T{}^1)$$

$$+ T_2(y_2{}^3 \cos\theta_T{}^2 \cos\psi_T{}^2 + y_2{}^1 \sin\theta_T{}^2)$$

$$+ T_3(y_3{}^3 \cos\theta_T{}^3 \cos\psi_T{}^3 + y_3{}^1 \sin\theta_T{}^3)$$

$$+ T_4(y_4{}^3 \cos\theta_T{}^3 \cos\psi_T{}^4 + y_4{}^1 \sin\theta_T{}^4)]$$

$$M_{T}{}^{3} = [T_{1}(y_{1}{}^{1} \cos \theta_{T}{}^{1} \sin \psi_{T}{}^{1} - y_{1}{}^{2} \cos \theta_{T}{}^{1} \cos \psi_{T}{}^{1})$$

$$+ T_{2}(y_{2}{}^{1} \cos \theta_{T}{}^{2} \sin \psi_{T}{}^{2} - y_{2}{}^{2} \cos \theta_{T}{}^{2} \cos \psi_{T}{}^{2})$$

$$+ T_{3}(y_{3}{}^{1} \cos \theta_{T}{}^{3} \sin \psi_{T}{}^{3} - y_{3}{}^{2} \cos \theta_{T}{}^{3} \cos \psi_{T}{}^{3})$$

$$+ T_{4}(y_{4}{}^{1} \cos \theta_{T}{}^{4} \sin \psi_{T}{}^{4} - y_{4}{}^{2} \cos \theta_{T}{}^{4} \cos \psi_{T}{}^{4})]$$

2.14 GRAVITY FORCES

Newton's law of gravitation states that every particle in the universe attracts every other particle with a force which is directly proportional to the product of the two masses and inversely proportional to the square of the distance between them, the direction of the force being in the line joining the two points (ref. 6); that is, the gravitational attraction of a particle of mass m_i toward a particle of mass m_j is

$$\frac{\lambda m_i m_j}{r_{ij}^2}$$

where λ is the gravitational constant and r_{ij} is the distance between the particles.

Since the law applies only to particles, the attraction exerted by bodies of finite size must be determined. However, because the potential is a scalar function and force a vector function, it is frequently more convenient to obtain the potential of the attracting mass first, and then determine the force associated with the given potential by using the well-known relation

$$F_g = \nabla V$$

Let the potential function be defined by the integral

$$V = \lambda \int \frac{dm}{\rho}$$

and consider the potential of the spherical shell shown in sketch (i).

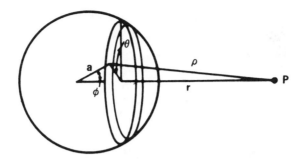

Sketch (i)

and

$$dm = \sigma a^2 t \sin \phi \, d\phi \, d\theta$$

where σ and t are the density and the thickness, respectively, and

$$\rho^2 = a^2 + r^2 - 2ar \cos \phi$$

which is independent of 0. On differentiating with respect to ρ, bearing in mind that a and r are constant, it is found that

$$\rho = ar \sin \phi \, \frac{d\phi}{d\rho}$$

Therefore

$$\frac{dm}{\rho} = \frac{\sigma at}{r} \, d\rho \int d\theta = \frac{2\pi at\sigma \, d\rho}{r}$$

and

$$V = \frac{\lambda 2\pi at\sigma}{r} \int_{r-a}^{r+a} d\rho = \frac{\lambda 4\pi a^2 t\sigma}{r} = \frac{\lambda M}{r}$$

From this result it follows that if M_s is the mass of a homogeneous shell and if the attracted point P lies outside the shell at a distance r from its center, the force of attraction of the shell on a unit mass at P is

$$F_g = \nabla V = -\frac{\lambda M_s}{r^2}$$

and is directed toward the center of the sphere. The resulting attraction of any number of such shells that are concentric is directed toward their common center, and its intensity is simply the sum of the intensities for the individual shells. Hence, if M is the mass of a sphere, its attraction upon an exterior point is

$$F_g = -\frac{\lambda M}{r^2}$$

It is not necessary that all the shells have the same volume density. It is sufficient that each shell separately shall be homogeneous. It is evident that a solid sphere which is homogeneous, or homogeneous in concentric layers, attracts a unit mass which is located at an exterior point, as though it were a particle of the same mass located at the center of the sphere.

If the acceleration due to gravity at the surface of a sphere of radius R is g, then the gravitational force attracting a body of mass m, located on the surface of the sphere, is

$$F_g = -\frac{\lambda Mm}{R^2} = -mg$$

and therefore

$$\lambda M = gR^2$$

It follows that the gravitational force attracting a body of mass m, which is located at a distance $r > R$ from the surface of the sphere, is given by the equation

$$F_g = -\frac{\lambda Mm}{r^2} = -mg\frac{R^2}{r^2}$$

The corresponding potential function is

$$V = \frac{\lambda Mm}{r} = \frac{mgR^2}{r}$$

It should be noted that this equation fails to give an accurate description of the influence of the Earth's gravitational field on bodies, such as satellites, in close Earth orbit. Due to lack of homogeneity and sphericity, the Earth's gravitational potential function consists of an infinite series of spherical harmonics. However, the assumption of a homogeneous, spherical Earth is adequate for most aeronautical applications. Indeed, it is frequently assumed, as the ancients did, that the Earth is flat. By assuming that the Earth is an indefinitely extended plane with a constant surface density, it is found that the acceleration due to gravity is independent of the distance from the Earth's surface. This can be seen as follows:

Consider a uniform circular disc of radius R (see sketch (j)), and surface density σ, then

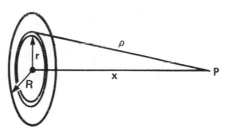

Sketch (j)

$$dm = \sigma 2\pi r \, dr$$

and

$$V = \lambda \int \frac{dm}{\rho} = \sigma 2\pi\lambda \int_{0}^{R} \frac{r \, dr}{\rho}$$

But

$$\rho^2 = r^2 + x^2$$

Therefore

$$\rho \, d\rho = r \, dr$$

and so

$$\frac{r \, dr}{\rho} = d\rho$$

Hence

$$V = 2\pi\sigma\lambda \int_{x}^{\sqrt{R^2+x^2}} d\rho = 2\pi\sigma\lambda\left(\sqrt{R^2 + x^2} - x\right)$$

The force of attraction of the disc on a unit mass on its axis is given by the equation

$$\frac{\partial V}{\partial x} = 2\pi\sigma\lambda\left(\frac{x}{\sqrt{R^2 + x^2}} - 1\right)$$

As the radius R tends to infinity, the first term on the right-hand side of this equation tends to zero, and the force of attraction on a unit mass assumes the form

$$\frac{\partial V}{\partial x} = -2\pi\sigma\lambda$$

Hence, the acceleration due to the gravitational attraction of an Earth which is assumed to be an indefinitely extended plane would be

$$g = 2\pi\sigma\lambda$$

and the gravitational force acting on a body of mass m would be

$$mg = 2\pi\sigma\lambda m = F_g$$

which is independent of the distance from the surface.

The gravitational force vector acting on an aircraft in flight will be assumed to have the value $m\bar{g}$, where m is the mass of the aircraft and $\bar{g}$ is the acceleration vector. The magnitude of g is assumed constant, which is tantamount to the assumption of a flat Earth.

The gravity vector is specified in an Earth-fixed system of axes, in which the coordinates $x_g{}^i$ are related to the body axes coordinate y^i by equations of the form

$$y^i = y^i(x_g{}^i, \psi_g, \theta_g, \phi_g) \qquad (2.14.1)$$

where ψ_g, θ_g, and ϕ_g are the Eulerian angles which relate the moving body axes to the set of Earth-fixed axes in which the gravity vector is specified. In accordance with aeronautical convention, these transformation equations represent the result of a rotation ψ_g about the y^3 body axis, followed by rotations θ_g and ϕ_g about the y^2 and y^3 axes, respectively. Hence, if it is assumed that the body axes and the Earth-fixed axes are initially coincident, equation (2.14.1) may be expressed in the following alternative forms

$$\begin{pmatrix} y^1 \\ y^2 \\ y^3 \end{pmatrix} = \begin{pmatrix} 1 & 0 & 0 \\ 0 & \cos\phi_g & \sin\phi_g \\ 0 & -\sin\phi_g & \cos\phi_g \end{pmatrix} \begin{pmatrix} \cos\theta_g & 0 & -\sin\theta_g \\ 0 & 1 & 0 \\ \sin\theta_g & 0 & \cos\theta_g \end{pmatrix} \begin{pmatrix} \cos\psi_g & \sin\psi_g & 0 \\ -\sin\psi_g & \cos\psi_g & 0 \\ 0 & 0 & 1 \end{pmatrix} \begin{pmatrix} x_g{}^1 \\ x_g{}^2 \\ x_g{}^3 \end{pmatrix}$$

$$\begin{pmatrix} y^1 \\ y^2 \\ y^3 \end{pmatrix} = \begin{pmatrix} \cos\theta_g\cos\psi_g & \cos\theta_g\sin\psi_g & -\sin\theta_g \\ -\cos\phi_g\sin\psi_g + \sin\phi_g\sin\theta_g\cos\psi_g & \cos\phi_g\cos\psi_g + \sin\phi_g\sin\theta_g\sin\psi_g & \sin\phi_g\cos\theta_g \\ \sin\phi_g\sin\psi_g + \cos\phi_g\sin\theta_g\cos\psi_g & -\sin\phi_g\cos\psi_g + \cos\phi_g\sin\theta_g\sin\psi_g & \cos\phi_g\cos\theta_g \end{pmatrix} \begin{pmatrix} x_g{}^1 \\ x_g{}^2 \\ x_g{}^3 \end{pmatrix}$$

$$(2.14.2)$$

Therefore

$$y^1 = x_g^1 \cos\theta_g \cos\psi_g + x_g^2 \cos\theta_g \sin\psi_g - x_g^3 \sin\theta_g$$

$$y^2 = x_g^1 (\sin\phi_g \sin\theta_g \cos\psi_g - \cos\phi_g \sin\psi_g)$$

$$+ x_g^2 (\cos\phi_g \cos\psi_g + \sin\phi_g \sin\theta_g \sin\psi_g + x_g^3 \sin\phi_g \cos\theta_g) \qquad (2.14.3)$$

$$y^3 = x_g^1 (\sin\phi_g \sin\psi_g + \cos\phi_g \sin\theta_g \cos\psi_g)$$

$$+ x_g^2 (\cos\phi_g \sin\theta_g \sin\psi_g - \sin\phi_g \cos\psi_g) + x_g^3 \cos\phi_g \cos\theta_g$$

Using the notation already established, these equations assume the more compact form

$$y^i = \frac{\partial y^i}{\partial x_g^\alpha} x_g^\alpha \qquad (2.14.4)$$

where $i, \alpha = 1, 2, 3$.

If a gravity force component in body axes be denoted by F_g^i and the gravity vector is assumed to coincide with the x_g^3 axis of the Earth-fixed system

$$F_g^i = \frac{\partial y^i}{\partial x_g^3} mg \qquad (2.14.5)$$

Equations (2.14.3) may now be used to evaluate the individual terms of equation (2.14.5). These are

$$F_g^1 = \frac{\partial y^1}{\partial x_g^3} mg = -mg \sin\theta_g$$

$$F_g^2 = \frac{\partial y^2}{\partial x_g^3} mg = mg \sin\phi_g \cos\theta_g \qquad (2.14.6)$$

$$F_g^3 = \frac{\partial y^3}{\partial x_g^3} mg = mg \cos\phi_g \cos\theta_g$$

2.15 SUMMATION OF FORCES AND MOMENTS

The total force acting on the aircraft is the sum of the aerodynamic, thrust, and gravity forces, that is

$$F^i = F_A{}^i + F_T{}^i + F_g{}^i \tag{2.15.1}$$

Hence

$$F^i = \left[\frac{\partial y^i}{\partial x^\alpha} f_0(\alpha) + \frac{\partial y^i}{\partial x^\alpha} \frac{\partial y^\tau}{\partial x^\beta} v^\tau f_{\overline{V}}(\alpha,\beta) + \frac{\partial y^i}{\partial x^\alpha} \frac{\partial y^\tau}{\partial x^\beta} p^\tau f_{\overline{\omega}}(\alpha,\beta) \right.$$
$$\left. + \frac{\partial y^i}{\partial x^\alpha} f_c(\alpha)\delta_{c_i} \right] + \frac{\partial y^i}{\partial x_n{}^1} T_n + \frac{\partial y^i}{\partial x_g{}^3} mg \tag{2.15.2}$$

Likewise, the total moment is the sum of the aerodynamic and thrust moments. These are

$$M^i = M_A{}^i + M_T{}^i \tag{2.15.3}$$

where

$$M^i = \left[\frac{\partial y^i}{\partial x^\alpha} m_0(\alpha) + \frac{\partial y^i}{\partial x^\alpha} \frac{\partial y^\tau}{\partial x^\beta} v^\tau m_{\overline{V}}(\alpha,\beta) + \frac{\partial y^i}{\partial x^\alpha} \frac{\partial y^\tau}{\partial x^\beta} p^\tau m_{\overline{\omega}}(\alpha,\beta) \right.$$
$$\left. + \frac{\partial y^i}{\partial x^\alpha} m_c(\alpha)\delta_{c_i} \right] + \left(y_n{}^j \frac{\partial Y_n{}^k}{\partial x_n{}^1} - y_n{}^k \frac{\partial Y_n{}^j}{\partial x_n{}^1} \right) T_n \tag{2.15.4}$$

and i, j, k are in cyclic order.

The equations of motion can now be formulated by invoking the principle of D'Alembert. This principle states that the external forces applied to a system must be balanced by the inertial forces. Therefore, we have merely to add to the aerodynamic, thrust, and gravity forces already obtained, the inertial forces $-F_I{}^i$. Similarly, the inertial moments $-M_I{}^i$ must be added to the aerodynamic and thrust moments to complete the description of the force and moment system.

When the inertial forces are added to the aerodynamic, thrust, and gravity forces in equation (2.15.1), D'Alembert's principle (ref. 7) requires that

$$F^i = F_A{}^i + F_T{}^i + F_g{}^i - F_I^i = 0 \tag{2.15.5}$$

This equation is usually written in the form

$$F_I^i = F_A{}^i + F_T{}^i + F_g{}^i \tag{2.15.6}$$

The forces F_I^i are equivalent to the rates of change of linear momentum. For bodies of constant mass m, the components are

$$F_I^i = m \frac{dv^i}{dt} = m\left(\frac{\partial v^i}{\partial t} + p^j v^k - p^k v^j\right) \tag{2.15.7}$$

where i, j, k are in cyclic order and where

$\dfrac{dv^i}{dt}$ rates of change of velocity components with respect to inertial space

$\dfrac{\partial v^i}{\partial t}$ rates of change of velocity components with respect to a set of moving body axes

v^i linear velocity components

p^j angular velocity components

When equations (2.15.2) and (2.15.7) are combined, we obtain for the force system

$$m\left(\frac{\partial v^i}{\partial t} + p^j v^k - p^k v^j\right) = \left[\frac{\partial y^i}{\partial x^\alpha} f_o(\alpha) + \frac{\partial y^i}{\partial x^\alpha}\frac{\partial y^\tau}{\partial x^\beta} v^\tau f_{\overline{V}}(\alpha,\beta) + \frac{\partial y^i}{\partial x^\alpha}\frac{\partial y^\tau}{\partial x^\beta} p^\tau f_{\overline{\omega}}(\alpha,\beta)\right.$$
$$\left. + \frac{\partial y^i}{\partial x^\alpha} f_c(\alpha)\delta_{c_i}\right] + \frac{\partial Y^i}{\partial x_n{}^1} T_n + \frac{\partial y^i}{\partial x_g{}^3} mg \tag{2.15.8}$$

where i, j, k are again in cyclic order.

Likewise, when the inertial moments $-M_I^i$ are added to the aerodynamic and thrust moments in equation (2.15.3), we have in accordance with D'Alembert's principle

$$M_A^i + M_T^i - M_I^i = 0 \tag{2.15.9}$$

or in the more familiar form

$$M_T^i = M_A^i + M_T^i \tag{2.15.10}$$

The moments M_I^i are equivalent to the rates of change of angular momentum. For a body with constant inertial components, these are

$$M_I^i = \frac{dH^i}{dt} = \frac{\partial H^i}{\partial t} + p^j H^k - p^k H^j \tag{2.15.11}$$

where i, j, k are in cyclic order and where

H^i components of angular momentum

$\dfrac{dH^i}{dt}$ rates of change of angular momentum components with respect to inertial space

$\dfrac{\partial H^i}{\partial t}$ rates of change of angular momentum components with respect to a set of moving body axes

Moreover

$$H^i = I^{i\alpha} p_\alpha = I_{i\alpha} p^\alpha \tag{2.15.12}$$

where $I_{i\alpha}$ are moments and products of inertia. Substitution of equation (2.15.12) in equation (2.15.11) yields

$$M_I^i = \frac{dH^i}{dt} = I_{i\alpha} \dot{p}^\alpha + I_{k\alpha} p^\alpha p^j - I_{j\alpha} p^\alpha p^k \tag{2.15.13}$$

In accordance with the summation convention, each term on the right-hand side of this equation must be summed on α.

The following definitions are required:

$$I_{11} = I_{y^1y^1} \quad ; \qquad I_{22} = I_{y^2y^2} \quad ; \qquad I_{33} = I_{y^3y^3}$$

$$I_{12} = -I_{y^1y^2}; \qquad I_{13} = -I_{y^1y^3}; \qquad I_{23} = -I_{y^2y^3}$$

These are the moments and products of inertia relative to the $y^1y^2y^3$ body axes. The above substitutions should only be made subsequent to the completion of summation on α.

When equations (2.15.4) and (2.15.13) are combined, we obtain

$$I_{i\alpha}\dot{p}^\alpha + I_{k\alpha}p^\alpha p^j - I_{j\alpha}p^\alpha p^k = \left[\frac{\partial y^i}{\partial x^\alpha} m_o(\alpha) + \frac{\partial y^i}{\partial x^\alpha}\frac{\partial y^\tau}{\partial x^\beta} v^\tau m_{\bar{v}}(\alpha,\beta) \right.$$

$$+ \frac{\partial y^i}{\partial x^\alpha}\frac{\partial y^\tau}{\partial x^\beta} p^\tau m_{\overline{\omega}}(\alpha,\beta) + \left. \frac{\partial y^i}{\partial x^\alpha} m_c(\alpha)\delta_{c_i} \right]$$

$$+ \left(y_n{}^j \frac{\partial Y_n{}^k}{\partial x_n{}^1} - y_n{}^k \frac{\partial Y_n{}^j}{\partial x_n{}^1} \right) T_n \qquad (2.15.14)$$

where $\alpha = 1,2,3$ and i,j,k are in cyclic order.

The solution of equations (2.15.8) and (2.15.14) yields the components v^i of linear velocity, and the components p^i of angular velocity. These components may be used to determine the geographical location of the aircraft and its spatial orientation.

2.16 SPATIAL ORIENTATION IN TERMS OF EULER ANGLES

The values of the angular velocity components p^i obtained by solving the equations of motion (2.15.8) and (2.15.14) may be used to determine the Euler angles ψ, θ, and ϕ, which relate the moving body axes to an Earth-fixed system. The equations relating the body angular rates p^i to the Euler angle rates may be obtained by considering sketch (k) and making the necessary transformations.

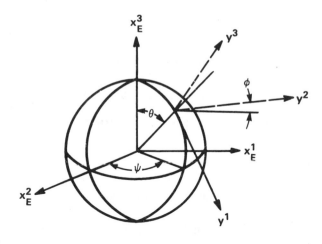

Sketch (k)

Consider a set of Earth-fixed axes with origin at the center of a sphere, and a set of moving axes having the same origin. The moving axes are subject to a rotation ψ about the y^3 axes, followed by a rotation θ about the y^2 axis and a rotation ϕ about the y^1 axis. The angular velocity components p^i of the moving body axes may be obtained as functions of ψ, θ, and ϕ by constructing a set of axes on the surface of the sphere, rather than at its center. This procedure produces a less cluttered diagram and simplifies the transformations.

The required relationships are most easily obtained by considering the contribution of each Euler angle rate to the angular velocity vector.

For a rotation ψ about the y^3 axis, the angular velocity components are

$$p^1 = 0 ; \quad p^2 = 0 ; \quad p^3 = \dot{\psi}$$

For a rotation ψ about the y^3 axis, followed by a rotation θ about the y^2 axis, the angular velocity components are

$$p^1 = -\dot{\psi} \sin \theta ; \quad p^2 = \dot{\theta} ; \quad p^3 = \dot{\psi} \cos \theta$$

Finally, for a rotation ψ about the y^3 axis, followed by a rotation θ about the y^2 axis and a rotation ϕ about the y^1 axis, the angular velocity components are

$$p^1 = \dot{\phi} - \dot{\psi} \sin \theta$$

$$p^2 = \dot{\theta} \cos \phi + \dot{\psi} \cos \theta \sin \phi$$

$$p^3 = \dot{\psi} \cos \theta \cos \phi - \dot{\theta} \sin \phi$$

In matrix notation, these equations assume the form

$$
\begin{pmatrix} p^1 \\ p^2 \\ p^3 \end{pmatrix} = \begin{pmatrix} 1 & 0 & -\sin \theta \\ 0 & \cos \phi & \cos \theta \sin \phi \\ 0 & -\sin \phi & \cos \theta \cos \phi \end{pmatrix} \begin{pmatrix} \dot{\phi} \\ \dot{\theta} \\ \dot{\psi} \end{pmatrix}
$$

and therefore

$$
\begin{pmatrix} \dot{\phi} \\ \dot{\theta} \\ \dot{\psi} \end{pmatrix} = \begin{pmatrix} 1 & \sin \phi \tan \theta & \cos \phi \tan \theta \\ 0 & \cos \phi & -\sin \phi \\ 0 & \sin \phi \sec \theta & \cos \phi \sec \theta \end{pmatrix} \begin{pmatrix} p^1 \\ p^2 \\ p^3 \end{pmatrix}
$$

or

$$\dot{\phi} = p^1 + \tan \theta (p^2 \sin \phi + p^3 \cos \phi)$$

$$\dot{\theta} = p^2 \cos \phi - p^3 \sin \phi$$

$$\dot{\psi} = (p^2 \sin \phi + p^3 \cos \phi)\sec \theta$$

When the values of p^i obtained by solving equations (2.15.8) and (2.15.14) are substituted in these equations, and the resulting equations integrated, the spatial orientation of the aircraft is determined.

The spatial orientation of the aircraft may also be obtained by considering the direction cosines relating the moving body axes to a set of Earth-fixed axes. Although the Euler angle equations are nonlinear, the direction cosine equations are linear. These may be obtained as described below.

2.17 SPATIAL ORIENTATION IN TERMS OF DIRECTION COSINES

Let an Earth-fixed reference system be determined by a triad of mutually orthogonal unit vectors $\hat{I}$, $\hat{J}$, $\hat{K}$, and let the moving body axes coincide initially with the unit vectors $\hat{i}$, $\hat{j}$, $\hat{k}$. These two systems of axes, which are assumed to have a common origin, are related by the matrix equation

$$\begin{pmatrix} \hat{I} \\ \hat{J} \\ \hat{K} \end{pmatrix} = \begin{pmatrix} d_{11} & d_{12} & d_{13} \\ d_{21} & d_{22} & d_{23} \\ d_{31} & d_{32} & d_{33} \end{pmatrix} \begin{pmatrix} \hat{i} \\ \hat{j} \\ \hat{k} \end{pmatrix}$$

where d_{ij} are the direction cosines.

Since the magnitudes and the directions of $\hat{I}, \hat{J}$, and $\hat{K}$ are constant, it follows that

$$\frac{d\hat{I}}{dt} = \frac{d\hat{J}}{dt} = \frac{d\hat{K}}{dt} = 0$$

As observed from an inertial reference frame, which is momentarily coincident with the moving body frame, the rates of change of these vectors are

$$\frac{d\hat{I}}{dt} = \frac{\partial\hat{I}}{\partial t} + \bar{\omega}x\hat{I} = 0$$

$$\frac{d\hat{J}}{dt} = \frac{\partial\hat{J}}{\partial t} + \bar{\omega}x\hat{J} = 0$$

$$\frac{d\hat{K}}{dt} = \frac{\partial\hat{K}}{\partial t} + \bar{\omega}x\hat{K} = 0$$

where

$\dfrac{\partial}{\partial t}$ denotes the rate of change with respect to the moving body axes

$\bar{\omega}x$ is the rate of change due to the rotation of the axes

Therefore

$$\frac{d\hat{I}}{dt} = (\dot{d}_{11}\hat{i} + \dot{d}_{12}\hat{j} + \dot{d}_{13}\hat{k}) + \bar{\omega}x(d_{11}\hat{i} + d_{12}\hat{j} + d_{13}\hat{k})$$

$$\frac{d\hat{J}}{dt} = (\dot{d}_{21}\hat{i} + \dot{d}_{22}\hat{j} + \dot{d}_{23}\hat{k}) + \bar{\omega}x(d_{21}\hat{i} + d_{22}\hat{j} + d_{23}\hat{k})$$

$$\frac{d\hat{K}}{dt} = (\dot{d}_{31}\hat{i} + \dot{d}_{32}\hat{j} + \dot{d}_{33}\hat{k}) + \bar{\omega}x(d_{31}\hat{i} + d_{32}\hat{j} + d_{33}\hat{k})$$

and

$$\bar{\omega} = p^1 i + p^2 j + p^3 k$$

Substitution of this value of $\bar{\omega}$ in the above equations yields the required differential equations for the direction cosines. These are

$$\dot{d}_{11} + p^2 d_{13} - p^3 d_{12} = 0$$

$$\dot{d}_{12} + p^3 d_{11} - p^1 d_{13} = 0$$

$$\dot{d}_{13} + p^1 d_{12} - p^2 d_{11} = 0$$

$$\dot{d}_{21} + p^2 d_{23} - p^3 d_{22} = 0$$

$$\dot{d}_{22} + p^3 d_{21} - p^1 d_{23} = 0$$

$$\dot{d}_{23} + p^1 d_{22} - p^2 d_{21} = 0$$

$$\dot{d}_{31} + p^2 d_{33} - p^2 d_{32} = 0$$

$$\dot{d}_{32} + p^3 d_{31} - p^1 d_{33} = 0$$

$$\dot{d}_{33} + p^1 d_{32} - p^2 d_{31} = 0$$

These are the differential equations for the direction cosines that relate the moving-body axes to the Earth-fixed axes. They may be written more compactly as follows:

$$d_{\alpha i} + d_{\alpha k} p^j - d_{\alpha j} p^k = 0 \qquad \alpha = 1, 2, 3$$

where i, j, k are in cyclic order.

Now that the linear velocity components and the Euler angles are known, it is possible to determine the geographical location of the aircraft in an Earth-fixed reference frame. The procedure is described below.

2.18 COORDINATES OF THE AIRCRAFT IN AN EARTH-FIXED REFERENCE FRAME

Equation (2.14.4) gives moving body axes components y^i as functions of the components $x_g{}^i$ in an Earth-fixed reference system, that is

$$y^i = \frac{\partial y^i}{\partial x_g{}^\alpha} x_g{}^\alpha$$

Since the superscripts i and α are dummy indices, this equation may be rewritten as follows:

$$y^\alpha = \frac{\partial y^\alpha}{\partial x_g{}^i} x_g{}^i$$

By solving this equation for the Earth-fixed coordinates $x_g{}^i$, we have

$$x_g{}^i = \frac{\partial x_g{}^i}{\partial y^\alpha} y^\alpha \tag{2.18.1}$$

Bearing in mind that

$$\frac{\partial y^i}{\partial x^j} = \frac{\partial x^j}{\partial y^i}$$

equation (2.14.3) may be used to obtain the partial differential coefficients required to formulate equation (2.18.1), and to determine the Earth-fixed coordinates. These are

$$x_g{}^1 = y^1 \cos \theta_g \cos \psi_g + y^2 (\sin \phi_g \sin \theta_g \cos \psi_g - \cos \phi_g \sin \psi_g)$$

$$+ y^3 (\sin \phi_g \sin \psi_g + \cos \phi_g \sin \theta_g \cos \psi_g) \tag{2.18.2}$$

$$x_g{}^2 = y^1 \cos \theta_g \sin \psi_g + y^2 (\sin \phi_g \sin \theta_g \sin \psi_g + \cos \phi_g \cos \psi_g)$$

$$+ y^3 (\cos \phi_g \sin \theta_g \sin \psi_g - \sin \phi_g \cos \psi_g) \tag{2.18.3}$$

$$x_g{}^3 = -y^1 \sin \theta_g + y^2 \sin \phi_g \cos \theta_g + y^3 \cos \phi_g \cos \theta_g \tag{2.18.4}$$

Let the linear velocity components in an Earth-fixed reference system be denoted by $\dot{X}_E{}^i$, then these components are related to the body axes components v^i by the equation

$$\dot{X}_E{}^i = \frac{\partial x_g{}^i}{\partial y^\alpha} v^\alpha \tag{2.18.5}$$

Evaluation of the differential coefficients from equations (2.18.2) through (2.18.4) and substitution in equation (2.18.5) yields

$$\dot{X}_E{}^1 = v^1 \cos \theta_g \cos \psi_g + v^2 (\sin \phi_g \sin \theta_g \cos \psi_g - \cos \phi_g \sin \psi_g)$$

$$+ v^3 (\sin \phi_g \sin \psi_g + \cos \phi_g \sin \theta_g \sin \theta_g \cos \psi_g)$$

$$\dot{X}_E{}^2 = v^1 \cos \theta_g \sin \psi_g + v^2 (\sin \phi_g \sin \theta_g \sin \psi_g + \cos \phi_g \cos \psi_g)$$

$$+ v^3 (\cos \phi_g \sin \theta_g \sin \psi_g - \sin \phi_g \cos \psi_g)$$

$$\dot{X}_E{}^3 = -v^1 \sin \theta_g + v^2 \sin \phi_g \cos \theta_g + v^3 \cos \phi_g \cos \theta_g$$

Integration of these equations gives the Earth-fixed coordinates of the aircraft at time t. These are

$$X_E{}^i = X^i{}_{Eo} + \int \dot{X}_E{}^i \, dt$$

where X^i_{Eo} are the initial values of the coordinates in the Earth-fixed reference system.

This completes the formulation of the simplified mathematical model of the aeronautical system considered, using linear aerodynamic theory. A more complete description of the system would include such items as the control loops that are interposed between the pilot's control levers and the various control surfaces. A description of these loops would entail discussions of linear and nonlinear control theory, and is beyond the scope of the present treatment.

It should be noted that the mathematical model has been formulated in such a way that specialized forms can be obtained by expanding the force and moment terms in accordance with the transformation laws established. Given the transformation laws and the system parameters, the model equations can be derived without further reference to the physics of the problem. The only operations required are differentiation and summation, and these can be performed either manually or by computer as previously indicated.

To demonstrate the feasibility of formulating mathematical models of aeronautical systems by algebraic computation, a mathematical model of a general aircraft has been formulated by computer in the final chapter. It will be seen that the interactive capability of the computer system used enhances the utility of the method by permitting the user to modify the formulation as he proceeds.

2.19 REFERENCES

1. Etkin, Bernard: Dynamics of Atmospheric Flight. John Wiley & Sons, Inc., 1972.

2. Gainer, Thomas G.; and Hoffman, Sherwood: Summary of Transformation Equations and Equations of Motion Used in Free-Flight and Wind-Tunnel Data Reductions and Analysis. NASA SP-3070, 1972.

3. Bond, E.; Auslander, M.; et al.: PL/I FORMAC Interpreter User's Reference Manual. IBM-Boston Programming Center, Cambridge, Mass., 1967.

4. Sokolnikoff, Ivan S.: Tensor Analysis; Theory and Applications. John Wiley & Sons, Inc., 1960.

5. McRuer, Duane; Ashkenas, Irving; and Graham, Dustan: Aircraft Dynamics and Automatic Control. Princeton University Press, 1973.

6. MacMillan, William D.: The Theory of the Potential. Dover Publications, Inc., 1958.

7. Goldstein, Herbert: Classical Mechanics. Addison-Wesley Publishing Company, Inc., 1959.

APPLICATIONS TO PARTICLE DYNAMICS

3.1 FORMULATION OF CHRISTOFFEL SYMBOLS USING COORDINATE TRANSFORMATION EQUATIONS

The importance of the Christoffel symbols (ref. 1) and their derivation in terms of either coordinate transformation equations or metric tensors was discussed in section 1.12. For reasons that will become apparent as we proceed, the equations of motion of a particle can be formulated in any curvilinear coordinate system once the Christoffel symbols are known. It is perhaps appropriate at this stage to review again the essential difference between the tensor method and the conventional approach, and the reason why a tensor formulation is so attractive. Conventionally, a vector is expressed in terms of its physical components and a corresponding set of unit base vectors. The tensor components of a vector are not, in general, the same as the physical components. Instead, they are components that obey transformation laws corresponding to their variance. The transformation laws for covariant and contra-variant vectors are given by equations (1.6.5) and (1.6.3), respectively. It may be noted that when the base vectors define an orthogonal Cartesian reference frame, the tensor components are the same as the physical components. As a consequence of the geometrical simplification inherent in the tensor method, the operations involved in obtaining derivatives and formulating the equations of mathematical physics in curvilinear coordinate reference frames are routine operations involving only summation and differentiation.

To illustrate the method of deriving the Christoffel symbols of the first kind from the coordinate transformation equations, consider the functional form given by equation (1.13.11) and the defining formula (1.13.14). The technique may be illus-trated by using the transformation from an orthogonal Cartesian reference frame to a curvilinear coordinate system in which x^i are cylindrical polar coordinates. If the curvilinear coordinate system is cylindrical polar, the Cartesian coordinates y^i are related to the curvilinear coordinates by the following transformation equations (fig. 3.1.1).

$$y^1 = x^1 \cos x^2$$

$$y^2 = x^1 \sin x^2$$

$$y^3 = x^3$$

(3.1.1)

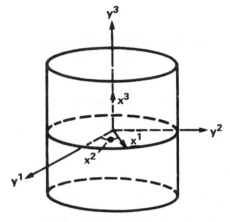

Figure 3.1.1.— Cylindrical coordinates.

The inverse transformation is given by

$$x^1 = \sqrt{(y^1)^2 + (y^2)^2}$$

$$x^2 = \tan^{-1}\left(\frac{y^2}{y^1}\right)$$

$$x^3 = y^3$$

(3.1.2)

By substitution from equations (3.1.1) and (3.1.2) in equations (1.13.14) and (1.13.15), the Christoffel symbols are obtained.

Equation (1.13.14) will be used to obtain the nonzero Christoffel symbols of the first kind. With the exception of the dummy index, the superscripts appearing on the right-hand side of equation (1.13.14) must correspond to those appearing on the Christoffel symbol. For example

$$[12,2] = \frac{\partial^2 y^\alpha}{\partial x^1 \partial x^2} \frac{\partial y^\alpha}{\partial x^2}$$

$$= \frac{\partial^2 y^1}{\partial x^1 \partial x^2} \frac{\partial y^1}{\partial x^2} + \frac{\partial^2 y^2}{\partial x^1 \partial x^2} \frac{\partial y^2}{\partial x^2} + \frac{\partial^2 y^3}{\partial x^1 \partial x^2} \frac{\partial y^3}{\partial x^2}$$

$$= -\sin x^2 (-x^1 \sin x^2) + \cos x^2 (x^1 \cos x^2)$$

$$= x^1 (\sin^2 x^2 + \cos^2 x^2) = x^1$$

Therefore

$$[12,2] = [21,2] = x^1$$

Likewise

$$[22,1] = \frac{\partial^2 y^\alpha}{\partial x^2 \partial x^2} \frac{\partial y^\alpha}{\partial x^1}$$

$$= \frac{\partial^2 y^1}{\partial x^2 \partial x^2} \frac{\partial y^1}{\partial x^1} + \frac{\partial^2 y^2}{\partial x^2 \partial x^2} \frac{\partial y^2}{\partial x^1} + \frac{\partial^2 y^3}{\partial x^2 \partial x^2} \frac{\partial y^3}{\partial x^1}$$

$$= -(x^1 \cos x^2)\cos x^2 - (x^1 \sin x^2)\sin x^2$$

$$= -x^1 (\cos^2 x^2 + \sin^2 x^2) = -x^1$$

The procedure for determining the Christoffel symbols of the first kind for a spherical polar coordinate system is the same as that used for a cylindrical polar coordinate system. In this case, however, the terms of equation (1.13.14) have to be obtained from a different set of coordinate transformation equations. The Cartesian coordinates y^i are related to the spherical polar coordinates x^i by the following transformation equations (fig. 3.1.2):

$$\left. \begin{array}{l} y^1 = x^1 \sin x^2 \cos x^3 \\ y^2 = x^1 \sin x^2 \sin x^3 \\ y^3 = x^1 \cos x^2 \end{array} \right\} \tag{3.1.3}$$

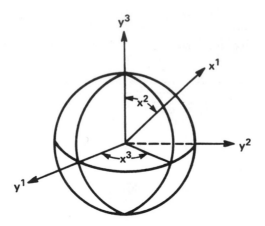

Figure 3.1.2.– Spherical coordinates.

The inverse transformation is given by

$$x^1 = \sqrt{(y^1)^2 + (y^2)^2 + (y^3)^2}$$

$$x^2 = \tan^{-1}\left(\frac{\sqrt{(y^1)^2 + (y^3)^2}}{y^3}\right) \qquad (3.1.4)$$

$$x^3 = \tan^{-1}\left(\frac{y^2}{y^1}\right)$$

By substitution from equation (3.1.3) in equation (1.13.14), the Christoffel symbols of the first kind are obtained. For the special case being considered, there are 6 nonzero symbols out of a total of 18. To illustrate, equation (1.13.14) will again be used to obtain the nonzero Christoffel symbols of the first kind. By substitution from equation (3.1.3) in the expanded form of equation (1.13.14), it is found that

$$[22,1] = \frac{\partial^2 y^1}{\partial x^2 \partial x^2}\frac{\partial y^1}{\partial x^1} + \frac{\partial^2 y^2}{\partial x^2 \partial x^2}\frac{\partial y^2}{\partial x^1} + \frac{\partial^2 y^3}{\partial x^2 \partial x^2}\frac{\partial y^3}{\partial x^1} = -x^1$$

$$[33,2] = \frac{\partial^2 y^1}{\partial x^3 \partial x^3}\frac{\partial y^1}{\partial x^2} + \frac{\partial^2 y^2}{\partial x^3 \partial x^3}\frac{\partial y^2}{\partial x^2} + \frac{\partial^2 y^3}{\partial x^3 \partial x^3}\frac{\partial y^3}{\partial x^2}$$

$$= -(x^1)^2 \sin x^2 \cos x^2$$

$$[12,2] = \frac{\partial^2 y^1}{\partial x^1 \partial x^2} \frac{\partial y^1}{\partial x^2} + \frac{\partial^2 y^2}{\partial x^1 \partial x^2} \frac{\partial y^2}{\partial x^2} + \frac{\partial^2 y^3}{\partial x^1 \partial x^2} \frac{\partial y^3}{\partial x^2} = x^1 = [21,2]$$

$$[13,3] = \frac{\partial^2 y^1}{\partial x^1 \partial x^3} \frac{\partial y^1}{\partial x^3} + \frac{\partial^2 y^2}{\partial x^1 \partial x^3} \frac{\partial y^2}{\partial x^3} + \frac{\partial^2 y^3}{\partial x^1 \partial x^3} \frac{\partial y^3}{\partial x^3}$$

$$= [31,3] = x^1 \sin^2 x^2$$

$$[33,1] = \frac{\partial^2 y^1}{\partial x^3 \partial x^3} \frac{\partial y^1}{\partial x^1} + \frac{\partial^2 y^2}{\partial x^3 \partial x^3} \frac{\partial y^2}{\partial x^1} + \frac{\partial^2 y^3}{\partial x^3 \partial x^3} \frac{\partial y^3}{\partial x^1}$$

$$= -x^1 \sin^2 x^2$$

$$[23,3] = \frac{\partial^2 y^1}{\partial x^2 \partial x^3} \frac{\partial y^1}{\partial x^3} + \frac{\partial^2 y^2}{\partial x^2 \partial x^3} \frac{\partial y^2}{\partial x^3} + \frac{\partial^2 y^3}{\partial x^2 \partial x^3} \frac{\partial y^3}{\partial x^3}$$

$$= (x^1)^2 \sin x^2 \cos x^2 = [32,3]$$

Equation (1.13.15) may be used to obtain the nonzero Christoffel symbols of the second kind. With the exception of the dummy index, the superscripts appearing on the right side of the equation (1.13.15) must correspond to those appearing in the Christoffel symbol. For example

$$\begin{Bmatrix} 1 \\ 22 \end{Bmatrix} = \frac{\partial^2 y^\alpha}{\partial x^2 \partial x^2} \frac{\partial x^1}{\partial y^\alpha}$$

By substitution from equations (3.1.3) and (3.1.4) in equation (1.13.15), all the Christoffel symbols are obtained. For the special case being considered there are 6 nonzero Christoffel symbols out of a total of 18. Of course, as indicated previously, the operation of obtaining Christoffel symbols from formula (1.13.15) and the use of the transformation equations could be performed by a computer programmed for this kind of operation. To illustrate, equation (1.13.15) will again be used to obtain the nonzero Christoffel symbols. By substitution from equations (3.1.3) and (3.1.4) in the expanded form of equation (1.13.15) it is found that

$$\begin{Bmatrix} 1 \\ 22 \end{Bmatrix} = \left(\frac{\partial^2 y^1}{\partial x^2 \partial x^2} \frac{\partial x^1}{\partial y^1} + \frac{\partial^2 y^2}{\partial x^2 \partial x^2} \frac{\partial x^1}{\partial y^2} + \frac{\partial^2 y^3}{\partial x^2 \partial x^2} \frac{\partial x^1}{\partial y^3} \right) = -x^1$$

$$\begin{Bmatrix} 2 \\ 33 \end{Bmatrix} = \left(\frac{\partial^2 y^1}{\partial x^3 \partial x^3} \frac{\partial x^2}{\partial y^1} + \frac{\partial^2 y^2}{\partial x^3 \partial x^3} \frac{\partial x^2}{\partial y^2} + \frac{\partial^2 y^3}{\partial x^3 \partial x^3} \frac{\partial x^2}{\partial y^3} \right) = -\sin x^2 \cos x^2$$

$$\begin{Bmatrix} 2 \\ 12 \end{Bmatrix} = \left(\frac{\partial^2 y^1}{\partial x^1 \partial x^2} \frac{\partial x^2}{\partial y^1} + \frac{\partial^2 y^2}{\partial x^1 \partial x^2} \frac{\partial x^2}{\partial y^2} + \frac{\partial^2 y^3}{\partial x^1 \partial x^2} \frac{\partial x^2}{\partial y^3} \right) = \begin{Bmatrix} 2 \\ 21 \end{Bmatrix} = \frac{1}{x^1}$$

$$\begin{Bmatrix} 3 \\ 13 \end{Bmatrix} = \left(\frac{\partial^2 y^1}{\partial x^1 \partial x^3} \frac{\partial x^3}{\partial y^1} + \frac{\partial^2 y^2}{\partial x^1 \partial x^3} \frac{\partial x^3}{\partial y^2} + \frac{\partial^2 y^3}{\partial x^1 \partial x^3} \frac{\partial x^3}{\partial y^3} \right) = \begin{Bmatrix} 3 \\ 31 \end{Bmatrix} = \frac{1}{x^1}$$

$$\begin{Bmatrix} 1 \\ 33 \end{Bmatrix} = \left(\frac{\partial^2 y^1}{\partial x^3 \partial x^3} \frac{\partial x^1}{\partial y^1} + \frac{\partial^2 y^2}{\partial x^3 \partial x^3} \frac{\partial x^1}{\partial y^2} + \frac{\partial^2 y^3}{\partial x^3 \partial x^3} \frac{\partial x^1}{\partial y^3} \right) = -x^1 \sin^2 x^2$$

$$\begin{Bmatrix} 3 \\ 23 \end{Bmatrix} = \left(\frac{\partial^2 y^1}{\partial x^2 \partial x^3} \frac{\partial x^3}{\partial y^1} + \frac{\partial^2 y^2}{\partial x^2 \partial x^3} \frac{\partial x^3}{\partial y^2} + \frac{\partial^2 y^3}{\partial x^2 \partial x^3} \frac{\partial x^3}{\partial y^3} \right) = \begin{Bmatrix} 3 \\ 32 \end{Bmatrix} = \cot x^2$$

These results are seen to agree with those obtained in equation (1.3.8).

3.2 METRIC TENSOR INPUTS

An alternative method of obtaining the Christoffel symbols is based on the use of the metric tensors. These were defined in terms of the scalar products of the base vectors and their reciprocals in equations (1.12.1) and (1.12.2), respectively. In general, it is more convenient to obtain these tensors in terms of the coefficients appearing in the fundamental quadratic form, equation (1.15.15), that is

$$ds^2 = g_{ij} \, dx^i \, dx^j \tag{3.2.1}$$

Substitution of the g_{ij} in the formulas of definition

$$[ij,k] = \frac{1}{2} \left(\frac{\partial g_{ik}}{\partial x^j} + \frac{\partial g_{jk}}{\partial x^i} - \frac{\partial g_{ij}}{\partial x^k} \right) \tag{3.2.2}$$

$$\begin{Bmatrix} i \\ jk \end{Bmatrix} = g^{i\alpha}[jk,\alpha] \tag{3.2.3}$$

yields the required set of Christoffel symbols. Given the metric tensors, the only operations required to formulate these symbols are partial differentiation, addition, and multiplication. These are routine operations that can be performed manually or executed with speed and efficiency on any computer equipped with a formula manipulation compiler. No multiplications are required to obtain the Christoffel symbols of the first kind. For example

$$[11,1] = \frac{1}{2}\left(\frac{\partial g_{11}}{\partial x^1} + \frac{\partial g_{11}}{\partial x^1} - \frac{\partial g_{11}}{\partial x^1}\right)$$

Therefore

$$[11,1] = \frac{1}{2}\frac{\partial g_{11}}{\partial x^1}$$

Since there is a repeated index in the definition of the Christoffel symbol of the second kind, a summation is required on that index. In a space of three dimensions

$$\left\{\begin{matrix} i \\ jk \end{matrix}\right\} = g^{i1}\,[jk,1] + g^{i2}\,[jk,2] + g^{i3}\,[jk,3]$$

Therefore

$$\left\{\begin{matrix} i \\ jk \end{matrix}\right\} = \frac{1}{2}\left[g^{i1}\left(\frac{\partial g_{j1}}{\partial x^k} + \frac{\partial g_{k1}}{\partial x^j} - \frac{\partial g_{jk}}{\partial x^1}\right)\right.$$

$$\left. + g^{i2}\left(\frac{\partial g_{j2}}{\partial x^k} + \frac{\partial g_{k2}}{\partial x^j} - \frac{\partial g_{jk}}{\partial x^2}\right) + g^{i3}\left(\frac{\partial g_{j3}}{\partial x^k} + \frac{\partial g_{k3}}{\partial x^j} - \frac{\partial g_{jk}}{\partial x^3}\right)\right] \qquad (3.2.4)$$

The determination of the fundamental quadratic form, equation (3.2.1), is no more difficult than finding the coordinate transformation equations (1.13.11) and (1.13.13). Indeed, it is often much simpler. For example, consider again the case of a problem being studied in a cylindrical polar coordinate system. In this system of coordinates, the square of an element of arc is given by the following equation:

$$ds^2 = (dx^1)^2 + (x^1\,dx^2)^2 + (dx^3)^2 \qquad (3.2.5)$$

Therefore

$$
\left.\begin{array}{l}
g_{11} = 1 \\[2mm]
g_{22} = (x^1)^2 \\[2mm]
g_{33} = 1
\end{array}\right\} \tag{3.2.6}
$$

where x^1 is the radial distance, x^2 is the angular displacement, and x^3 is the axial displacement. In this case the coordinate transformation equations assume the form

$$
\left.\begin{array}{l}
y^1 = x^1 \cos x^2 \\[2mm]
y^2 = x^1 \sin x^2 \\[2mm]
y^3 = x^3
\end{array}\right\} \tag{3.2.7}
$$

Likewise, the square of an element of arc in a spherical coordinate system is given by

$$
ds^2 = (dx^1)^2 + (x^1 \, dx^2)^2 + (x^1 \sin x^2 \, dx^3)^2 \tag{3.2.8}
$$

where x^2 is the polar angle and x^3 is the longitude.

$$
\left.\begin{array}{l}
g_{11} = 1 \\[2mm]
g_{22} = (x_1)^2 \\[2mm]
g_{33} = (x^1 \sin x^2)^2
\end{array}\right\} \tag{3.2.9}
$$

The corresponding coordinate transformation equations are given by equation (3.1.3).

In these two cases, the determination of the metric tensors from the square of the line element is no more difficult than finding the coordinate transformation equations. More importantly, if the functions $g_{ij}(x)$ are such that the system of equations (1.13.12) has no solution, then no admissible transformation of coordinates exists, which reduces equation (3.2.1) to the Pythagorean form. In this case, the manifold is nonEuclidean and the use of coordinate transformation equations as inputs will fail.

The manual derivation of the Christoffel symbols of the first kind for a cylindrical polar coordinate system would proceed as follows: Referring to equation (3.2.6), it is seen that g_{22} is the only metric tensor component that is a function of a coordinate, the coordinate x^1. In view of this, it is clear that the only nonzero Christoffel symbols of the first kind are those in which the indices assume the following values:

$$[12,2] \; ; \quad [21,2] \; ; \quad [22,1]$$

When evaluated, these yield

$$[12,2] = \frac{1}{2}\left(\frac{\partial g_{12}}{\partial x^2} + \frac{\partial g_{22}}{\partial x^1} - \frac{\partial g_{12}}{\partial x^2}\right) = x^1$$

In view of the symmetry of the metric tensors

$$g_{ij} = g_{ji}$$

and the definition of the Christoffel symbols of the first kind

$$[ij,k] = \frac{1}{2}\left(\frac{\partial g_{ik}}{\partial x^j} + \frac{\partial g_{jk}}{\partial x^i} - \frac{\partial g_{ij}}{\partial x^k}\right)$$

It follows that

$$[ij,k] = [ji,k]$$

and therefore

$$[21,2] = [12,2] = x^1$$

Or by direct evaluation

$$[21,2] = \frac{1}{2}\left(\frac{\partial g_{22}}{\partial x^1} + \frac{\partial g_{12}}{\partial x^2} - \frac{\partial g_{21}}{\partial x^2}\right) = x^1$$

The remaining symbol is

$$[22,1] = \frac{1}{2}\left(\frac{\partial g_{21}}{\partial x^2} + \frac{\partial g_{21}}{\partial x^2} - \frac{\partial g_{22}}{\partial x^1}\right) = -x^1$$

Hence, the only nonzero Christoffel symbols of the first kind are

$$[21,2] = [12,2] = x^1$$

and

$$[22,1] = -x^1$$

In this case it is much quicker to formulate the Christoffel symbols of the first kind manually than by digital computer. Nevertheless, a computer program will be written to mechanize the formulation and to prepare the reader for more complex cases to follow. A computer program and the corresponding output would assume the following form if the coordinate system were cylindrical polar.

```
LET(G(1,1)=1);

LET(G(2,2)=(X(1))**2);

LET(G(3,3)=1);

PRINT_OUT(G(1,1) G(2,2) G(3,3));

PUT SKIP(5);

DO I=1 TO 3 BY 1;

DO J=1 TO 3 BY 1;

DO K=1 TO 3 BY 1;

LET(I="I");

LET(J="J");

LET(K="K");
```

```
LET(D(I,J,K)=(1/2)*DERIV(G(I,K),X(J)));

LET(E(J,K,I)=(1/2)*DERIV(G(J,K),X(I)));

LET(F(I,J,K)=(1/2)*DERIV(G(I,J),X(K)));

LET(C(I,J,K)=(D(I,J,K)+E(J,K,I)-F(I,J,K)));

PRINT=OUT(C(I,J,K));

END;

END;

END;
```

The output from this program follows:

$G(1,1) = 1$	$C(2,1,1) - 0$
$G(2,2) = X(1)^2$	$C(2,1,2) = X(1)$
$G(3,3) = 1$	$C(2,1,3) = 0$
$C(1,1,1) = 0$	$C(2,2,1) = -X(1)$
$C(1,1,2) = 0$	$C(2,2,2) = 0$
$C(1,1,3) = 0$	$C(2,2,3) = 0$
$C(1,2,1) = 0$	$C(2,3,1) = 0$
$C(1,2,2) = X(1)$	$C(2,3,2) = 0$
$C(1,2,3) = 0$	$C(2,3,3) = 0$
$C(1,3,1) = 0$	$C(3,1,1) = 0$
$C(1,3,2) = 0$	$C(3,1,2) = 0$
$C(1,3,3) = 0$	$C(3,1,3) = 0$

$$C(3,2,1) = 0$$

$$C(3,2,2) = 0$$

$$C(3,2,3) = 0$$

$$C(3,3,1) = 0$$

$$C(3,3,2) = 0$$

$$C(3,3,3) = 0$$

Proceeding next to the formulation of the Christoffel symbols of the first kind for a spherical polar coordinate system and noting from equation (3.2.9) that g_{11} is a constant, g_{22} a function of x^1, and g_{33} a function of x^1 and x^2, it follows that the only nonzero Christoffel symbols are those with indices as follows:

$$[22,1] \; ; \quad [21,2] \; ; \quad [12,2]$$

$$[33,1] \; ; \quad [31,3] \; ; \quad [13,3]$$

$$[33,2] \; ; \quad [32,3] \; ; \quad [23,3]$$

Moreover, the general property already established for the Christoffel symbols

$$[ij,k] = [ji,k]$$

can be used to reduce the number of independent Christoffel symbols from nine to six. These are

$$[22,1] \; ; \quad [33,1] \; ; \quad [33,2]$$

$$[21,2] = [12,2]$$

$$[31,3] = [13,3]$$

$$[32,3] = [23,3]$$

Using again the defining formula, equation (3.2.2), we obtain

$$[22,1] = \frac{1}{2}\left(\frac{\partial g_{21}}{\partial x^2} + \frac{\partial g_{21}}{\partial x^2} - \frac{\partial g_{22}}{\partial x^1}\right) = -x^1$$

$$[21,2] = \frac{1}{2}\left(\frac{\partial g_{22}}{\partial x^1} + \frac{\partial g_{12}}{\partial x^2} - \frac{\partial g_{21}}{\partial x^2}\right) = x^1 = [12,2]$$

$$[33,1] = \frac{1}{2}\left(\frac{\partial g_{31}}{\partial x^3} + \frac{\partial g_{31}}{\partial x^3} - \frac{\partial g_{33}}{\partial x^1}\right) = -x^1 \sin^2 x^2$$

$$[31,3] = \frac{1}{2}\left(\frac{\partial g_{33}}{\partial x^1} + \frac{\partial g_{13}}{\partial x^3} - \frac{\partial g_{31}}{\partial x^3}\right) = x^1 \sin^2 x^2 = [13,3]$$

$$[33,2] = \frac{1}{2}\left(\frac{\partial g_{32}}{\partial x^3} + \frac{\partial g_{32}}{\partial x^3} - \frac{\partial g_{33}}{\partial x^2}\right) = -(x^1)^2 \sin x^2 \cos x^2$$

$$[32,3] = \frac{1}{2}\left(\frac{\partial g_{33}}{\partial x^2} + \frac{\partial g_{23}}{\partial x^3} - \frac{\partial g_{32}}{\partial x^3}\right) = (x^1)^2 \sin x^2 \cos x^2 = [23,3]$$

An advantage of a computer formulation is that once a program is written it can be used to derive the Christoffel symbols in any orthogonal curvilinear coordinate system of interest, the only requirement being that the metric tensor inputs are appropriate to the coordinate system being used. For example, when the same program is used to formulate the Christoffel symbols of the first kind, in a spherical polar coordinate system, we obtain

```
LET(G(1,1)=1);

LET(G(2,2)=X(1)**2);

LET(G(3,3)=(X(1)*SIN(X(2)))**2);

PRINT_OUT(G(1,1);G(2,2);G(3,3));

PUT SKIP(5);

DO I=1 TO 3 BY 1;
```

```
LET(I="I");
DO J=1 TO 3 BY 1;
DO K=1 TO 3 BY 1;
LET(J="J");
LET(K="K");
LET(D(I,J,K)=(1/2)*DERIV(G(I,K),X(J)));
LET(E(J,K,I)=(1/2)*DERIV(G(J,K),X(I)));
LET(F(I,J,K)=(1/2)*DERIV(G(I,J),X(K)));
LET(C(I,J,K)=(D(I,J,K)+E(J,K,I)-F(I,J,K)));
PRINT_OUT(C(I,J,K));
END;
END;
END;
```

The output from this program follows:

$G(I,1) = 1$

$G(2,2) = X(1)^2$

$G(3,3) = SIN^2 (X(2)) X(1)^2$

$C(1,1,1) = 0$

$C(1,1,2) = 0$

$C(1,1,3) = 0$

$C(1,2,1) = 0$

C(1,2,2) = X(1)

C(1,2,3) = 0

C(1,3,1) = 0

C(1,3,2) = 0

C(1,3,3) = SIN^2 (X(2)) X(1)

C(2,1,1) = 0

C(2,1,2) = X(1)

C(2,1,3) = 0

C(2,2,1) = -X(1)

C(2,2,2) = 0

C(2,2,3) - 0

C(2,3,1) = 0

C(2,3,2) = 0

C(2,3,3) = COS (X(2)) SIN(X(2) X(1)^2

C(3,1,1) = 0

C(3,1,2) = 0

C(3,1,3) = SIN^2 (X(2)) X(1)

C(3,2,1) = 0

C(3,2,2) = 0

C(3,2,3) = COS (X(2)) SIN(X(2)) X(1)^2

C(3,3,1) = -SIN^2 (X(2)) X(1)

$$C(3,3,2) = -COS (X(2)) SIN (X(2)) X(1)^2$$

$$C(3,3,3) = 0$$

3.3 THE VELOCITY VECTOR

Three methods of obtaining the metric tensors have been indicated: one of these uses the method of vector calculus; another uses the known differential coefficients from the coordinate transformation equations; and the method described in the preceding section uses the coefficients of the fundamental quadratic form. Since the coordinate transformation method is more adaptable to digital logic than the vector method, it can be used for all Euclidian applications. However, a formulation using metric tensor coefficients can be used for Euclidian and nonEuclidian applications.

Given the Christoffel symbols, it is seen that there are two forms for the intrinsic or absolute derivative of a vector. Equation (1.12.19) gives the intrinsic derivative in terms of the contravariant components; and equation (1.13.33) gives the same in terms of the covariant components. Either of these equations may be used. However, to avoid the necessity of transforming covariant components into contravariant components, and vice versa, it is better to match the formula to the variance of the vectors. In the course of the analysis, it will become evident what the variance of the vectors is. For example, the variance of the differential elements can be determined as follows: the differential elements dy^i in the y coordinate system are related to the elements dx^j in the x coordinate system by the following equation:

$$dy^i = \frac{\partial y^i}{\partial x^j} \, dx^j \tag{3.3.1}$$

By comparing this equation with equation (1.6.3), it is seen that the differential elements are the components of a contravariant vector. Likewise, equation (3.3.1) shows that the components of velocity in the y coordinate system are related to those in the x coordinate system by the equation

$$\frac{dy^i}{dt} = \frac{\partial y^i}{\partial x^j} \frac{dx^j}{dt}$$

That is,

$$V^i(y) = \frac{\partial y^i}{\partial x^j} U^j(x) \qquad (3.3.2)$$

where $V^i(y)$ are the velocity components in the y coordinate system, and $U^j(x)$ are the velocity components in the x coordinate system. Comparison of equation (3.3.2) with equation (1.6.3) shows that the components of the velocity vector also obey the contravariant transformation law. To obtain the velocity vector from equation (1.12.19), the position vector $\bar{r}$ is substituted for the vector $\bar{A}$, that is

$$\bar{A} = A^i \bar{a}_i = \bar{r} \qquad (3.3.3)$$

Hence, in a cylindrical polar coordinate system

$$A^1 = x^1, \quad A^2 = 0, \quad A^3 = x^3 \qquad (3.3.4)$$

By substitution of these values in equation (1.13.4), the velocity vector is obtained as follows

$$\bar{V} = \frac{dx^1}{dt} \bar{a}_1 + \begin{Bmatrix} 2 \\ 1\,2 \end{Bmatrix} x^1 \frac{dx^2}{dt} \bar{a}_2 + \frac{dx^3}{dt} \bar{a}_3$$

When the appropriate value of the Christoffel symbol is substituted from equation (1.13.3), the tensor components of the velocity vector are given by

$$\bar{V} = \frac{dx^1}{dt} \bar{a}_1 + \frac{dx^2}{dt} \bar{a}_2 + \frac{dx^3}{dt} \bar{a}_3$$

that is,

$$\bar{V} = \frac{dx^i}{dt} \bar{a}_i \qquad (3.3.5)$$

In order to reduce equation (3.3.5) to the conventional form, where the physical components of velocity are associated with a set of unit base vectors, equation (1.9.1) may be used to express the base vectors in unitary form. In this form the velocity $\bar{V}$ is given by

$$\overline{V} = \frac{dx^1}{dt}\,\hat{a}_1 + \left(x^1\,\frac{dx^2}{dt}\right)\hat{a}_2 + \frac{dx^3}{dt}\,\hat{a}_3 \qquad (3.3.6)$$

If the coordinate x^1 is identified with the radial distance r, the coordinate x^2 with the polar angle θ, and the coordinate x^3 with the axial displacement z, the equation for the velocity in a cylindrical polar coordinate system assumes the familiar form

$$\overline{V} = \frac{dr}{dt}\,\hat{a}_1 + \left(r\,\frac{d\theta}{dt}\right)\hat{a}_2 + \frac{dz}{dt}\,\hat{a}_3$$

where $\hat{a}_1, \hat{a}_2$, and $\hat{a}_3$ are a triad of mutually orthogonal unit vectors in the directions of increasing r, θ, and z, respectively.

In a spherical polar coordinate system, the vector $\overline{A}$ has the following components:

$$A^1 = x^1, \quad A^2 = A^3 = 0$$

When these values are substituted in equation (1.13.9), the velocity vector in this coordinate system is given by

$$\overline{V} = \frac{dx^1}{dt}\,\bar{a}_1 + \left\{\begin{matrix}2\\12\end{matrix}\right\}x^1\,\frac{dx^2}{dt}\,\bar{a}_2 + \left\{\begin{matrix}3\\13\end{matrix}\right\}x^1\,\frac{dx^3}{dt}\,\bar{a}_3 \qquad (3.3.7)$$

Again, by substitution of the Christoffel symbols from equation (1.13.8), the velocity vector may be expressed in terms of its tensor components and a corresponding set of base vectors as follows:

$$\overline{V} = \frac{dx^1}{dt}\,\bar{a}_1 + \frac{dx^2}{dt}\,\bar{a}_2 + \frac{dx^3}{dt}\,\bar{a}_3$$

that is,

$$\overline{V} = \frac{dx^i}{dt}\,\bar{a}_i \qquad (3.3.8)$$

From a comparison of equations (3.3.5) and (3.3.8), it is seen that when expressed in terms of its tensor components, the velocity vector has the same form in both coordinate systems. This is true, in general, since by definition

$$\overline{V} = \frac{d\overline{r}}{dt} = \frac{\partial \overline{r}}{\partial x^i} \frac{dx^i}{dt} \qquad (3.3.9)$$

And substitution from equation (1.12.9) in equation (3.3.9) gives

$$\overline{V} = \frac{dx^i}{dt} \, \overline{a}_i$$

Of course, the physical components of velocity are different, as can be seen when the base vectors are reduced to unitary form. By substitution from equation (1.13.6) in equation (3.3.7), the velocity vector may be expressed in terms of its physical components and a set of unit base vectors as follows:

$$\overline{V} = \frac{dx^1}{dt} \, \hat{a}_1 + \left(x^1 \frac{dx^2}{dt} \right) \hat{a}_2 + \left(x^1 \sin x^2 \frac{dx^3}{dt} \right) \hat{a}_3 \qquad (3.3.10)$$

When the coordinate x^1 is identified with the radial distance r, the coordinate x^2 with the polar angle θ, and the coordinate x^3 with the azimuth angle ψ, the equation for $\overline{V}$ assumes the more familiar form

$$\overline{V} = \frac{dr}{dt} \hat{a}_1 + \left(r \frac{d\theta}{dt} \right) \hat{a}_2 + \left(r \sin \theta \frac{d\psi}{dt} \right) \hat{a}_3 \qquad (3.3.11)$$

where $\hat{a}_1$, $\hat{a}_2$, and $\hat{a}_3$ are a triad of mutually orthogonal unit vectors in the directions of increasing r, θ, and ψ, respectively.

3.4 THE ACCELERATION VECTOR

If the acceleration vector were required, the velocity vector $\overline{V}$ would be substituted for the vector $\overline{A}$ in the equation for the intrinsic derivative:

$$\overline{A} = \overline{V} = \frac{dx^i}{dt} \, \overline{a}_i \qquad (3.4.1)$$

Hence, in a general curvilinear coordinate system, the acceleration vector is given by

$$\frac{d\overline{V}}{dt} = \left(\frac{dV^i}{dt} + \begin{Bmatrix} i \\ jk \end{Bmatrix} V^j \frac{dx^k}{dt} \right) \bar{a}_i \qquad (3.4.2)$$

By substitution from equation (3.4.1) in equation (3.4.2), the acceleration vector may be written in the following alternative form:

$$\frac{d\overline{V}}{dt} = \left(\frac{d^2 x^i}{dt^2} + \begin{Bmatrix} i \\ jk \end{Bmatrix} \frac{dx^j}{dt} \frac{dx^k}{dt} \right) \bar{a}_i \qquad (3.4.3)$$

This equation gives the acceleration in any coordinate system, provided the Christoffel symbols are appropriate to the coordinate system chosen to describe the problem.

In a three-dimensional cylindrical polar coordinate system, equation (3.4.2) reduces to the form given by equation (1.13.4) when the vector $\overline{V}$ is substituted for the vector $\overline{A}$. Likewise, in a three-dimensional spherical polar coordinate system, equation (3.4.2) reduces to the form given by equation (1.13.9) when the vector $\overline{V}$ is substituted for the vector $\overline{A}$. If equation (1.13.9) is used to obtain the acceleration vector, the tensor components of velocity, rather than the physical components, must always be used. The tensor components of velocity are given by equation (3.4.1). These are

$$A^1 = \frac{dx^1}{dt} \ , \qquad A^2 = \frac{dx^2}{dt} \ , \qquad A^3 = \frac{dx^3}{dt} \qquad (3.4.4)$$

Substituting these values in equation (3.4.3) gives the acceleration in terms of spherical polar coordinates

$$\frac{d\overline{V}}{dt} = \left(\frac{d^2 x^1}{dt^2} + \begin{Bmatrix} 1 \\ 22 \end{Bmatrix} \frac{dx^2}{dt} \frac{dx^2}{dt} + \begin{Bmatrix} 1 \\ 33 \end{Bmatrix} \frac{dx^3}{dt} \frac{dx^3}{dt} \right) \bar{a}_1$$

$$+ \left(\frac{d^2 x^2}{dt^2} + 2 \begin{Bmatrix} 2 \\ 12 \end{Bmatrix} \frac{dx^1}{dt} \frac{dx^2}{dt} + \begin{Bmatrix} 2 \\ 33 \end{Bmatrix} \frac{dx^3}{dt} \frac{dx^3}{dt} \right) \bar{a}_2$$

$$+ \left(\frac{d^2 x^3}{dt^2} + 2 \begin{Bmatrix} 3 \\ 13 \end{Bmatrix} \frac{dx^1}{dt} \frac{dx^3}{dt} + 2 \begin{Bmatrix} 3 \\ 23 \end{Bmatrix} \frac{dx^2}{dt} \frac{dx^3}{dt} \right) \bar{a}_3 \qquad (3.4.5)$$

By substitution of the Christoffel symbols from equation (1.13.8) in equation (3.4.5), the acceleration vector may be expressed in terms of its tensor components and associated base vectors as follows:

$$\frac{d\overline{V}}{dt} = \left[\frac{d^2x^1}{dt^2} - x^1 \left(\frac{dx^2}{dt}\right)^2 - x^1 \sin x^2 \left(\frac{dx^3}{dt}\right)^2 \right] \bar{a}_1$$

$$+ \left[\frac{d^2x^2}{dt^2} + \frac{2}{x^1} \frac{dx^1}{dt} \frac{dx^2}{dt} - \sin x^2 \cos x^2 \left(\frac{dx^3}{dt}\right)^2 \right] \bar{a}_2$$

$$+ \left(\frac{d^2x^3}{dt^2} + \frac{2}{x^1} \frac{dx^1}{dt} \frac{dx^3}{dt} + 2 \cot x^2 \frac{dx^2}{dt} \frac{dx^3}{dt} \right) \bar{a}_3 \qquad (3.4.6)$$

The corresponding physical components of the acceleration vector are obtained when the base vectors are expressed in terms of unit vectors in accordance with equation (1.13.6). When appropriate substitutions are made, equation (3.4.6) gives

$$\frac{d\overline{V}}{dt} = \left[\frac{d^2x^1}{dt^2} - x^1 \left(\frac{dx^2}{dt}\right)^2 - x^1 \left(\sin x^2 \frac{dx^3}{dt}\right)^2 \right] \hat{a}_1$$

$$+ \left[x^1 \frac{d^2x^2}{dt^2} + 2\frac{dx^1}{dt} \frac{dx^2}{dt} - x^1 \sin x^2 \cos x^2 \left(\frac{dx^3}{dt}\right)^2 \right] \hat{a}_2$$

$$+ \left(x^1 \sin x^2 \frac{d^2x^3}{dt^2} + 2 \sin x^2 \frac{dx^1}{dt} \frac{dx^3}{dt} + 2x^1 \cos x^2 \frac{dx^2}{dt} \frac{dx^3}{dt} \right) \hat{a}_3$$

3.5 EQUATIONS OF MOTION IN A GENERAL CURVILINEAR COORDINATE SYSTEM

In using tensor methods to derive equations of motion, it is again important to remember that the acceleration and force vectors must always be expressed in terms of their tensor components rather than their physical components. Hence, the two sides of every equation must balance with respect to their covariant or contravariant properties before applying Newton's second law of motion. In this connection it is worth noting that, although the acceleration vector is expressed in contravariant form in equation (3.4.3), the force vector may appear in the form of a covariant vector. The force vector assumes the covariant form when it appears as the gradient of a scalar point function. This occurs in the equations of motion of a space vehicle which, in addition to the thrust force, is subject to gravitational forces. If the gravitational forces are expressed in the form of the gradient of a gravitational potential function, the force vector is

$$\bar{F} = \nabla\phi + \bar{T} \tag{3.5.1}$$

where ϕ is the gravitational potential function, which may include the influence of oblateness and extraterrestrial gravitational forces, and $\bar{T}$ is the thrust vector.

The gradient of a scalar point function assumes the form

$$\nabla\phi = \frac{\partial\phi}{\partial x^i} \, \bar{a}^i \tag{3.5.2}$$

The use of the reciprocal base vector $\bar{a}^i$ in equation (3.5.2) is justified by the following considerations: the components of the gradient of the gravitational potential function in the y coordinate system are related to those in the x coordinate system by the following equation:

$$\frac{\partial\phi}{\partial y^i} = \frac{\partial\phi}{\partial x^j} \frac{\partial x^j}{\partial y^i}$$

or

$$F_i(y) = \frac{\partial x^j}{\partial y^i} \, F_j(x) \tag{3.5.3}$$

where

$$F_i(y) = \frac{\partial\phi}{\partial y^i}$$

$$F_j(x) = \frac{\partial\phi}{\partial x^j}$$

Therefore, the transformation of the components of the gradient vector from the x coordinate system to the y coordinate system obeys the covariant transformation law as defined in equation (1.4.6).

The equation of motion of a point mass which is subject to gravitational and thrust forces is obtained by combining equations (3.4.3) and (3.5.1):

$$M\left(\frac{d^2 x^i}{dt^2} + \left\{\begin{matrix} i \\ jk \end{matrix}\right\} \frac{dx^j}{dt}\frac{dx^k}{dt}\right)\bar{a}_i = \nabla\phi + \bar{T} \qquad (3.5.4)$$

where M is the mass.

It is seen that the acceleration components represented by the left-hand side of this equation are all contravariant. The thrust vector, on the other hand, is usually given in terms of its physical components, and as already indicated in equation (3.5.3), the gravitational forces assume the form of covariant vectors. To have a force system compatible with the accelerations, it is necessary to convert all the force terms to the contravariant form. The potential gradient function may be converted to contravariant form with the aid of equation (1.7.6). From equations (3.5.2) and (1.7.6)

$$\nabla\phi = \frac{\partial\phi}{\partial x^j}\,\bar{a}^j = g^{ij}\frac{\partial\phi}{\partial x^j}\,\bar{a}_i \qquad (3.5.5)$$

The thrust vector may be expressed in the following alternative forms:

$$\bar{T} = T^i\bar{a}_i = \tau^i\hat{a}_i$$

where T^i are the contravariant components of the thrust vector, and τ^i are the corresponding physical components. The physical components of the thrust vector are related to the contravariant components by equation (1.9.2)

$$T^i = \frac{1}{\sqrt{g_{(ii)}}}\,\tau^i \qquad (3.5.6)$$

By substitution from equations (3.5.5) and (3.5.6) in equation (3.5.4), the equation of motion assumes the following form

$$M\left(\frac{d^2x^i}{dt^2} + \left\{\begin{matrix} i \\ jk \end{matrix}\right\}\frac{dx^j}{dt}\frac{dx^k}{dt}\right)\bar{a}_i = \left[g^{ij}\frac{\partial\phi}{\partial x^j} + \frac{\tau^i}{\sqrt{g_{(ii)}}}\right]\bar{a}_i$$

Therefore

$$M\left(\frac{d^2x^i}{dt^2} + \left\{\begin{matrix} i \\ jk \end{matrix}\right\}\frac{dx^j}{dt}\frac{dx^k}{dt}\right) = g^{ij}\frac{\partial\phi}{\partial x^j} + \frac{\tau^i}{\sqrt{g_{(ii)}}} \tag{3.5.7}$$

When the expression for the gravitational forces is expanded in a general three-dimensional coordinate system, equation (3.5.7) becomes

$$M\left(\frac{d^2x^i}{dt^2} + \left\{\begin{matrix} i \\ jk \end{matrix}\right\}\frac{dx^j}{dt}\frac{dx^k}{dt}\right) = g^{i1}\frac{\partial\phi}{\partial x^1} + g^{i2}\frac{\partial\phi}{\partial x^2} + g^{i3}\frac{\partial\phi}{\partial x^3} + \frac{\tau^i}{\sqrt{g_{(ii)}}} \tag{3.5.8}$$

However, in a rectangular coordinate system

$$g^{ij} = 0 \quad \text{for} \quad i \neq j$$

and

$$g^{(ii)} = 1/g_{(ii)}$$

where the parentheses imply suspension of the summation convention.

Substituting these values in equation (3.5.8) gives for orthogonal systems

$$M\left(\frac{d^2x^i}{dt^2} + \left\{\begin{matrix} i \\ jk \end{matrix}\right\}\frac{dx^j}{dt}\frac{dx^k}{dt}\right) = \frac{1}{g_{(ii)}}\frac{\partial\phi}{\partial x^i} + \frac{\tau^i}{\sqrt{g_{(ii)}}} \tag{3.5.9}$$

Equation (3.5.9) may be rewritten as follows:

$$M\left[g_{(ii)}\frac{d^2x^i}{dt^2} + g_{(ii)}\left\{\begin{matrix} i \\ jk \end{matrix}\right\}\frac{dx^j}{dt}\frac{dx^k}{dt}\right] = \frac{\partial\phi}{\partial x^i} + \sqrt{g_{(ii)}}\,\tau^i$$

or in the alternative form

$$M \left[g_{(ii)} \frac{d^2 x^i}{dt^2} + [jk,i] \frac{dx^j}{dt} \frac{dx^k}{dt} \right] = \frac{\partial \phi}{\partial x^i} + \sqrt{g_{(ii)}} \, \tau^i \qquad (3.5.10)$$

The derivation of equation (3.5.10) may, at first sight, seem to involve an unnecessary degree of complexity. However, when it is realized that this equation is valid in all orthogonal curvilinear coordinate systems, the effort expended will be seen to be worthwhile. Although in using the conventional approach, the equations have to be reformulated each time a new set of coordinates is considered, the tensor equation requires no such modification. Hence, the expenditure of time and effort, which has to be made repeatedly when the formulation is conventional, is avoided when the tensor equation is used.

Equation (3.5.10) may be formulated in terms of either coordinate transformation equations or metric tensors. When a particular application suggests a formulation in terms of coordinate transformation equations, the metric tensors and the Christoffel symbols may be replaced by their equivalents from equations (1.13.12) and (1.13.14), respectively.

When these substitutions are made, equation (3.5.10) assumes the following modified form:

$$M \left\{ \left[\frac{\partial y^\alpha}{\partial x^{(i)}} \frac{\partial y^\alpha}{\partial x^{(i)}} \right] \frac{d^2 x^i}{dt^2} + \left(\frac{\partial^2 y^\alpha}{\partial x^j \partial x^k} \frac{\partial y^\alpha}{\partial x^i} \right) \frac{dx^j}{dt} \frac{dx^k}{dt} \right\} = \frac{\partial \phi}{\partial x^i} + \sqrt{\frac{\partial y^\alpha}{\partial x^{(i)}} \frac{\partial y^\alpha}{\partial x^{(i)}}} \, \tau^i$$

$$(3.5.11)$$

As indicated previously, a repeated index implies summation with respect to that index. An exception to this rule occurs when repeated indices are enclosed in parentheses. Parentheses around an index imply that the summation convention is to be suspended for that index. This means that for each value of the index i, equation (3.5.11) must be summed on α, j, and k. For example, when equation (3.5.11) is summed on α, it appears as follows:

$$M \left\{ \left[\frac{\partial y^1}{\partial x^{(i)}} \frac{\partial y^1}{\partial x^{(i)}} + \frac{\partial y^2}{\partial x^{(i)}} \frac{\partial y^2}{\partial x^{(i)}} + \frac{\partial y^3}{\partial x^{(i)}} \frac{\partial y^3}{\partial x^{(i)}} \right] \frac{d^2 x^i}{dt^2} \right.$$

$$\left. + \left(\frac{\partial^2 y^1}{\partial x^j \partial x^k} \frac{\partial y^1}{\partial x^i} + \frac{\partial^2 y^2}{\partial x^j \partial x^k} \frac{\partial y^2}{\partial x^i} + \frac{\partial^2 y^3}{\partial x^j \partial x^k} \frac{\partial y^3}{\partial x^i} \right) \frac{dx^j}{dt} \frac{dx^k}{dt} \right\}$$

$$= \frac{\partial \phi}{\partial x^i} + \sqrt{\frac{\partial y^1}{\partial x^{(i)}} \frac{\partial y^1}{\partial x^{(i)}} + \frac{\partial y^2}{\partial x^{(i)}} \frac{\partial y^2}{\partial x^{(i)}} + \frac{\partial y^3}{\partial x^{(i)}} \frac{\partial y^3}{\partial x^{(i)}}} \, \tau^i \qquad (3.5.12)$$

The left side of this equation must also be summed on j and k. When each of these indices is permitted to take the values 1, 2, 3, in turn, equation (3.5.12) assumes the following form:

$$
M\left\{\left[\frac{\partial y^1}{\partial x^{(i)}}\frac{\partial y^1}{\partial x^{(i)}} + \frac{\partial y^2}{\partial x^{(i)}}\frac{\partial y^2}{\partial x^{(i)}} + \frac{\partial y^3}{\partial x^{(i)}}\frac{\partial y^3}{\partial x^{(i)}}\right]\frac{d^2 x^i}{dt^2}\right.
$$

$$
+ \left(\frac{\partial^2 y^1}{\partial x^1 \partial x^1}\frac{\partial y^1}{\partial x^i} + \frac{\partial^2 y^2}{\partial x^1 \partial x^1}\frac{\partial y^2}{\partial x^i} + \frac{\partial^2 y^3}{\partial x^1 \partial x^1}\frac{\partial y^3}{\partial x^i}\right)\frac{dx^1}{dt}\frac{dx^1}{dt}
$$

$$
+ \left(\frac{\partial^2 y^1}{\partial x^1 \partial x^2}\frac{\partial y^1}{\partial x^i} + \frac{\partial^2 y^2}{\partial x^1 \partial x^2}\frac{\partial y^2}{\partial x^i} + \frac{\partial^2 y^3}{\partial x^1 \partial x^2}\frac{\partial y^3}{\partial x^i}\right)\frac{dx^1}{dt}\frac{dx^2}{dt}
$$

$$
+ \left(\frac{\partial^2 y^1}{\partial x^1 \partial x^3}\frac{\partial y^1}{\partial x^i} + \frac{\partial^2 y^2}{\partial x^1 \partial x^3}\frac{\partial y^2}{\partial x^i} + \frac{\partial^2 y^3}{\partial x^1 \partial x^3}\frac{\partial y^3}{\partial x^i}\right)\frac{dx^1}{dt}\frac{dx^3}{dt}
$$

$$
+ \left(\frac{\partial^2 y^1}{\partial x^2 \partial x^1}\frac{\partial y^1}{\partial x^i} + \frac{\partial^2 y^2}{\partial x^2 \partial x^1}\frac{\partial y^2}{\partial x^i} + \frac{\partial^2 y^3}{\partial x^2 \partial x^1}\frac{\partial y^3}{\partial x^i}\right)\frac{dx^2}{dt}\frac{dx^1}{dt}
$$

$$
+ \left(\frac{\partial^2 y^1}{\partial x^2 \partial x^2}\frac{\partial y^1}{\partial x^i} + \frac{\partial^2 y^2}{\partial x^2 \partial x^2}\frac{\partial y^2}{\partial x^i} + \frac{\partial^2 y^3}{\partial x^2 \partial x^2}\frac{\partial y^3}{\partial x^i}\right)\frac{dx^2}{dt}\frac{dx^2}{dt}
$$

$$
+ \left(\frac{\partial^2 y^1}{\partial x^2 \partial x^3}\frac{\partial y^1}{\partial x^i} + \frac{\partial^2 y^2}{\partial x^2 \partial x^3}\frac{\partial y^2}{\partial x^i} + \frac{\partial^2 y^3}{\partial x^2 \partial x^3}\frac{\partial y^3}{\partial x^i}\right)\frac{dx^2}{dt}\frac{dx^3}{dt}
$$

$$
+ \left(\frac{\partial^2 y^1}{\partial x^3 \partial x^1}\frac{\partial y^1}{\partial x^i} + \frac{\partial^2 y^2}{\partial x^3 \partial x^1}\frac{\partial y^2}{\partial x^i} + \frac{\partial^2 y^3}{\partial x^3 \partial x^1}\frac{\partial y^3}{\partial x^i}\right)\frac{dx^3}{dt}\frac{dx^1}{dt}
$$

$$
+ \left(\frac{\partial^2 y^1}{\partial x^3 \partial x^2}\frac{\partial y^1}{\partial x^i} + \frac{\partial^2 y^2}{\partial x^3 \partial x^2}\frac{\partial y^2}{\partial x^i} + \frac{\partial^2 y^3}{\partial x^3 \partial x^2}\frac{\partial y^3}{\partial x^i}\right)\frac{dx^3}{dt}\frac{dx^2}{dt}
$$

$$
+ \left.\left(\frac{\partial^2 y^1}{\partial x^3 \partial x^3}\frac{\partial y^1}{\partial x^i} + \frac{\partial^2 y^2}{\partial x^3 \partial x^3}\frac{\partial y^2}{\partial x^i} + \frac{\partial^2 y^3}{\partial x^3 \partial x^3}\frac{\partial y^3}{\partial x^i}\right)\frac{dx^3}{dt}\frac{dx^3}{dt}\right\}
$$

$$
= \frac{\partial \phi}{\partial x^{(i)}} + \sqrt{\frac{\partial y^1}{\partial x^{(i)}}\frac{\partial y^1}{\partial x^{(i)}} + \frac{\partial y^2}{\partial x^{(i)}}\frac{\partial y^2}{\partial x^{(i)}} + \frac{\partial y^3}{\partial x^{(i)}}\frac{\partial y^3}{\partial x^{(i)}}}\,\tau^i
$$

$$(3.5.13)$$

The form of this equation is well suited to routine non-numeric computer operations. The large number of terms appearing in equation (3.5.13) is due to the generality of the equation, which is applicable to any space of three dimensions. Hence, to obtain the equations of motion in any system of coordinates by using transformation equation inputs, the only information required is the special form of equation (1.4.4), relating that system of coordinates to the orthogonal Cartesian coordinates y^i.

3.6 COMPUTER DERIVATIONS OF EQUATIONS OF MOTION OF A PARTICLE

Equation (3.5.11) can be mechanized to formulate the equations of motion of a particle in any orthogonal curvilinear coordinate system requested by the user. The key statement in the mechanization program utilizes the analytic differentiation routine. This statement must be written according to the rules of analytic differentiation specified in the user's manual. In the case of computers equipped with formula manipulation compilers, the statement corresponding to equation (3.5.11) would assume the following form (ref. 2):

```
M*(((DERIV(Y(A),X(I)))**2)*A(I)

+(DERIV(Y(A),X(J),X(K)))*(DERIV(Y(A),X(I)))*V(J)*V(K))

= DPHI(I) + SQRT((DERIV(Y(A),X(I)))**2)*T(I)         (3.6.1)
```

where

$$A(I) = \frac{d^2 x^i}{dt^2} \quad ; \qquad T(I) = \tau^i \tag{3.6.2}$$

$$V(J) = \frac{dx^j}{dt} \quad ; \qquad V(K) = \frac{dx^k}{dt} \tag{3.6.3}$$

$$Y(I) = y^i \quad ; \qquad Y(A) = y^\alpha \tag{3.6.4}$$

A simple program using equation (3.6.1) and supporting statements to formulate the equation of motion of a particle will require as input, the coordinate transformation equations corresponding to the coordinate systems being considered; the following are examples.

3.6.1 Spherical Polar Coordinates

Consider a transformation of coordinates specifying the relation between the spherical polar coordinates x^i and the orthogonal Cartesian coordinates y^i. In this case, equation (1.4.4) becomes

$$\text{Input} = \begin{cases} y^1 = x^1 \sin x^2 \cos x^3 \\[2mm] y^2 = x^1 \sin x^2 \sin x^3 \\[2mm] y^3 = x^1 \cos x^2 \end{cases}$$

The corresponding output is

$$M\left[\frac{d^2 x^1}{dt^2} - x^1 \left(\frac{dx^2}{dt}\right)^2 - x^1 \left(\sin x^2 \frac{dx^3}{dt}\right)^2\right] = \frac{\partial \phi}{\partial x^1} + \tau^1$$

$$M\left[(x^1)^2 \frac{d^2 x^2}{dt^2} + 2x^1 \frac{dx^1}{dt}\frac{dx^2}{dt} - (x^1)^2 \sin x^2 \cos x^2 \left(\frac{dx^3}{dt}\right)^2\right] = \frac{\partial \phi}{\partial x^2} + x^1 \tau^2$$

$$M\left[(x^1 \sin x^2)^2 \frac{d^2 x^3}{dt^2} + 2x^1 \sin^2 x^2 \frac{dx^1}{dt}\frac{dx^3}{dt} + 2(x^1)^2 \sin x^2 \cos x^2 \frac{dx^2}{dt}\frac{dx^3}{dt}\right]$$

$$= \frac{\partial \phi}{\partial x^3} + x^1 \sin x^2 \tau^3$$

Because of its generality, equation (3.5.13) is applicable in all coordinate systems. Therefore, to obtain the equations of motion in any other coordinate system, all that is required is to supply the computer with the appropriate coordinate transformation equations.

3.6.2 Cylindrical Polar Coordinates

As a further illustration of the procedure involved, consider the equations of motion in a cylindrical polar system of coordinates. In this case, the coordinate transformation equations are

$$y^1 = x^1 \cos x^2$$

$$y^2 = x^1 \sin x^2$$

$$y^3 = x^3$$

When these coordinate transformation equations were used to evaluate the terms of equation (3.5.13), the following output was obtained.

$$M\left(\frac{d^2 x^1}{dt^2} - x^1 \frac{dx^2}{dt}\frac{dx^2}{dt}\right) = \frac{\partial \phi}{\partial x^1} + \tau^1$$

$$M\left[(x^1)^2 \frac{d^2 x^2}{dt^2} + 2x^1 \frac{dx^1}{dt}\frac{dx^2}{dt}\right] = \frac{\partial \phi}{\partial x^2} + x^1 \tau^2$$

$$M\left(\frac{d^2 x^3}{dt^2}\right) = \frac{\partial \phi}{\partial x^2} + \tau^3$$

3.6.3 Prolate Spheroidal Coordinates

Another interesting system of orthogonal curvilinear coordinates is the prolate spheroidal coordinates. Coordinate surfaces are obtained by rotating a family of confocal ellipses and hyperbolas about their major axes. Rotating these conic sections gives rise to a system of prolate spheroids and hyperboloids of two sheets. A family of planes through the axis of rotation completes the system of orthogonal surfaces. The curvilinear coordinate systems generated by the curves of intersection of these surfaces are useful in certain quantum mechanical problems. The transformation equations relating this system of coordinates to the orthogonal Cartesian system are as follows (ref. 3):

$$y^1 = a \sinh x^1 \sin x^2 \cos x^3$$

$$y^2 = a \sinh x^1 \sin x^2 \sin x^3$$

$$y^3 = a \cosh x^1 \cos x^2$$

To obtain the equations of motion relative to a prolate spheroidal system of coordinates, these transformation equations were substituted for equation (1.4.4) in the computer program. The equations of motion were obtained as follows:

$$M\left[a^2(\sin^2 x^2 + \sinh^2 x^1)\frac{d^2x^1}{dt^2} + 2a^2 \sin x^2 \cos x^2 \frac{dx^1}{dt}\frac{dx^2}{dt}\right.$$

$$+ a^2 \sinh x^1 \cosh x^1 \frac{dx^1}{dt}\frac{dx^1}{dt} - a^2 \sinh x^1 \cosh x^1 \frac{dx^2}{dt}\frac{dx^2}{dt}$$

$$\left. - a^2 \sin^2 x^2 \sinh x^1 \cosh x^1 \frac{dx^2}{dt}\frac{dx^3}{dt}\right] = a\sqrt{\sin^2 x^2 + \sinh^2 x^1}\,\tau^1 + \frac{\partial \phi}{\partial x^1}$$

$$M\left[a^2(\sin^2 x^2 + \sinh^2 x^1)\frac{d^2x^1}{dt^2} - a^2 \sin x^2 \cos x^2 \frac{dx^1}{dt}\frac{dx^1}{dt}\right.$$

$$+ 2a^2 \sinh x^1 \cosh x^1 \frac{dx^1}{dt}\frac{dx^2}{dt} + a^2 \sin x^2 \cos x^2 \frac{dx^2}{dt}\frac{dx^2}{dt}$$

$$\left. - a^2 \sin x^2 \cos x^2 \sinh^2 x^1 \frac{dx^3}{dt}\frac{dx^3}{dt}\right] = a\left(\sqrt{\sin^2 x^2 + \sinh^2 x^1}\right)\tau^2 + \frac{\partial \phi}{\partial x^2}$$

$$M\left[a^2 \sin^2 x^2 \sinh^2 x^1 \frac{d^2x^3}{dt^3} + 2a^2 \sin^2 x^2 \sinh x^1 \cosh x^1 \frac{dx^1}{dt}\frac{dx^3}{dt}\right.$$

$$\left. + 2a^2 \sin x^2 \cos x^2 \sinh^2 x^1 \frac{dx^2}{dt}\frac{dx^3}{dt}\right] = a \sin x^2 \sinh x^1 \tau^3 + \frac{\partial \phi}{\partial x^3}$$

3.6.4 Oblate Spheroidal Coordinates

Confocal ellipses and hyperbolas rotated about their minor axes generate the oblate spheroids and hyperboloids of one sheet. These surfaces, together with a family of planes through the axis of rotation, constitute a family of orthogonal surfaces. The curvilinear coordinate systems generated by the curves of intersection of these surfaces are called oblate spheroidal coordinates. Oblate spheroids are sometimes referred to as planetary ellipsoids, because the Earth and the planet Jupiter are approximately of this form. The transformation equations relating this system of coordinates to the orthogonal Cartesian system are as follows (ref. 3):

$$y^1 = a \cosh x^1 \sin x^2 \cos x^3$$

$$y^2 = a \cosh x^1 \sin x^2 \sin x^3$$

$$y^3 = a \sinh x^1 \cos x^2$$

These transformation equations again take the place of equation (1.4.4) in the computer program. The equations of motion relative to a system of oblate spheroidal coordinates were obtained in the following form:

$$M\left[a^2(\sinh^2 x^1 + \cos^2 x^2)\frac{d^2 x^1}{dt^2} + a^2(\sinh x^1 \cosh x^1)\frac{dx^1}{dt}\frac{dx^1}{dt} \right.$$

$$- 2a^2 \cos x^2 \sin x^2 \frac{dx^1}{dt}\frac{dx^2}{dt} - a^2 \sinh x^1 \cosh x^1 \frac{dx^2}{dt}\frac{dx^2}{dt}$$

$$\left. - a^2 \cosh x^1 \sinh x^1 \sin^2 x^2 \frac{dx^3}{dt}\frac{dx^3}{dt} \right] = a\left(\sqrt{\sinh^2 x^1 + \cos^2 x^2}\right)\tau^1 + \frac{\partial \phi}{\partial x^1}$$

$$M\left[a^2(\sinh^2 x^1 + \cos^2 x^2)\frac{d^2 x^2}{dt^2} + a^2 \sin x^2 \cos x^2 \frac{dx^1}{dt}\frac{dx^1}{dt} \right.$$

$$+ 2a^2 \sinh x^1 \cosh x^1 \frac{dx^1}{dt}\frac{dx^2}{dt} - a^2 \sin x^2 \cos x^2 \frac{dx^2}{dt}\frac{dx^2}{dt}$$

$$\left. - a^2 \cosh x^1 \sin x^2 \cos x^2 \frac{dx^3}{dt}\frac{dx^3}{dt} \right] = a\sqrt{\sinh^2 x^1 + \cos^2 x^2}\ \tau^2 + \frac{\partial \phi}{\partial x^2}$$

$$M\left(a^2 \cosh^2 x^1 \sin^2 x^2 \frac{d^2 x^3}{dt^2} + 2a^2 \sinh x^1 \cosh x^1 \sin^2 x^2 \frac{dx^1}{dt}\frac{dx^3}{dt} \right.$$

$$\left. + 2a^2 \cosh^2 x^1 \sin x^2 \cos x^2 \frac{dx^2}{dt}\frac{dx^3}{dt} \right) = a \cosh x^1 \sin x^2 \tau^3 + \frac{\partial \phi}{\partial x^3}$$

The preceding technique for formulating equations of motion by symbolic mathematical computation is based on the use of coordinate transformation equations. However, in many cases it will be convenient to use the metric tensors, rather than the coordinate transformation equations. When this procedure is adopted, the necessity for evaluating equation (1.13.12) is eliminated and equation (3.5.10) is modified as follows:

$$
M\left\{g_{(ii)}\frac{d^2x^1}{dt^2}+\frac{1}{2}\left[\left(\frac{\partial g_{1i}}{\partial x^1}+\frac{\partial g_{1i}}{\partial x^1}-\frac{\partial g_{11}}{\partial x^i}\right)\frac{dx^1}{dt}\frac{dx^1}{dt}+\left(\frac{\partial g_{1i}}{\partial x^2}+\frac{\partial g_{1i}}{\partial x^1}\right.\right.\right.
$$

$$
\left.-\frac{\partial g_{12}}{\partial x^i}\right)\frac{dx^1}{dt}\frac{dx^2}{dt}+\left(\frac{\partial g_{1i}}{\partial x^3}+\frac{\partial g_{3i}}{\partial x^1}-\frac{\partial g_{13}}{\partial x^i}\right)\frac{dx^1}{dt}\frac{dx^3}{dt}\right]+\frac{1}{2}\left[\left(\frac{\partial g_{2i}}{\partial x^1}+\frac{\partial g_{1i}}{\partial x^2}\right.\right.
$$

$$
\left.-\frac{\partial g_{21}}{\partial x^i}\right)\frac{dx^2}{dt}\frac{dx^1}{dt}+\left(\frac{\partial g_{2i}}{\partial x^2}+\frac{\partial g_{2i}}{\partial x^2}-\frac{\partial g_{22}}{\partial x^i}\right)\frac{dx^2}{dt}\frac{dx^2}{dt}+\left(\frac{\partial g_{2i}}{\partial x^3}+\frac{\partial g_{3i}}{\partial x^2}\right.
$$

$$
\left.-\frac{\partial g_{23}}{\partial x^i}\right)\frac{dx^2}{dt}\frac{dx^3}{dt}\right]+\frac{1}{2}\left[\left(\frac{\partial g_{3i}}{\partial x^1}+\frac{\partial g_{1i}}{\partial x^3}-\frac{\partial g_{31}}{\partial x^i}\right)\frac{dx^3}{dt}\frac{dx^1}{dt}+\left(\frac{\partial g_{3i}}{\partial x^2}+\frac{\partial g_{2i}}{\partial x^3}\right.\right.
$$

$$
\left.\left.\left.-\frac{\partial g_{32}}{\partial x^i}\right)\frac{dx^3}{dt}\frac{dx^2}{dt}+\left(\frac{\partial g_{3i}}{\partial x^3}+\frac{\partial g_{3i}}{\partial x^3}-\frac{\partial g_{33}}{\partial x^i}\right)\frac{dx^3}{dt}\frac{dx^3}{dt}\right]\right\}=\frac{\partial\phi}{\partial x^{(i)}}+\sqrt{g_{(ii)}}\,\tau^i
$$

If this form is used to derive the equations of motion of a particle in cylindrical polar or spherical polar coordinate systems, the inputs to the computer program would be given by equations (3.2.6) and (3.2.9), respectively, rather than by equations (3.1.1) and (3.1.3).

Consider again the problem of formulating the equations of motion of a particle in a cylindrical polar coordinate system. In this case the metric tensors are

$$g_{11} = 1$$

$$g_{22} = (x^1)^2$$

$$g_{33} = 1$$

The following simple program consisting of two separate DO loops may be used to formulate the equations of motion of a particle in a cylindrical polar coordinate system

```
LET(G(1,1)=1);

LET(G(2,2)=X(1)**2);

LET(G(3,3)=1);

PRINT_OUT(G(1,1);G(2,2);G(3,3));

PUT SKIP(5);

DO I=1 TO 3 BY 1;

LET(I="I");

DO J=1 TO 3 BY 1;

LET(J="J");

DO K=1 TO 3 BY 1;

LET(K="K");

LET(D(I,J,K)=(1/2)*DERIV(G(I,K),X(J)));

LET(E(J,K,I)=(1/2)*DERIV(G(J,K),X(I)));

LET(F(I,J,K)=(1/2)*DERIV(G(I,J),X(K)));

LET(C(I,J,K)=(D(I,J,K)+E(J,K,I)-F(I,J,K)));

PRINT_OUT(G(I,J,K));

END;

END;

END;

PUT SKIP(3);

DO I=1 TO 3 BY 1;

LET(I="I");
```

```
LET(ST(I)=0);

DO J=1 TO 3 BY 1;

LET(J="J");

DO K=1 TO 3 BY 1;

LET(K="K");

LET(ST(I)=ST(I)+C(J,K,I)*V(J)*V(K));

LET(A(I)=((DPHI(I)+SQRT(G(I,1))*T(I))/M-ST(I))/G(I,1));

END;

END;

PRINT_OUT(A(I));

PUT SKIP(3);

END;
```

The output from this program is as follows:

$G(1,1) = 1$	$C(1,2,3) = 0$
$G(2,2) = X(1)^2$	$C(1,3,1) = 0$
$G(3,3) = 1$	$C(1,3,2) = 0$
	$C(1,3,3) = 0$
$C(1,1,1) = 0$	$C(2,1,1) = 0$
$C(1,1,2) = 0$	$C(2,1,2) = X(1)$
$C(1,1,3) = 0$	$C(2,1,3) = 0$
$C(1,2,1) = 0$	$C(2,2,1) = -X(1)$
$C(1,2,2) = X(1)$	$C(2,2,2) = 0$

C(2,2,3) = 0   C(3,2,1) = 0

C(2,3,1) = 0   C(3,2,2) = 0

C(2,3,2) = 0   C(3,2,3) = 0

C(2,3,3) = 0   C(3,3,1) = 0

C(3,1,1) = 0   C(3,3,2) = 0

C(3,1,2) = 0   C(3,3,3) = 0

C(3,1,3) = 0

A(1) = X(1) V(2)^2 + (DPHI(1) + T(1))/M

A(2) = (-2X(1) V(1) V(2) + (DPHI(2) + X(1) T(2))/M)/X(1)^2

A(3) = (DPHI(3) + T(3))/M

In interpreting these output statements, it should be noted that

$$C(I,J,K) = [ij,k]$$

where $[ijk]$ are the Christoffel symbols of the first kind, and

$$A(I) = \frac{d^2 x^i}{dt^2} \quad ; \qquad DPHI(I) = \frac{\partial \phi}{\partial x^i}$$

$$V(I) = \frac{dx^i}{dt} \quad ; \qquad M = \text{Mass}$$

Also, $T(I)$ is the ith component of the thrust vector.

In terms of conventional mathematical symbolism, these equations are

$$M\left(\frac{d^2 x^1}{dt^2} - x^1 \frac{dx^2}{dt} \frac{dx^2}{dt}\right) = \frac{\partial \phi}{\partial x^1} + \tau^1$$

$$M\left[(x^1)^2 \frac{d^2 x^2}{dt^2} + 2x^1 \frac{dx^1}{dt} \frac{dx^2}{dt}\right] = \frac{\partial \phi}{\partial x^2} + x^1 \tau^2$$

$$M\left(\frac{d^2 x^3}{dt^2}\right) = \frac{\partial \phi}{\partial x^3} + \tau^3$$

The same program may be used to formulate the Christoffel symbols of the first kind and the equations of motion of a particle in a spherical coordinate system, provided the metric tensors from equation (3.2.9) are used as input.

With this change, the program and the output appear as follows:

```
LET(G(1,1)=1);

LET(G(2,2)=X(1)**2);

LET(G(3,3)=X(1)*SIN(X(2)))**2);

PRINT_OUT(G(1,1);G(2,2);G(3,3));

PUT SKIP(5);

DO I=1 TO 3 BY 1;

LET(I="I");

DO J=1 TO 3 BY 1;

LET(J="J");

DO K=1 TO 3 BY 1;

LET(K="K");

LET(D(I,J,K)=(1/2)*DERIV(G(I,K),X(J)));

LET(E(J,K,I)=(1/2)*DERIV(G(J,K),X(I)));

LET(F(I,J,K)=(1/2)*DERIV(G(I,J),X(K)));

LET(C(I,J,K)=(D(I,J,K)+E(J,K,I)-F(I,J,K)));
```

```
PRINT_OUT(C(I,J,K));
END;
END;
END;
PUT SKIP(3);
DO I=1 TO 3 BY 1;
LET(I="I");
LET(ST(I)=0);
DO J=1 TO 3 BY 1;
LET(J="J");
DO K=1 TO 3 BY 1;
LET(K="K");
LET(ST(I)=ST(I)+C(J,K,I)*V(J)*V(K));
LET(A(I)=((DPHI(I)+SQRT(G(I,I))*T(I))/M-ST(I))/G(I,I));
END;
END;
PRINT_OUT(A(I));
PUT SKIP(3);
END;
```

$G(1,1) = 1$

$G(2,2) = X(1)^2$

$G(3,3) = SIN^2 (X(2)) X(1)^2$

C(1,1,1) = 0

C(1,1,2) = 0

C(1,1,3) = 0

C(1,2,1) = 0

C(1,2,2) = X(1)

C(1,2,3) = 0

C(1,3,1) = 0

C(1,3,2) = 0

C(1,3,3) = SIN^2 (X(2)) X(1)

C(2,1,1) = 0

C(2,1,2) = X(1)

C(2,1,3) = 0

C(2,2,1) = -X(1)

C(2,2,2) = 0

C(2,2,3) = 0

C(2,3,1) = 0

C(2,3,2) = 0

C(2,3,3) = COS (X(2)) SIN (X(2)) X(1)^2

C(3,1,1) = 0

C(3,1,2) = 0

C(3,1,3) = SIN^2 (X(2)) X(1)

C(3,2,1) = 0

C(3,2,2) = 0

C(3,2,3) = COS (X(2)) SIN (X(2)) X(1)²

C(3,3,1) = -SIN² (X(2)) X(1)

C(3,3,2) = -COS (X(2)) SIN (X(2)) X(1)²

C(3,3,3) = 0

A(1) = X(1) V(2)² + SIN² (X(2)) X(1) V(3)² + (DPHI(1) + T(1))/M

A(2) = (-2X(1) V(1) V(2) + COS (X(2)) SIN (X(2)) X(1)² V(3)²
+ (DPHI(2) + X(1) T(2))/M)/X(1)²

A(3) = (-2 COS (X(2)) SIN (X(2)) X(1)² V(3) V(2)
- 2 SIN² (X(2)) X(1) V(1) V(3) + (DPHI(3)
+ SIN (X(2)) X(1) T(3))/M)/(SIN² (X(2)) X(1)²)

In conventional notation, these equations assume the more familiar form

$$M\left[\frac{d^2x^1}{dt^2} - x^1\left(\frac{dx^2}{dt}\right)^2 - x^1\left(\sin x^2 \frac{dx^3}{dt}\right)^2\right] = \frac{\partial\phi}{\partial x^1} + \tau^1$$

$$M\left[(x^1)^2 \frac{d^2x^2}{dt^2} + 2x^1 \frac{dx^1}{dt}\frac{dx^2}{dt} - (x^1)^2 \sin x^2 \cos x^2 \left(\frac{dx^3}{dt}\right)^2\right] = \left(\frac{\partial\phi}{\partial x^2} + x^1\tau^2\right)$$

$$M\left[(x^1 \sin x^2)^2 \frac{d^2x^3}{dt^2} + 2x^1 \sin^2 x^2 \frac{dx^1}{dt}\frac{dx^3}{dt}\right.$$
$$\left. + 2(x^1)^2 \sin x^2 \cos x^2 \frac{dx^2}{dt}\frac{dx^3}{dt}\right] = \left(\frac{\partial\phi}{\partial x^3} + x^1 \sin x^2\tau^3\right)$$

The same procedure can be employed to obtain the equations of motion of a particle in any other coordinate system provided the fundamental quadratic form is known. For example, the technique has been used to derive the equations of motion of a particle relative to a prolate spheroidal system of coordinates. In this form, the equations are useful in certain quantum-mechanical problems.

3.7 OBSERVATIONS

It is seen that a digital computer can be used to facilitate the formulation of the equations of motion of a particle in any curvilinear coordinate system of interest. The simplification inherent in the tensor method is again evident. Equations (3.5.10) and (3.5.11) are applicable in all three-dimensional systems of coordinates. With these equations, the user has a choice of two methods: (1) a formulation based on metric tensor inputs and (2) a formulation based on coordinate transformation equation inputs. As in the case of aeronautical applications, it should be observed that, in each case, the only operations involved are summation and symbolic differentiation.

3.8 ILLUSTRATIVE EXAMPLES

The following illustrations and applications of the equations derived in preceding sections are designed to provide readers with some physical insight and an opportunity to reexamine the equations when they are expressed in more familiar symbology.

The use of indices, such as superscripts and subscripts, is advantageous from the point of view of symbolic mathematical computation. However, some readers may feel more comfortable with the equations when the following substitutions are made (see sketch (1)):

$$\left. \begin{array}{l} x^1 = r \\[6pt] x^2 = \theta \\[6pt] x^3 = \phi \end{array} \right\} \quad \text{polar coordinates}$$

$$\left.\begin{array}{l} x^1 = r \\[6pt] x^2 = \theta \\[6pt] x^3 = z \end{array}\right\} \quad \text{cylindrical coordinates}$$

Sketch (l)

Equations (3.5.10), and subsequent equations, give the components of acceleration in any curvilinear coordinate system. In terms of the more familiar (r, θ, ϕ) coordinates these components are

$$f_r = (\ddot{r} - r\dot{\theta}^2 - r \sin^2 \theta \, \dot{\phi}^2)$$

$$f_\theta = (r\ddot{\theta} + 2\dot{r}\dot{\theta} - r\dot{\phi}^2 \sin \theta \cos \theta)$$

$$f_\phi = (r\ddot{\phi} \sin \theta + 2\dot{r}\dot{\phi} \sin \theta + 2r\dot{\phi}\dot{\theta} \cos \theta)$$

3.8.1 *Motion on the Surface of a Sphere*

For the special case of motion on the surface of a sphere of radius a, these equations take a simpler form.

The acceleration toward the center is

$$a(\dot{\theta}^2 + \sin^2 \theta \dot{\phi}^2)$$

Acceleration along a meridian curve is

$$a(\ddot{\theta} - \dot{\phi}^2 \sin \theta \cos \theta)$$

The acceleration perpendicular to a meridian plane is

$$a(\ddot{\phi} \sin \theta + 2\dot{\phi}\dot{\theta} \cos \theta)$$

3.8.2 *Motion of a Particle on a Right Circular Cone (sketch (m))*

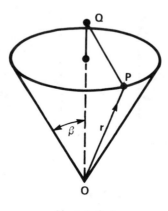

Sketch (m)

For this kind of motion, we substitute $\theta = \beta$, and obtain for the acceleration along a generator OP

$$(\ddot{r} - r\dot{\phi}^2 \sin^2 \beta)$$

Along the inward normal PQ, the acceleration is

$$(r\dot{\phi}^2 \sin \beta \cos \beta)$$

and the acceleration normal to a meridian plane is

$$(r\ddot{\phi} \sin \beta + 2\dot{r}\dot{\phi} \sin \beta)$$

3.8.3 Motion of a Particle in a Central Force Field Varying Inversely as the Square of the Radial Distance From the Center

3.8.3.1 Polar orbits. The equations of motion of a particle describing a polar orbit are obtained by substituting $\phi = \alpha = $ constant, and by applying the appropriate force function, that is

$$(\ddot{r} - r\dot{\theta}^2) = -\frac{\mu}{r^2}$$
$$(r\ddot{\theta} + 2\dot{r}\dot{\theta}) = 0$$

where μ is the force per unit mass. The second of these equations may be rewritten as follows:

$$\frac{1}{r} \frac{d}{dt} (r^2 \dot{\theta}) = 0$$

Therefore

$$r^2 \dot{\theta} = h$$

where h is the constant angular momentum of the particle. By substituting for $\dot{\theta}$ in the first equation we obtain

$$\left(\ddot{r} - \frac{h^2}{r^3}\right) = -\frac{\mu}{r^2}$$

A useful form of solution is obtained by making the substitution $u = 1/r$, which gives

$$\frac{d}{dt} = hu^2 \frac{d}{d\theta}$$

Hence

$$\dot{r} = -h \frac{du}{d\theta}$$

and

$$\ddot{r} = \frac{d}{dt}\left(-h \frac{du}{d\theta}\right) = -h^2 u^2 \frac{d^2 u}{d\theta^2}$$

Therefore

$$\left(\ddot{r} - \frac{h^2}{r^3}\right) = \left(-h^2 u^2 \frac{d^2 u}{d\theta^2} - h^2 u^3\right) = -\mu u^2$$

or

$$\left(\frac{d^2 u}{d\theta^2} + u\right) = \frac{\mu}{h^2}$$

This well-known equation has a solution of the form

$$u = \frac{\mu}{h^2} + D \cos(\theta - \tilde{\omega}) = \frac{1}{r}$$

where D and $\tilde{\omega}$ are constants of integration.
   This equation may be rewritten as

$$\frac{h^2}{\mu r} = 1 + \frac{Dh^2}{\mu} \cos(\theta - \tilde{\omega})$$

which is recognized as the polar equation of a conic with focus as origin. The semilatus rectum is l, where

$$l = \frac{h^2}{\mu}$$

In terms of this relationship the polar equation of the conic, when expanded, is (see sketch (n))

$$\frac{l}{r} = 1 + A \cos\theta + B \sin\theta$$

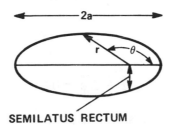

SEMILATUS RECTUM

Sketch (n)

To determine the integration constants, it is noted that when

$$\theta = \frac{\pi}{2} , \quad r = l$$

Therefore, $B = 0$ and $l/r = (1 + A \cos\theta)$.

The constant A is determined by noting that when

$$\theta = 0 , \quad r = a(1 - e)$$

where a and e are the semimajor axis and the eccentricity, respectively. Substitution of these values and use of the known relationship

$$l = a(1 - e^2)$$

yields the value

$$A = e$$

and gives the equation of the orbit in the form

$$\frac{l}{r} = 1 + e \cos \theta$$

which is the typical form of the equation of a conic.

   3.8.3.2 *Equatorial orbits.* Equatorial orbits are obtained by letting $\theta = \pi/2$, and subjecting the particle to the same force function, that is,

$$\ddot{r} - r\dot{\phi}^2 = -\frac{\mu}{r^2}$$

$$r\ddot{\phi} + 2\dot{r}\dot{\phi} = 0$$

Solving these equations leads to a solution of the same form as the preceding case, that is,

$$\frac{l}{r} = 1 + e \cos \phi$$

   3.8.3.3 *Kepler's second law.* Let dA be the element of shaded area in sketch (o) below.

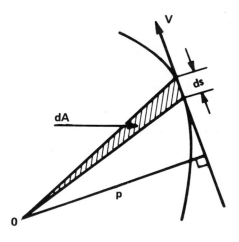

Sketch (o)

Then

$$dA = \frac{1}{2} p \, dS$$

where p is the perpendicular distance to the tangent vector and dS is an element of orbital arc. It follows that

$$\frac{dA}{dt} = \frac{1}{2}p\,\frac{dS}{dt} = \frac{1}{2}pV$$

From the second orbital equation $r^2\dot{\theta} = r^2\dot{\phi} = h$, which is the constant angular momentum of the particle. Moreover, the angular momentum of the particle is also equal to pV. Hence

$$r^2\dot{\theta} = r^2\dot{\phi} = h = pV$$

Therefore

$$\frac{dA}{dt} = \frac{1}{2}pV = \frac{1}{2}h$$

This is the mathematical expression of Kepler's second law (ref. 4), which states: the radius drawn from the center of force to a planet describes equal areas in equal times.

3.8.3.4 *Kepler's third law.* Kepler's third law follows immediately from this result. If P be the planetary period, then

$$P = \frac{\pi ab}{h/2} = \frac{2\pi ab}{h}$$

but

$$h^2 = \frac{\mu b^2}{a}$$

or

$$h = b\sqrt{\frac{\mu}{a}}$$

Therefore

$$P = \frac{2\pi a^{3/2}}{\sqrt{\mu}}$$

and

$$P^2 = \frac{4\pi^2 a^3}{\mu}$$

This is Kepler's third law which states that the square of the periodic times of different planets is proportional to the cubes of the semimajor axes of their orbits (ref. 5).

3.8.4 Motion on the Surface of a Cone

As an example of the use of cylindrical coordinates, consider the motion of a heavy particle of mass m on the surface of a smooth, right circular cone, with axis vertical and vertex downward (sketch (p)). By resolving along a generator, we obtain

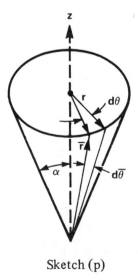

Sketch (p)

$$m(\ddot{r} - r\dot{\theta}^2)\sin\alpha + \ddot{z}\cos\alpha = -mg\cos\alpha$$

and
$$z = r \cot \alpha$$

Therefore
$$(\ddot{r} - r\dot{\theta}^2)\sin \alpha + \frac{\ddot{r} \cos^2 \alpha}{\sin \alpha} = -g \cos \alpha$$

or
$$\ddot{r} - r\dot{\theta}^2 \sin^2 \alpha = -g \sin \alpha \cos \alpha$$

$$r d\theta = \bar{r} d\bar{\theta} \quad \text{and} \quad r = \bar{r} \sin \alpha$$

As demonstrated in a previous example, this equation may be rewritten as

$$\frac{d^2 u}{d\theta^2} + u \sin^2 \alpha = \frac{g \sin \alpha \cos \alpha}{h^2 u^2}$$

This is the differential equation of the projection of the path on a horizontal plane. Moreover, since
$$d\theta = d\bar{\theta} \operatorname{cosec} \alpha ; \quad u = \bar{u} \operatorname{cosec} \alpha$$

Making these substitutions in the equation of the projection of the path on a horizontal plane yields

$$\frac{d^2 \bar{u}}{d\bar{\theta}^2} + \bar{u} = \frac{g \sin^2 \alpha \cos \alpha}{h^2 \bar{u}^2}$$

Hence, if the cone be developed into a plane, it is seen that the orbit of the path on the surface will be the same as would be produced by a particle moving in a plane under the action of a constant central force.

3.9 REFERENCES

1. Sokolnikoff, Ivan S.: Tensor Analysis; Theory and Applications. John Wiley & Sons, Inc., 1960.

2. Bond, E.; Auslander, M.; et al.: PL/I FORMAC Interpreter User's Reference Manual. IBM — Boston Programming Center, Cambridge, Mass., 1967.

3. Margenaw, Henry; and Murphy, George M.: The Mathematics of Physics and Chemistry. D. Van Nostrand Company, Inc., 1964.

4. Green, Stanley L.: Dynamics. University Tutorial Press, London, 1948.

5. Moulton, F. R.: An Introduction to Celestial Mechanics. Second ed. The Macmillan Company, 1959.

4

APPLICATIONS TO FLUID MECHANICS

4.1 FORMULATION OF THE NAVIER-STOKES EQUATIONS AND THE CONTINUITY EQUATION

The Navier-Stokes equations form the basis of the whole science of fluid mechanics (ref. 1). The technique described in preceding chapters can be used to facilitate the derivation of these equations and the corresponding continuity equation. In fact, any equation or system of equations that can be expressed in tensor form is amenable to formulation by the methods described. Again, it will be seen that in order to formulate the equations describing the flow over a given surface, it is only necessary to know the metric tensors for that particular surface, since all essential metric properties of the surface are completely determined by this tensor. The metric tensors may be obtained from the fundamental quadratic form, which is an expression for the square of the distance between two adjacent points on a surface, or in terms of coordinate transformation equations. If a formulation in terms of coordinate transformation equations is adopted, the methods described in section 1.13 may be used. For reasons that will become apparent as we proceed, the relationships expressed in equations (1.13.12), (1.13.14), and (1.13.15) are adequate to our needs.

A form of the Navier-Stokes equations of motion of a compressible viscous fluid, which is valid in all curvilinear coordinate systems, is (ref. 2)

$$
\begin{aligned}
\frac{\partial V^i}{\partial t} = \nu g^{jk} &\left[\frac{\partial^2 V^i}{\partial x^j \partial x^k} + \begin{Bmatrix} i \\ \alpha k \end{Bmatrix} \frac{\partial V^\alpha}{\partial x^j} + \begin{Bmatrix} i \\ \alpha j \end{Bmatrix} \frac{\partial V^\alpha}{\partial x^k} - \begin{Bmatrix} \alpha \\ jk \end{Bmatrix} \frac{\partial V^i}{\partial x^\alpha} + \left(\frac{\partial}{\partial x^k} \begin{Bmatrix} i \\ \alpha j \end{Bmatrix} \right. \right. \\
&\left. \left. + \begin{Bmatrix} i \\ mk \end{Bmatrix} \begin{Bmatrix} m \\ \alpha j \end{Bmatrix} - \begin{Bmatrix} i \\ \alpha m \end{Bmatrix} \begin{Bmatrix} m \\ jk \end{Bmatrix} \right) V^\alpha \right] - \frac{g^{ij}}{\rho} \frac{\partial p}{\partial x^j} - V^j \left(\frac{\partial V^i}{\partial x^j} + \begin{Bmatrix} i \\ \alpha j \end{Bmatrix} V^\alpha \right) \\
&+ \frac{\nu}{3} g^{ij} \frac{\partial}{\partial x^j} \left(\frac{\partial V^k}{\partial x^k} + \begin{Bmatrix} k \\ \alpha k \end{Bmatrix} V^\alpha \right) + \bar{F}^i
\end{aligned}
\tag{4.1.1}
$$

In this equation, the summation convention is assumed. That is to say, if in any term an index occurs twice, the term is to be summed with respect to that index for all admissible values of the index.

If the body forces are assumed to be known, it is seen that the Navier-Stokes equations involve five unknowns: $V^i(x,t)$, $(i = 1,2,3)$, $p(x,t)$, and $\rho(x,t)$. To complete the system, two more equations are added. One of these is the equation of state that relates the pressure and the density. It may be written as follows:

$$p = p(\rho) \tag{4.1.2}$$

The other equation expresses the principle of conservation of mass and assumes the form (ref. 1)

$$\frac{d\rho}{dt} + \rho\left(\frac{\partial V^k}{\partial x^k} + \begin{Bmatrix} k \\ \alpha k \end{Bmatrix} V^\alpha\right) = 0 \tag{4.1.3}$$

Furthermore, if the process is not isothermal it is necessary to make use of the energy equation, which draws up a balance between mechanical and thermal energy and furnishes a differential equation for the temperature distribution. However, to simplify and clarify the exposition, the flow will be assumed to be incompressible and viscous. Hence, this equation will not be included here. Moreover, equation (4.1.2) will not be required in this case, and equation (4.1.3) will assume the simpler form (ref. 3)

$$\left(\frac{\partial V^k}{\partial x^k} + \begin{Bmatrix} k \\ \alpha k \end{Bmatrix} V^\alpha\right) = 0 \tag{4.1.4}$$

For orthogonal coordinate systems

$$g^{ij} = 0 \quad \text{for} \quad i \neq j \tag{4.1.5}$$

and

$$g^{(ii)} = \frac{1}{g_{(ii)}} \tag{4.1.6}$$

where parentheses around a repeated index imply suspension of the summation convention for that particular index. With these simplifications, equation (4.1.1) may be rewritten as follows:

$$\frac{\partial V^i}{\partial t} = vg^{jj} \left[\frac{\partial^2 V^i}{\partial x^j \partial x^j} + 2 \begin{Bmatrix} i \\ \alpha j \end{Bmatrix} \frac{\partial V^\alpha}{\partial x^j} - \begin{Bmatrix} \alpha \\ jj \end{Bmatrix} \frac{\partial V^i}{\partial x^\alpha} + \left(\frac{\partial}{\partial x^j} \begin{Bmatrix} i \\ \alpha j \end{Bmatrix} + \begin{Bmatrix} i \\ mj \end{Bmatrix} \begin{Bmatrix} m \\ \alpha j \end{Bmatrix} \right. \right.$$

$$\left. \left. - \begin{Bmatrix} i \\ \alpha m \end{Bmatrix} \begin{Bmatrix} m \\ jj \end{Bmatrix} \right) V^\alpha \right] - \frac{1}{g_{(ii)}} \frac{1}{\rho} \frac{\partial p}{\partial x^i} - V^j \left(\frac{\partial V^i}{\partial x^j} + \begin{Bmatrix} i \\ \alpha j \end{Bmatrix} V^\alpha \right) + \bar{F}^i \qquad (4.1.7)$$

where the kinematic viscosity v is defined by the equation

$$v = \frac{\mu}{\rho} \qquad (4.1.8)$$

It is instructive to dwell on this rather complicated equation for a moment and examine its meaning and the meaning of the individual terms and coefficients. This is the equation of incompressible, viscous flow in which the density ρ and the kinematic viscosity v are assumed to be constants. It represents the three equations of motion obtained by invoking the principle of conservation of momentum. Likewise, the principle of conservation of mass yields equation (4.1.4). With known body forces F^i, there are, therefore, four equations for the three unknown velocity components V^i and the pressure p. Since the coefficients are metric tensors or Christoffel symbols, which are functions of the metric tensors, the formulation of these equations in any particular coordinate system depends only on the specification of the metric tensors. For a formulation in terms of coordinate transformation equations, the curvilinear coordinates x^j are assumed to be related to an orthogonal Cartesian triad y^i by the following coordinate transformation equations:

$$\left. \begin{aligned} y^i &= y^i(x^1 x^2 x^3) \\ i &= 1,2,3 \end{aligned} \right\} \qquad (4.1.9)$$

This transformation is assumed to be reversible and one-to-one. Hence

$$x^i = x^i(y^1 y^2 y^3) \qquad (4.1.10)$$

In terms of these functional relationships, the metric tensors and the Christoffel symbols are defined as follows:

$$g_{ij} = \frac{\partial y^\alpha}{\partial x^i} \frac{\partial y^\alpha}{\partial x^j} \qquad (4.1.11)$$

$$\begin{Bmatrix} i \\ jk \end{Bmatrix} = \left(\frac{\partial^2 y^\beta}{\partial x^j \partial x^k} \frac{\partial x^i}{\partial y^\beta} \right)$$ (4.1.12)

Before substituting these relationships in equation (4.1.7), the following simplification will be helpful

$$g_{(ii)} \begin{Bmatrix} i \\ jk \end{Bmatrix} = \left[\frac{\partial y^\alpha}{\partial x^{(i)}} \frac{\partial y^\alpha}{\partial x^{(i)}} \right] \left[\frac{\partial^2 y^\beta}{\partial x^j \partial x^k} \frac{\partial x^i}{\partial y^\beta} \right]$$

Therefore

$$g_{(ii)} \begin{Bmatrix} i \\ jk \end{Bmatrix} = \frac{\partial^2 y^\beta}{\partial x^j \partial x^k} \delta_\beta{}^\alpha \frac{\partial y^\alpha}{\partial x^i}$$

where

$$\delta_\beta^\alpha = \begin{matrix} 0 & \text{for} & \alpha \neq \beta \\ 1 & \text{for} & \alpha = \beta \end{matrix}$$

Hence

$$g_{(ii)} \begin{Bmatrix} i \\ jk \end{Bmatrix} = \left(\frac{\partial^2 y^\alpha}{\partial x^j \partial x^k} \frac{\partial y^\alpha}{\partial x^i} \right)$$ (4.1.13)

The expression on the right-hand side of equation (4.1.13) defines the Christoffel symbol of the first kind in terms of coordinate transformation equations. The Christoffel symbol of the first kind was defined in terms of metric tensors in equation (1.12.3), that is

$$[ij,k] = \frac{1}{2} \left(\frac{\partial g_{ik}}{\partial x^j} + \frac{\partial g_{jk}}{\partial x^i} - \frac{\partial g_{ij}}{\partial x^k} \right)$$

Substitution from equations (4.1.11) through (4.1.13) in equation (4.1.7) gives

$$\rho\left[\frac{\partial y^\beta}{\partial x^{(i)}}\frac{\partial y^\beta}{\partial x^{(i)}}\left(\frac{\partial V^i}{\partial t}+V^j\frac{\partial V^i}{\partial x^j}\right)+\left(\frac{\partial^2 y^m}{\partial x^\alpha \partial x^j}\frac{\partial y^m}{\partial x^i}\right)V^\alpha V^j\right]$$

$$=\left[\frac{\partial y^\beta}{\partial x^{(i)}}\frac{\partial y^\beta}{\partial x^{(i)}}\right]F^i-\frac{\partial p}{\partial x^i}+\mu\left[\frac{\partial y^k}{\partial x^{(j)}}\frac{\partial y^k}{\partial x^{(j)}}\right]^{-1}\left(\left[\frac{\partial y^\beta}{\partial x^{(i)}}\frac{\partial y^\beta}{\partial x^{(i)}}\right]\frac{\partial^2 V^i}{\partial x^j \partial x^j}\right.$$

$$+2\left[\frac{\partial^2 y^m}{\partial x^\alpha \partial x^j}\frac{\partial y^m}{\partial x^i}\right]\frac{\partial V^\alpha}{\partial x^j}-\left[\frac{\partial y^\beta}{\partial x^{(i)}}\frac{\partial y^\beta}{\partial x^{(i)}}\right]\left[\frac{\partial^2 y^m}{\partial x^j \partial x^j}\frac{\partial x^\alpha}{\partial y^m}\right]\frac{\partial V^i}{\partial x^\alpha}$$

$$+\left\{\left[\frac{\partial y^\beta}{\partial x^{(i)}}\frac{\partial y^\beta}{\partial x^{(i)}}\right]\frac{\partial}{\partial x^j}\left(\frac{\partial^2 y^m}{\partial x^\alpha \partial x^j}\frac{\partial x^i}{\partial y^m}\right)+\left(\frac{\partial^2 y^\gamma}{\partial x^r \partial x^j}\frac{\partial y^\gamma}{\partial x^i}\right)\left(\frac{\partial^2 y^m}{\partial x^\alpha \partial x^j}\frac{\partial x^r}{\partial y^m}\right)\right.$$

$$\left.\left.-\left(\frac{\partial^2 y^\gamma}{\partial x^\alpha \partial x^r}\frac{\partial y^\gamma}{\partial x^i}\right)\left(\frac{\partial^2 y^m}{\partial x^j \partial x^j}\frac{\partial x^r}{\partial y^m}\right)\right\}V^\alpha\right) \qquad (4.1.14)$$

Equation (4.1.14) represents three tensor equations of motion of an incompressible viscous fluid relative to any orthogonal curvilinear coordinate system. Since the derivation merely requires the determination of the partial differential coefficients of y with respect to x, and of x with respect to y, the form of this equation is well suited to routine nonnumeric computer operations. Moreover, the only inputs required are the coordinate transformation equations that are given in equations (4.1.9) and (4.1.10). The required inputs can be reduced further by elimination of the partial differential coefficients of x with respect to y from equation (4.1.14). This can be done by using the following relationship, which is valid in all orthogonal coordinate systems:

$$\frac{\partial^2 y^\alpha}{\partial x^i \partial x^k}\frac{\partial x^j}{\partial y^\alpha}=\left(\frac{\partial y^\beta}{\partial x^j}\frac{\partial y^\beta}{\partial x^j}\right)^{-1}\frac{\partial^2 y^\alpha}{\partial x^i \partial x^k}\frac{\partial y^\alpha}{\partial x^j} \qquad (4.1.15)$$

By substitution from equation (4.1.15) in equation (4.1.14), a form of the Navier-Stokes equation is obtained whose derivation in any coordinate system depends only on equation (4.1.9). This is

$$\rho\left\{\left[\frac{\partial y^\beta}{\partial x^{(i)}}\frac{\partial y^\beta}{\partial x^{(i)}}\right]\left(\frac{\partial V^i}{\partial t}+V^j\frac{\partial V^i}{\partial x^j}\right)+\left(\frac{\partial^2 y^m}{\partial x^\alpha \partial x^j}\frac{\partial y^m}{\partial x^i}\right)V^\alpha V^j\right\}$$

$$=\left[\frac{\partial y^\beta}{\partial x^{(i)}}\frac{\partial y^\beta}{\partial x^{(i)}}\right]F^i-\frac{\partial p}{\partial x^i}+\mu\left[\frac{\partial y^k}{\partial x^{(j)}}\frac{\partial y^k}{\partial x^{(j)}}\right]^{-1}\left(\left[\frac{\partial y^\beta}{\partial x^{(i)}}\frac{\partial y^\beta}{\partial x^{(i)}}\right]\frac{\partial^2 V^i}{\partial x^j \partial x^j}\right.$$

$$+2\left(\frac{\partial^2 y^m}{\partial x^\alpha \partial x^j}\frac{\partial y^m}{\partial x^i}\right)\frac{\partial V^\alpha}{\partial x^j}-\left[\frac{\partial y^\beta}{\partial x^{(i)}}\frac{\partial y^\beta}{\partial x^{(i)}}\right]\left[\frac{\partial y^\tau}{\partial x^{(\alpha)}}\frac{\partial y^\tau}{\partial x^{(\alpha)}}\right]^{-1}\left(\frac{\partial^2 y^m}{\partial x^j \partial x^j}\frac{\partial y^m}{\partial x^\alpha}\right)\frac{\partial V^i}{\partial x^\alpha}$$

$$+\left\{\left[\frac{\partial y^\beta}{\partial x^{(i)}}\frac{\partial y^\beta}{\partial x^{(i)}}\right]\frac{\partial}{\partial x^j}\left[\frac{\partial y^\beta}{\partial x^{(i)}}\frac{\partial y^\beta}{\partial x^{(i)}}\right]^{-1}\left(\frac{\partial^2 y^m}{\partial x^\alpha \partial x^j}\frac{\partial y^m}{\partial x^i}\right)\right.$$

$$+\left(\frac{\partial^2 y^\gamma}{\partial x^r \partial x^j}\frac{\partial y^\gamma}{\partial x^i}\right)\left[\frac{\partial y^\tau}{\partial x^{(r)}}\frac{\partial y^\tau}{\partial x^{(r)}}\right]^{-1}\left(\frac{\partial^2 y^m}{\partial x^\alpha \partial x^j}\frac{\partial y^m}{\partial x^r}\right)$$

$$\left.\left.-\left(\frac{\partial^2 y^\gamma}{\partial x^\alpha \partial x^r}\frac{\partial y^\tau}{\partial x^i}\right)\left[\frac{\partial y^\tau}{\partial x^{(r)}}\frac{\partial y^\tau}{\partial x^{(r)}}\right]^{-1}\left(\frac{\partial^2 y^m}{\partial x^j \partial x^j}\frac{\partial y^m}{\partial x^r}\right)\right\}V^\alpha\right) \qquad (4.1.16)$$

4.2 THE PHYSICAL FORM OF THE NAVIER-STOKES EQUATIONS

Because equation (4.1.16) is a tensor equation, all the velocity and force components occurring in this equation are tensor components. This form of the equation is well suited to theoretical studies. However, in practical applications, it is the physical components that are of interest (ref. 1). The tensor components of a vector, or first-order tensor, are related to its physical components as follows (ref. 2):

$$V^i=\frac{v^i}{\sqrt{\dfrac{\partial y^\beta}{\partial x^{(i)}}\dfrac{\partial y^\beta}{\partial x^{(i)}}}} \qquad (4.2.1)$$

$$F^i=\frac{f^i}{\sqrt{\dfrac{\partial y^\beta}{\partial x^{(i)}}\dfrac{\partial y^\beta}{\partial x^{(i)}}}} \qquad (4.2.2)$$

where V^i and v^i are the tensor and physical components, respectively, of the velocity vector $\bar{V}$; and F^i, f^i are the tensor and physical components, respectively, of the force vector $\bar{F}$. By substitution from equations (4.2.1) and (4.2.2) in equation (4.1.16), the physical form of the Navier-Stokes equations is obtained

$$
\begin{aligned}
\rho &\left\{ \left[\frac{\partial y^\beta}{\partial x^{(i)}} \frac{\partial y^\beta}{\partial x^{(i)}} \right] \left[\frac{\partial}{\partial t} \frac{v^i}{\sqrt{\frac{\partial y^\beta}{\partial x^{(i)}} \frac{\partial y^\beta}{\partial x^{(i)}}}} + \frac{v^j}{\sqrt{\frac{\partial y^\beta}{\partial x^{(j)}} \frac{\partial y^\beta}{\partial x^{(j)}}}} \frac{\partial}{\partial x^j} \frac{v^i}{\sqrt{\frac{\partial y^\beta}{\partial x^{(i)}} \frac{\partial y^\beta}{\partial x^{(i)}}}} \right] \right. \\
&\left. + \left(\frac{\partial^2 y^m}{\partial x^\alpha \partial x^j} \frac{\partial y^m}{\partial x^i} \right) \frac{v^\alpha}{\sqrt{\frac{\partial y^\beta}{\partial x^{(\alpha)}} \frac{\partial y^\beta}{\partial x^{(\alpha)}}}} \frac{v^j}{\sqrt{\frac{\partial y^\beta}{\partial x^{(j)}} \frac{\partial y^\beta}{\partial x^{(j)}}}} \right\} \\
&= \sqrt{\frac{\partial y^\beta}{\partial x^{(i)}} \frac{\partial y^\beta}{\partial x^{(i)}}} \, f^i - \frac{\partial p}{\partial x^i} + \mu \left[\frac{\partial y^k}{\partial x^{(j)}} \frac{\partial y^k}{\partial x^{(j)}} \right] \left(\left[\frac{\partial y^\beta}{\partial x^{(i)}} \frac{\partial y^\beta}{\partial x^{(i)}} \right] \frac{\partial^2}{\partial x^j \partial x^j} \left[\frac{v^i}{\sqrt{\frac{\partial y^\beta}{\partial x^{(i)}} \frac{\partial y^\beta}{\partial x^{(i)}}}} \right. \right. \\
&\left. + 2 \left(\frac{\partial^2 y^m}{\partial x^\alpha \partial x^j} \frac{\partial y^m}{\partial x^i} \right) \frac{\partial}{\partial x^j} \frac{v^\alpha}{\sqrt{\frac{\partial y^\beta}{\partial x^{(\alpha)}} \frac{\partial y^\beta}{\partial x^{(\alpha)}}}} \right] \\
&- \left[\frac{\partial y^\beta}{\partial x^{(i)}} \frac{\partial y^\beta}{\partial x^{(i)}} \right] \left[\frac{\partial y^\tau}{\partial x^{(\alpha)}} \frac{\partial y^\tau}{\partial x^{(\alpha)}} \right]^{-1} \left(\frac{\partial^2 y^m}{\partial x^j \partial x^j} \frac{\partial y^m}{\partial x^\alpha} \right) \frac{\partial}{\partial x^\alpha} \frac{v^i}{\sqrt{\frac{\partial y^\beta}{\partial x^{(i)}} \frac{\partial y^\beta}{\partial x^{(i)}}}} \\
&+ \left\{ \left[\frac{\partial y^\beta}{\partial x^{(i)}} \frac{\partial y^\beta}{\partial x^{(i)}} \right] \frac{\partial}{\partial x^j} \left[\frac{\partial y^\beta}{\partial x^{(i)}} \frac{\partial y^\beta}{\partial x^{(i)}} \right]^{-1} \left(\frac{\partial^2 y^m}{\partial x^\alpha \partial x^j} \frac{\partial y^m}{\partial x^i} \right) \right. \\
&+ \left(\frac{\partial^2 y^\gamma}{\partial x^r \partial x^j} \frac{\partial y^\gamma}{\partial x^i} \right) \left[\frac{\partial y^\tau}{\partial x^{(r)}} \frac{\partial y^\tau}{\partial x^{(r)}} \right]^{-1} \left(\frac{\partial^2 y^m}{\partial x^\alpha \partial x^j} \frac{\partial y^m}{\partial x^r} \right) \\
&\left. \left. - \left(\frac{\partial^2 y^\gamma}{\partial x^\alpha \partial x^r} \frac{\partial y^\gamma}{\partial x^i} \right) \left[\frac{\partial y^\tau}{\partial x^{(r)}} \frac{\partial y^\tau}{\partial x^{(r)}} \right]^{-1} \left(\frac{\partial^2 y^m}{\partial x^j \partial x^j} \frac{\partial y^m}{\partial x^r} \right) \right\} \frac{v^\alpha}{\sqrt{\frac{\partial y^\beta}{\partial x^{(\alpha)}} \frac{\partial y^\beta}{\partial x^{(\alpha)}}}} \right)
\end{aligned}
\tag{4.2.3}
$$

Because the index i can assume the values 1, 2, or 3, equation (4.2.3) represents three equations. If the forces f^i are assumed given, there are only three equations for the four variables v^i and p. However, as indicated previously, another equation can be obtained from the continuity condition. Satisfaction of this condition requires that the physical components of $\bar{V}$ satisfy the equation

$$\frac{\partial}{\partial x^k}\left[\frac{v^k}{\sqrt{\frac{\partial y^\beta}{\partial x^{(k)}}\frac{\partial y^\beta}{\partial x^{(k)}}}}\right] + \left[\frac{\partial y^\alpha}{\partial x^{(k)}}\frac{\partial y^\alpha}{\partial x^{(k)}}\right]^{-1}\left(\frac{\partial^2 y^\tau}{\partial x^k \partial x^j}\frac{\partial y^\tau}{\partial x^k}\right)\frac{v^j}{\sqrt{\frac{\partial y^\beta}{\partial x^{(j)}}\frac{\partial y^\beta}{\partial x^{(j)}}}} = 0 \quad (4.2.4)$$

The form of equations (4.2.3) and (4.2.4) is such that the only computer input required is the transformation equation expressing the Cartesian coordinates y^i as functions of the curvilinear coordinates x^i. The inverse transformation is no longer required.

Although these equations are complicated, the only computer operations required to formulate them are summation and symbolic differentiation. A program that consists of a few statements for controlling symbolic differentiation, supported by a simple computational algorithm for exploiting the summation convention, was demonstrated in chapter 2. The methods employed there can be used with equal facility to formulate equations (4.2.3) and (4.2.4). Moreover, once these equations are programmed and made available in a program library, the researcher need only specify the surface or boundary that determines the flow.

Those who do not have access to a digital computer will find that a manual formulation has certain advantages. Indeed, even in those cases where a digital computer is available, it is recommended that at least some of the formulations be performed manually. By doing so, the user will be made aware of the inherent simplicity of the method, which involves nothing more than repeated partial differentiation. A point to be noted is that once the general formulation is available, the specialized form can be obtained without further reference to the physics of the problem. The nature of the problem is such that given a set of transformation equations, the required model can be formulated by anyone who can differentiate.

To demonstrate again the inherent simplicity of the method, and the fact that the formulation of complex models only requires a series of repeated differentiations, one of the equations (4.1.16) will be evaluated in a cylindrical polar coordinate frame of reference, where

$$y^1 = x^1 \cos x^2 \\ y^2 = x^1 \sin x^2 \\ y^3 = x^3 \quad \Bigg\} \tag{4.2.5}$$

The first factor in equation (4.1.16) is

$$\frac{\partial y^\beta}{\partial x^{(i)}} \frac{\partial y^\beta}{\partial x^{(i)}}$$

It will be recalled that parentheses around an index implies suspension of the summation convention for that index. Bearing this in mind we sum this factor on β, remembering to keep the i constant, that is

$$\left[\frac{\partial y^\beta}{\partial x^{(i)}} \frac{\partial y^\beta}{\partial x^{(i)}} \right] = \left[\frac{\partial y^1}{\partial x^{(i)}} \right]^2 + \left[\frac{\partial y^2}{\partial x^{(i)}} \right]^2 + \left[\frac{\partial y^3}{\partial x^{(i)}} \right]^2$$

And for $i = 1$

$$\left[\frac{\partial y^\beta}{\partial x^{(i)}} \frac{\partial y^\beta}{\partial x^{(i)}} \right]_{i=1} = \left(\frac{\partial y^1}{\partial x^1} \right)^2 + \left(\frac{\partial y^2}{\partial x^2} \right)^2 + \left(\frac{\partial y^3}{\partial x^1} \right)^2 \tag{4.2.6}$$

From equation (4.2.5), we have

$$\frac{\partial y^1}{\partial x^1} = \cos x^1 \quad ; \qquad \frac{\partial y^2}{\partial x^1} = \sin x^1 \quad ; \qquad \frac{\partial y^3}{\partial x^1} = 0$$

Substitution of these values in equation (4.2.6) yields

$$\frac{\partial y^\beta}{\partial x^1} \frac{\partial y^\beta}{\partial x^1} = (\cos^2 x^1 + \sin^2 x^1) = 1 \tag{4.2.7}$$

Therefore, for $i = 1$, the first term of equation (4.1.16) is

$$\rho \left(\frac{\partial y^\beta}{\partial x^1} \frac{\partial y^\beta}{\partial x^1} \right) \frac{\partial V^1}{\partial t} = \rho \frac{\partial V^1}{\partial t} \tag{4.2.8}$$

To this must be added

$$V^j \frac{\partial V^1}{\partial x^j} = \left(V^1 \frac{\partial V^1}{\partial x^1} + V^2 \frac{\partial V^1}{\partial x^2} + V^3 \frac{\partial V^1}{\partial x^3} \right)$$

and the first two terms are

$$\rho \left(\frac{\partial V^1}{\partial t} + V^1 \frac{\partial V^1}{\partial x^1} + V^2 \frac{\partial V^1}{\partial x^2} + V^3 \frac{\partial V^1}{\partial x^3} \right)$$

The third term involves a summation on m, α, and j. Summing first on m gives

$$\left(\frac{\partial^2 y^m}{\partial x^\alpha \partial x^j} \frac{\partial y^m}{\partial x^i} \right)_{i=1} V^\alpha V^j = \left(\frac{\partial^2 y^1}{\partial x^\alpha \partial x^j} \frac{\partial y^1}{\partial x^1} + \frac{\partial^2 y^2}{\partial x^\alpha \partial x^j} \frac{\partial y^2}{\partial x^1} + \frac{\partial^2 y^3}{\partial x^\alpha \partial x^j} \frac{\partial y^3}{\partial x^1} \right) V^\alpha V^j$$

From equations (4.2.5), it is seen that y^3 is independent of x^1. Hence, the last term on the right-hand side of this equation vanishes, and we are left with

$$\left(\frac{\partial^2 y^m}{\partial x^\alpha \partial x^j} \frac{\partial y^m}{\partial x^1} \right) V^\alpha V^j = \left(\frac{\partial^2 y^1}{\partial x^\alpha \partial x^j} \frac{\partial y^1}{\partial x^1} + \frac{\partial^2 y^2}{\partial x^\alpha \partial x^j} \frac{\partial y^2}{\partial x^1} \right) V^\alpha V^j$$

Summing next on α, and noting that terms involving $\alpha = 3$ vanish, we obtain

$$\left[\left(\frac{\partial^2 y^1}{\partial x^1 \partial x^j} \frac{\partial y^1}{\partial x^1} + \frac{\partial^2 y^2}{\partial x^1 \partial x^j} \frac{\partial y^2}{\partial x^1} \right) V^1 V^j + \left(\frac{\partial^2 y^1}{\partial x^2 \partial x^j} \frac{\partial y^1}{\partial x^1} + \frac{\partial^2 y^2}{\partial x^2 \partial x^j} \frac{\partial y^2}{\partial x^1} \right) V^2 V^j \right]$$

Summing next on j, we find that the coefficient of $(V^1)^2$ is zero since

$$\frac{\partial^2 y^1}{\partial x^1 \partial x^1} = \frac{\partial^2 y^2}{\partial x^1 \partial x^1} = 0$$

The coefficient of $V^1 V^2$ is also zero because

$$\left(\frac{\partial^2 y^1}{\partial x^1 \partial x^2} \frac{\partial y^1}{\partial x^1} + \frac{\partial^2 y^2}{\partial x^1 \partial x^2} \frac{\partial y^2}{\partial x^1} \right) = (-\sin x^2 \cos x^2 + \cos x^2 \sin x^2) = 0$$

and the coefficient of $V^2 V^1$ is zero for the same reason. Hence, the only nonzero coefficient is the coefficient of $V^2 V^2$, which is

$$\left(\frac{\partial^2 y^1}{\partial x^2 \partial x^2} \frac{\partial y^1}{\partial x^1} + \frac{\partial^2 y^2}{\partial x^2 \partial x^2} \frac{\partial y^2}{\partial x^1} \right) V^2 V^2 = -x^1 V^2 V^2$$

and when expanded, the first three terms of equation (4.1.16) yield

$$\rho \left[\frac{\partial V^1}{\partial t} + V^1 \frac{\partial V^1}{\partial x^1} + V^2 \frac{\partial V^1}{\partial x^2} + V^3 \frac{\partial V^1}{\partial x^3} - x^1 (V^2)^2 \right] \qquad (4.2.9)$$

Proceeding next to the right-hand side of equation (4.1.16), but still retaining the value $i = 1$, we find on using the value already obtained for

$$\frac{\partial y^\beta}{\partial x^1} \frac{\partial y^\beta}{\partial x^1}$$

in equation (4.2.7), that the body force and pressure gradient components are

$$F^1 - \frac{\partial p}{\partial x^1} \qquad (4.2.10)$$

The first viscosity term on the right-hand side of equation (4.1.16) is

$$\mu \left[\frac{\partial y^k}{\partial x^{(j)}} \frac{\partial y^k}{\partial x^{(j)}} \right] \left(\frac{\partial y^\beta}{\partial x^1} \frac{\partial y^\beta}{\partial x^1} \right)^{-1} \frac{\partial^2 V^1}{\partial x^j \partial x^j}$$

Because the term

$$\frac{\partial y^\beta}{\partial x^1} \frac{\partial y^\beta}{\partial x^1}$$

has already been evaluated and found to be equal to unity, the first viscous term reduces to

$$\mu \left[\frac{\partial y^k}{\partial x^{(j)}} \frac{\partial y^k}{\partial x^{(j)}} \right]^{-1} \frac{\partial^2 V^1}{\partial x^j \partial x^j}$$

When expanded, these factors give rise to the following three terms:

$$\mu\left[\left(\frac{\partial y^k}{\partial x^1}\frac{\partial y^k}{\partial x^1}\right)^{-1}\frac{\partial^2 V^1}{\partial x^1\partial x^1}+\left(\frac{\partial y^k}{\partial x^2}\frac{\partial y^k}{\partial x^2}\right)^{-1}\frac{\partial^2 V^1}{\partial x^2\partial x^2}+\left(\frac{\partial y^k}{\partial x^3}\frac{\partial y^k}{\partial x^3}\right)^{-1}\frac{\partial^2 V^1}{\partial x^3\partial x^3}\right]$$

Because the index k is a dummy index, it can be replaced by any other index. In particular, it can be replaced by the index β, that is

$$\left(\frac{\partial y^k}{\partial x^1}\frac{\partial y^k}{\partial x^1}\right)=\left(\frac{\partial y^\beta}{\partial x^1}\frac{\partial y^\beta}{\partial x^1}\right)=1$$

and the first viscosity component is

$$\mu\frac{\partial^2 V^1}{\partial x^1\partial x^1}$$

To evaluate the coefficient of the second term

$$\frac{\partial y^k}{\partial x^2}\frac{\partial y^k}{\partial x^2}$$

it is necessary to refer again to equation (4.2.5), which gives

$$\frac{\partial y^k}{\partial x^2}\frac{\partial y^k}{\partial x^2}=(x^1)^2 \tag{4.2.10a}$$

Therefore, the second viscosity component is

$$\frac{\mu}{(x^1)^2}\frac{\partial^2 V^1}{\partial x^2\partial x^2}$$

Likewise, the coefficient of the last term is

$$\left(\frac{\partial y^k}{\partial x^3}\frac{\partial y^k}{\partial x^3}\right)^{-1}=1 \tag{4.2.10b}$$

and the last term is simply

$$\mu \frac{\partial^2 V^1}{\partial x^3 \partial x^3}$$

Combining these three components yields the first viscosity term as follows:

$$\mu \left[\frac{\partial^2 V^1}{\partial x^1 \partial x^1} + \frac{1}{(x^1)^2} \frac{\partial^2 V^1}{\partial x^2 \partial x^2} + \frac{\partial^2 V^1}{\partial x^3 \partial x^3} \right] \qquad (4.2.11)$$

The next viscosity term is

$$2\mu \left[\frac{\partial y^k}{\partial x^{(j)}} \frac{\partial y^k}{\partial x^{(j)}} \right]^{-1} \frac{\partial^2 y^m}{\partial x^\alpha \partial x^j} \frac{\partial y^m}{\partial x^1} \frac{\partial V^\alpha}{\partial x^j} \qquad (4.2.12)$$

When this expression is summed on j, we obtain

$$\left[2\mu \left(\frac{\partial y^k}{\partial x^1} \frac{\partial y^k}{\partial x^1} \right)^{-1} \left(\frac{\partial^2 y^m}{\partial x^\alpha \partial x^1} \frac{\partial y^m}{\partial x^1} \right) \frac{\partial V^\alpha}{\partial x^1} + 2\mu \left(\frac{\partial y^k}{\partial x^2} \frac{\partial y^k}{\partial x^2} \right)^{-1} \left(\frac{\partial^2 y^m}{\partial x^\alpha \partial x^2} \frac{\partial y^m}{\partial x^1} \right) \frac{\partial V^\alpha}{\partial x^2} \right.$$

$$\left. + 2\mu \left(\frac{\partial y^k}{\partial x^3} \frac{\partial y^k}{\partial x^3} \right)^{-1} \left(\frac{\partial^2 y^m}{\partial x^\alpha \partial x^3} \frac{\partial y^m}{\partial x^1} \right) \frac{\partial V^\alpha}{\partial x^3} \right]$$

It has been shown that

$$\frac{\partial y^k}{\partial x^1} \frac{\partial y^k}{\partial x^1} = 1 \quad ; \qquad \frac{\partial y^k}{\partial x^2} \frac{\partial y^k}{\partial x^2} = (x^1)^2$$

and

$$\frac{\partial y^k}{\partial x^3} \frac{\partial y^k}{\partial x^3} = 1$$

Therefore, equation (4.2.12) reduces to

$$2\mu \left[\left(\frac{\partial^2 y^m}{\partial x^\alpha \partial x^1} \frac{\partial y^m}{\partial x^1} \right) \frac{\partial V^\alpha}{\partial x^1} + \frac{1}{(x^1)^2} \left(\frac{\partial^2 y^m}{\partial x^\alpha \partial x^2} \frac{\partial y^m}{\partial x^1} \right) \frac{\partial V^\alpha}{\partial x^2} + \left(\frac{\partial^2 y^m}{\partial x^\alpha \partial x^3} \frac{\partial y^m}{\partial x^1} \right) \frac{\partial V^\alpha}{\partial x^3} \right]$$

The next step in the formulation is a summation on m which yields

$$2\mu\left[\left(\frac{\partial^2 y^1}{\partial x^\alpha \partial x^1}\frac{\partial y^1}{\partial x^1} + \frac{\partial^2 y^2}{\partial x^\alpha \partial x^1}\frac{\partial y^2}{\partial x^1} + \frac{\partial^2 y^3}{\partial x^\alpha \partial x^1}\frac{\partial y^3}{\partial x^1}\right)\frac{\partial V^\alpha}{\partial x^1}\right.$$

$$+ \frac{1}{(x^1)^2}\left(\frac{\partial^2 y^1}{\partial x^\alpha \partial x^2}\frac{\partial y^1}{\partial x^1} + \frac{\partial^2 y^2}{\partial x^\alpha \partial x^2}\frac{\partial y^2}{\partial x^1} + \frac{\partial^2 y^3}{\partial x^\alpha \partial x^1}\frac{\partial y^3}{\partial x^1}\right)\frac{\partial V^\alpha}{\partial x^2}$$

$$+ \left.\left(\frac{\partial^2 y^1}{\partial x^\alpha \partial x^3}\frac{\partial y^1}{\partial x^1} + \frac{\partial^2 y^2}{\partial x^\alpha \partial x^3}\frac{\partial y^2}{\partial x^1} + \frac{\partial^2 y^3}{\partial x^\alpha \partial x^3}\frac{\partial y^3}{\partial x^1}\right)\frac{\partial V^\alpha}{\partial x^3}\right]$$

Consider the coefficient of $\partial V^\alpha/\partial x^1$ in this expression, and sum on α. For $\alpha = 1$, each term of this coefficient is zero since

$$\frac{\partial^2 y^1}{\partial x^1 \partial x^1} = \frac{\partial^2 y^2}{\partial x^1 \partial x^1} = \frac{\partial^2 y^3}{\partial x^1 \partial x^1} = 0$$

For $\alpha = 2$, we have

$$\frac{\partial^2 y^1}{\partial x^2 \partial x^1}\frac{\partial y^1}{\partial x^1} = -\sin x^2 \cos x^2$$

and

$$\frac{\partial^2 y^2}{\partial x^2 \partial x^1}\frac{\partial y^2}{\partial x^1} = \sin x^2 \cos x^2$$

Because y^3 is independent of x^1 and x^2, the last term is zero, that is

$$\frac{\partial^2 y^3}{\partial x^2 \partial x^1}\frac{\partial y^3}{\partial x^1} = 0$$

Hence, for $\alpha = 2$, the coefficient of $\partial V^\alpha/\partial x^1$ is zero.

For $\alpha = 3$, all terms of this coefficient vanish because y^1 and y^2 are independent of x^3, and y^3 is independent of x^1. Therefore, the coefficient of $\partial V^\alpha/\partial x^1$ is zero for all α.

Proceeding next to the coefficient of $\partial V^{\alpha}/\partial x^2$, and summing for $\alpha = 1$, we note that the sum of the terms vanishes. This is because the coefficient of $\partial V^{\alpha}/\partial x^1$ for $\alpha = 2$, which has been shown to be zero, is the same as the coefficient of $\partial V^{\alpha}/\partial x^2$ for $\alpha = 1$.

For $\alpha = 2$

$$\frac{\partial^2 y^1}{\partial x^2 \partial x^2} \frac{\partial y^1}{\partial x^1} = -x^1 \cos^2 x^2$$

and

$$\frac{\partial^2 y^2}{\partial x^2 \partial x^2} \frac{\partial y^2}{\partial x^1} = -x^1 \sin^2 x^2$$

Because y^3 is independent of x^1 and x^2, the last term of this coefficient vanishes, and the coefficient of $\partial V^{\alpha}/\partial x^2$ is

$$\frac{1}{(x^1)^2} (-x^1 \cos^2 x^2 - x^1 \sin^2 x^2) = -\frac{1}{x^1}$$

Recalling that because y^1 and y^2 are independent of x^3 and y^3 is independent of x^1, the coefficient of $\partial V^{\alpha}/\partial x^2$ is zero for $\alpha = 3$. For the same reason, the coefficient of $\partial V^{\alpha}/\partial x^3$ is zero for all values of α. It follows that when the viscosity term (4.2.12) is summed on α, j, k, and m, the result is simply

$$-\left(\frac{2\mu}{x^1} \frac{\partial V^{\alpha}}{\partial x^2} \right)$$

Combining the results obtained so far, we have for $i = 1$

$$\rho \left[\frac{\partial V^1}{\partial t} + V^1 \frac{\partial V^1}{\partial x^1} + V^2 \frac{\partial V^1}{\partial x^2} + V^3 \frac{\partial V^1}{\partial x^3} - x^1 (V^2)^2 \right]$$

$$= F^1 \left[-\frac{\partial p}{\partial x^1} + \mu \frac{\partial^2 V^1}{\partial x^1 \partial x^1} + \frac{1}{(x^1)^2} \frac{\partial^2 V^1}{\partial x^2 \partial x^2} + \frac{\partial^2 V^1}{\partial x^3 \partial x^3} - \frac{2}{x^1} \frac{\partial V^2}{\partial x^2} \right] \qquad (4.2.13)$$

The next viscous term to be formulated is

$$-\mu\left[\frac{\partial y^k}{\partial y^{(j)}}\;\frac{\partial y^k}{\partial y^{(j)}}\right]^{-1}\left(\frac{\partial y^\beta}{\partial x^1}\;\frac{\partial y^\beta}{\partial x^1}\right)\left[\frac{\partial y^\tau}{\partial x^{(\alpha)}}\;\frac{\partial y^\tau}{\partial x^{(\alpha)}}\right]^{-1}\left(\frac{\partial^2 y^m}{\partial x^j\partial x^j}\;\frac{\partial y^m}{\partial x^\alpha}\right)\frac{\partial V^1}{\partial x^\alpha}$$

In spite of its forbidding appearance, this term can be formulated easily if one uses the results of the previous work. A little practice with expressions of this type will show that it is not always necessary to carry out all the summations specified for the general case. By referring to the second viscous term considered (4.2.12), it will be observed that in the subsequent summations, the only nonzero value of the factor

$$\frac{\partial^2 y^m}{\partial x^\alpha \partial x^j}\;\frac{\partial y^m}{\partial x^1}$$

occurred when the indices assumed the following values: $\alpha = 2; j = 2$.

When formulated in terms of the given transformation equations and summed on m, this factor was shown to satisfy the equation

$$\frac{\partial^2 y^m}{\partial x^2 \partial x^2}\;\frac{\partial y^m}{\partial x^1} = -x^1$$

It follows that the only indices that need to be considered are $\alpha = 1; j = 2$. Moreover, it has been shown that for $j = 2$

$$\frac{\partial y^k}{\partial x^2}\;\frac{\partial y^k}{\partial x^2} = (x^1)^2$$

and

$$\left(\frac{\partial y^\beta}{\partial x^1}\;\frac{\partial y^\beta}{\partial x^1}\right) = \left(\frac{\partial y^\tau}{\partial x^1}\;\frac{\partial y^\tau}{\partial x^1}\right) = 1$$

and this viscous term reduces to

$$-\mu\;\frac{1}{(x^1)^2}\;(-x^1)\;\frac{\partial V^1}{\partial x^1} = \frac{\mu}{x^1}\;\frac{\partial V^1}{\partial x^1}$$

When we add this component to the previously obtained results (4.2.13), we have

$$\rho \left[\frac{\partial V^1}{\partial t} + V^1 \frac{\partial V^1}{\partial x^1} + V^2 \frac{\partial V^1}{\partial x^2} + V^3 \frac{\partial V^1}{\partial x^3} - x^1 (V^2)^2 \right]$$

$$= F^1 - \frac{\partial p}{\partial x^1} + \mu \left[\frac{\partial^2 V^1}{\partial x^1 \partial x^1} + \frac{1}{(x^1)^2} \frac{\partial^2 V^1}{\partial x^2 \partial x^2} + \frac{\partial^2 V^1}{\partial x^3 \partial x^3} - \frac{2}{x^1} \frac{\partial V^2}{\partial x^2} + \frac{1}{x^1} \frac{\partial V^1}{\partial x^1} \right]$$

$$(4.2.14)$$

Although the remaining three terms look equally forbidding, all the factors except one can be evaluated by inspection.

Consider first the term

$$\mu \left[\frac{\partial y^k}{\partial x^{(j)}} \frac{\partial y^k}{\partial x^{(j)}} \right]^{-1} \left(\frac{\partial y^\beta}{\partial x^1} \frac{\partial y^\beta}{\partial x^1} \right) \frac{\partial}{\partial x^j} \left(\frac{\partial y^\beta}{\partial x^1} \frac{\partial y^\beta}{\partial x^1} \right)^{-1} \left(\frac{\partial^2 y^m}{\partial x^\alpha \partial x^j} \frac{\partial y^m}{\partial x^1} \right) V^\alpha$$

We have already determined that

$$\frac{\partial y^\beta}{\partial x^1} \frac{\partial y^\beta}{\partial x^1} = 1$$

and therefore

$$\frac{\partial}{\partial x^j} \left(\frac{\partial y^\beta}{\partial x^1} \frac{\partial y^\beta}{\partial x^1} \right) = 0$$

Hence, this term vanishes for all values of the indices.

The next term to be considered is

$$\mu \left[\frac{\partial y^k}{\partial x^{(j)}} \frac{\partial y^k}{\partial x^{(j)}} \right]^{-1} \left(\frac{\partial^2 y^\gamma}{\partial x^r \partial x^j} \frac{\partial y^\gamma}{\partial x^1} \right) \left(\frac{\partial y^\tau}{\partial x^{(r)}} \frac{\partial y^\tau}{\partial x^{(r)}} \right)^{-1} \left(\frac{\partial^2 y^m}{\partial x^\alpha \partial x^j} \frac{\partial y^m}{\partial x^r} \right) V^\alpha$$

The factor

$$\frac{\partial^2 y^\gamma}{\partial x^r \partial x^j} \frac{\partial y^\gamma}{\partial x^1}$$

has been shown to be zero unless the indices assume the values $r = 2, j = 2$, in which case it satisfies the equation

$$\frac{\partial^2 y^\gamma}{\partial x^2 \partial x^2} \frac{\partial y^\gamma}{\partial x^1} = -\frac{1}{x^1}$$

With these indices, the other factors are

$$\left(\frac{\partial y^k}{\partial x^2} \frac{\partial y^k}{\partial x^2}\right) = \left(\frac{\partial y^\tau}{\partial x^2} \frac{\partial y^\tau}{\partial x^2}\right) = \frac{1}{(x^1)^2}$$

$$\left(\frac{\partial^2 y^\gamma}{\partial x^2 \partial x^2} \frac{\partial y^\gamma}{\partial x^1}\right) = -x^1$$

and this viscous term assumes the simpler form

$$-\frac{\mu}{(x^1)^3}\left(\frac{\partial^2 y^m}{\partial x^\alpha \partial x^2} \frac{\partial y^m}{\partial x^2}\right) V^\alpha$$

The only nonzero value for this factor occurs when the index $\alpha = 1$. In this case

$$\frac{\partial^2 y^m}{\partial x^1 \partial x^2} \frac{\partial y^m}{\partial x^2} = x^1$$

and this viscous component is simply

$$-\frac{\mu}{(x^1)^2} V^1$$

When this component is added to the previously obtained results (4.2.14), we have

$$\rho\left[\frac{\partial V^1}{\partial t} + V^1\frac{\partial V^1}{\partial x^1} + V^2\frac{\partial V^1}{\partial x^2} + V^3\frac{\partial V^1}{\partial x^3} - x^1(V^2)^2\right]$$

$$= F^1 - \frac{\partial p}{\partial x^1} + \mu\left[\frac{\partial^2 V^1}{\partial x^1\partial x^1} + \frac{1}{(x^1)^2}\frac{\partial^2 V^1}{\partial x^2\partial x^2} + \frac{\partial^2 V^1}{\partial x^3\partial x^3}\right.$$

$$\left. - \frac{2}{x^1}\frac{\partial V^2}{\partial x^2} + \frac{1}{x^1}\frac{\partial V^1}{\partial x^1} - \frac{V^1}{(x^1)^2}\right] \tag{4.2.15}$$

Again, by inspection we find that the last term

$$\mu\left[\frac{\partial y^k}{\partial x^{(j)}}\frac{\partial y^k}{\partial x^{(j)}}\right]^{-1}\left(\frac{\partial^2 y^\gamma}{\partial x^\alpha\partial x^r}\frac{\partial y^\gamma}{\partial x^1}\right)\left[\frac{\partial y^\tau}{\partial x^{(r)}}\frac{\partial y^\tau}{\partial x^{(r)}}\right]^{-1}\left(\frac{\partial^2 y^m}{\partial x^j\partial x^j}\frac{\partial y^m}{\partial x^r}\right)V^\alpha = 0$$

This term vanishes because the only nonzero value of the factor

$$\frac{\partial^2 y^m}{\partial x^j\partial x^j}\frac{\partial y^m}{\partial x^r}$$

occurs when $j = 2$, $r = 1$, and for $r = 1$, the factor

$$\left(\frac{\partial^2 y^\gamma}{\partial x^\alpha\partial x^r}\frac{\partial y^\gamma}{\partial x^1}\right)_{r=1} = 0$$

Hence, this component vanishes for all values of the indices, and equation (4.2.15) is the complete equation for $i = 1$. The remaining two equations of (4.1.16) can be formulated in the same manner.

Equation (4.2.15) is a tensor equation, that is, all the force and velocity components are tensor components. A formulation in terms of physical components is obtained by transforming all tensor components to physical components. From equation (4.2.1)

$$V^1 = \frac{v^1}{\sqrt{\dfrac{\partial y^\beta}{\partial x^1}\dfrac{\partial y^\beta}{\partial x^1}}}$$

By substitution from equation (4.2.7) in this equation, we obtain

$$V^1 = v^1 \tag{4.2.16}$$

Again, from equation (4.2.1), we have

$$V^2 = \frac{v^2}{\sqrt{\dfrac{\partial y^\beta}{\partial x^2} \dfrac{\partial y^\beta}{\partial x^2}}}$$

Substitution from equation (4.2.10a) in this equation gives

$$V^2 = \frac{v^2}{x^1} \tag{4.2.17}$$

The component V^3 is given by

$$V^3 = \frac{v^3}{\sqrt{\dfrac{\partial y^\beta}{\partial x^3} \dfrac{\partial y^\beta}{\partial x^3}}}$$

and from equation (4.2.10b)

$$\frac{\partial y^\beta}{\partial x^3} \frac{\partial y^\beta}{\partial x^3} = 1$$

Hence

$$V^3 = v^3 \tag{4.2.18}$$

The force component F^1 is subject to the transformation (4.2.2), that is

$$F^1 = \frac{f_1}{\sqrt{\dfrac{\partial y^\beta}{\partial x^1} \dfrac{\partial y^\beta}{\partial x^1}}}$$

Substituting again from equation (4.2.7), we find that

$$F^1 = f^1 \tag{4.2.19}$$

Substitution from equations (4.2.16) through (4.2.19) in equation (4.2.15) gives the physical form of this equation as follows:

$$\rho\left[\frac{\partial v^1}{\partial t} + v^1\frac{\partial v^1}{\partial x^1} + \frac{v^2}{x^1}\frac{\partial v^1}{\partial x^2} + v^3\frac{\partial v^1}{\partial x^3} - \frac{(v^2)^2}{x^1}\right]$$

$$= f^1 - \frac{\partial p}{\partial x^1} + \mu\left[\frac{\partial^2 v^1}{\partial x^1\partial x^1} + \frac{1}{(x^1)^2}\frac{\partial^2 v^1}{\partial x^2\partial x^2} + \frac{\partial^2 v^1}{\partial x^3\partial x^3}\right.$$

$$\left. - \frac{2}{(x^1)^2}\frac{\partial v^2}{\partial x^2} + \frac{1}{x^1}\frac{\partial v^1}{\partial x^1} - \frac{v^1}{(x^1)^2}\right] \tag{4.2.20}$$

A formulation in terms of metric tensor inputs will be described next. Specifically, attention will be focused on the formulation of the Christoffel symbols. Once the user knows how to derive these symbols, he can formulate the Navier-Stokes equations and the corresponding continuity equation in any coordinate system of interest.

4.3 FORMULATION OF THE CHRISTOFFEL SYMBOLS OF THE SECOND KIND

From the formulas of definition, equations (1.12.3) and (1.12.4), the Christoffel symbols are

$$[ij,k] = \frac{1}{2}\left(\frac{\partial g_{ik}}{\partial x^j} + \frac{\partial g_{jk}}{\partial x^i} - \frac{\partial g_{ij}}{\partial x^k}\right)$$

$$\left\{\begin{matrix} k \\ ij \end{matrix}\right\} = g^{kl}[ij,l]$$

Therefore

$$\begin{Bmatrix} k \\ ij \end{Bmatrix} = g^{k1}[ij,1] + g^{k2}[ij,2] + g^{k3}[ij,3] \qquad (4.3.1)$$

In orthogonal coordinate systems $l = k$, and in this case

$$\begin{Bmatrix} k \\ ij \end{Bmatrix} = g^{kk}[ij,k]$$

Therefore

$$\begin{Bmatrix} k \\ ij \end{Bmatrix} = g^{11}[ij,1] + g^{22}[ij,2] + g^{33}[ij,3] \qquad (4.3.2)$$

A simple program designed to formulate the Christoffel symbols of the first and second kinds, by using the metric tensors as input, would assume the following form if the coordinate system were cylindrical polar. Equations (3.2.6) give the metric tensors for this particular coordinate system (ref. 4).

```
LET(G(1,1)=1);
LET(G(2,2)=X(1)**2);
LET(G(3,3)=1);
PRINT_OUT(G(1,1);G(2,2);G(3,3));
PUT SKIP(5);
LET(H(1,1)=1);
LET(H(1,2)=0);
LET(H(1,3)=0);
LET(H(2,1)=0);
LET(H(2,2)=1/G(2,2));
```

```
LET(H(2,3)=0);

LET(H(3,1)=0);

LET(H(3,2)=0);

LET(H(3,3)=1/G(3,3));

PRINT_OUT(H(1,1);H(2,2);H(3,3));

PUT SKIP(3);

DO I=1 TO 3 BY 1;

LET(I="I");

DO J=1 TO 3 BY 1;

LET(J="J");

DO K=1 TO 3 BY 1;

LET(K="K");

LET(D(I,J,K)=(1/2)*DERIV(G(I,K),X(J)));

LET(E(J,K,I)=(1/2)*DERIV(G(J,K),X(I)));

LET(F(I,J,K)=(1/2)*DERIV(G(I,J),X(K)));

LET(C(I,J,K)=(D(I,J,K)+E(J,K,I)-F(I,J,K)));

PRINT_OUT(C(I,J,K));

END;

END;

END;

PUT SKIP(3);
```

```
DO I=1 TO 3 BY 1;
LET(I="I");
DO J=1 TO 3 BY 1;
LET(J="J");
DO K=1 TO 3 BY 1;
LET(K="K");
LET(T(I,J,K)=0);
DO N=1 TO 3 BY 1;
LET(N="N");
LET(CI(I,J,K)=H(N,K)*C(I,J,N));
LET(T(I,J,K)=T(I,J,K)+CT(I,J,K));
END;
PRINT_OUT(T(I,J,K));
END;
END;
END;
```

The output from this program is as follows:

$G(1,1) = 1$	$H(1,1) = 1$
$\mathring{G}(2,2) = X(1)^2$	$H(2,2) = 1/X(1)^2$
$G(3,3) = 1$	$H(3,3) = 1$

C(1,1,1) = 0

C(1,1,2) = 0

C(1,1,3) = 0

C(1,2,1) = 0

C(1,2,2) = X(1)

C(1,2,3) = 0

C(1,3,1) = 0

C(1,3,2) = 0

C(1,3,3) = 0

C(2,1,1) - 0

C(2,1,2) = X(1)

C(2,1,3) = 0

C(2,2,1) = -X(1)

C(2,2,2) = 0

C(2,2,3) = 0

C(2,3,1) = 0

C(2,3,2) = 0

C(2,3,3) = 0

C(3,1,1) = 0

C(3,1,2) = 0

C(3,1,3) = 0

C(3,2,1) = 0

C(3,2,2) = 0

C(3,2,3) = 0

C(3,3,1) = 0

C(3,3,2) = 0

C(3,3,3) = 0

T(1,1,1) = 0

T(1,1,2) = 0

T(1,1,3) = 0

T(1,2,1) = 0

T(1,2,2) = 1/X(1)

T(1,2,3) = 0

T(1,3,1) = 0

T(1,3,2) = 0

T(1,3,3) = 0

T(2,1,1) = 0

T(2,1,2) = 1/X(1)

T(2,1,3) = 0

T(2,2,1) = -X(1)

T(2,2,2) = 0

T(2,2,3) = 0

T(2,3,1) = 0	T(3,2,1) = 0
T(2,3,2) = 0	T(3,2,2) = 0
T(2,3,3) = 0	T(3,2,3) = 0
T(3,1,1) = 0	T(3,3,1) = 0
T(3,1,2) = 0	T(3,3,2) = 0
T(3,1,3) = 0	T(3,3,3) = 0

In this program and output, $C(I,J,K)$ denotes a Christoffel symbol of the first kind and $T(I,J,K)$ is a Christoffel symbol of the second kind.

When formulated in this system of coordinates, the Navier-Stokes equations and the continuity equation are in a form suitable for studying flow through pipes. Moreover, in physiological applications, the equations may be used to study hemodynamic problems involving flow through distensible arteries. The equations can be formulated by specifying the metric tensors, deriving the Christoffel symbols of the second kind and proceeding as indicated in equations (4.1.4) and (4.1.7). Mechanization of these operations led to the following results:

G(1,1) = 1	T(1,1,1) = 0
G(2,2) = X(1)^2	T(1,1,2) = 0
G(3,3) = 1	T(1,1,3) = 0
	T(1,2,1) = 0
H(1,1) = 1	T(1,2,2) = 1/X(1)
H(2,2) = 1/X(1)^2	T(1,2,3) = 0
H(3,3) = 1	T(1,3,1) = 0
	T(1,3,2) = 0
	T(1,3,3) = 0

$$T(2,1,1) = 0 \qquad\qquad T(3,1,1) = 0$$

$$T(2,1,2) = 1/X(1) \qquad\qquad T(3,1,2) = 0$$

$$T(2,1,3) = 0 \qquad\qquad T(3,1,3) = 0$$

$$T(2,2,1) = -X(1) \qquad\qquad T(3,2,1) = 0$$

$$T(2,2,2) = 0 \qquad\qquad T(3,2,2) = 0$$

$$T(2,2,3) = 0 \qquad\qquad T(3,2,3) = 0$$

$$T(2,3,1) = 0 \qquad\qquad T(3,3,1) = 0$$

$$T(2,3,2) = 0 \qquad\qquad T(3,3,2) = 0$$

$$T(2,3,3) = 0 \qquad\qquad T(3,3,3) = 0$$

$$A(1) = F(1) - U(1) \; NU/X(1)^2 + U(1)^{(1\ 2)} \cdot (X(2),X(2)) \; NU/X(1)^2$$

$$- 2U(2)^{(1)} \cdot (X(2)) \; NU/X(1)^2 + U(1)^{(1)} \cdot (X(1)) \; NU/X(1)$$

$$+ U(1)^{(1\ 2)} \cdot (X(3), X(3)) \; NU + U(1)^{(1\ 2)} \cdot (X(1), X(1)) \; NU$$

$$- U(1)^{(1)} \cdot (X(1)) \; U(1) - U(1)^{(1)} \cdot (X(2)) \; U(2)/X(1)$$

$$- U(1)^{(1)} \cdot (X(3)) \; U(3) - P^{(1)} \cdot (X(1))/PHO + U(2)^2/X(1)$$

$$A(2) = F(2) - U(2) \; NU/X(1)^2 + 2U(1)^{(1)} \cdot (X(2)) \; NU/X(1)^2$$

$$+ U(2)^{(1\ 2)} \cdot (X(2), X(2)) \; NU/X(1)^2 + U(2)^{(1)}$$

$$\cdot (X(1)) \; NU/X(1) + U(2)^{(1\ 2)} \cdot (X(3), X(3)) \; NU + U(2)^{(1\ 2)}$$

$$\cdot (X(1), X(1)) \; NU - U(2) \; U(1)/X(1) - U(2)^{(1)} \cdot (X(1)) \; U(1)$$

$$- U(2)^{(1)} \cdot (X(2)) \; U(2)/X(1) - U(2)^{(1)} \cdot (X(3)) \; U(3) - P^{(1)}$$

$$\cdot (X(2))/X(1) \; RHO$$

A(3) = F(3) + U(3)$^{(1\ 2)}$. (X(2), X(2)) NU/X(1)^2 + U(3)$^{(1)}$

$\quad$. (X(1)) NU/X(1) + U(3)$^{(1\ 2)}$. (X(3), X(3)) NU + U(3)$^{(1\ 2)}$

$\quad$. (X(1), X(1)) NU - U(3)$^{(1)}$. (X(1)) U(1) - U(3)$^{(1)}$

$\quad$. (X(2)) U(2)/X(1) - U(3)$^{(1)}$. (X(3)) U(3) - P$^{(1)}$

$\quad$. (X(3))/RHO

DRHO = -U(1) RHO/X(1) - U(2)$^{(1)}$. (X(2)) RHO/X(1) - U(1)$^{(1)}$

$\quad$. (X(1)) RHO - U(3)$^{(1)}$. (X(3)) RHO

Using the definitions on page 250, these equations are interpreted as follows:

$$\rho\left[\frac{\partial v^1}{\partial t} + v^1\frac{\partial v^1}{\partial x^1} + \frac{v^2}{x^1}\frac{\partial v^1}{\partial x^2} - \frac{(v^2)^2}{x^1} + v^3\frac{\partial v^1}{\partial x^3}\right]$$

$$= f^1 - \frac{\partial p}{\partial x^1} + \mu\left[\frac{\partial^2 v^1}{\partial x^1 \partial x^1} + \frac{1}{x^1}\frac{\partial v^1}{\partial x^1} - \frac{v^1}{(x^1)^2} + \frac{1}{(x^1)^2}\frac{\partial^2 v^1}{\partial x^2 \partial x^2}\right.$$

$$\left. - \frac{2}{(x^1)^2}\frac{\partial v^2}{\partial x^2} + \frac{\partial^2 v^1}{\partial x^3 \partial x^3}\right] \qquad (4.3.3)$$

$$\rho\left(\frac{\partial v^2}{\partial t} + v^1\frac{\partial v^2}{\partial x^1} + \frac{v^2}{x^1}\frac{\partial v^2}{\partial x^2} + \frac{v^1 v^2}{x^1} + v^3\frac{\partial v^2}{\partial x^3}\right)$$

$$= f^2 - \frac{1}{x^1}\frac{\partial p}{\partial x^2} + \mu\left[\frac{\partial^2 v^2}{\partial x^1 \partial x^1} + \frac{1}{x^1}\frac{\partial v^2}{\partial x^1} - \frac{v^2}{(x^1)^2} + \frac{1}{(x^1)^2}\frac{\partial^2 v^2}{\partial x^2 \partial x^2}\right.$$

$$\left. + \frac{2}{(x^1)^2}\frac{\partial v^1}{\partial x^2} + \frac{\partial^2 v^2}{\partial x^3 \partial x^3}\right] \qquad (4.3.4)$$

$$\rho\left(\frac{\partial v^3}{\partial t} + v^1 \frac{\partial v^3}{\partial x^1} + \frac{v^2}{x^1} \frac{\partial v^3}{\partial x^2} + v^3 \frac{\partial v^3}{\partial x^3}\right)$$

$$= f^3 - \frac{\partial p}{\partial x^3} + \mu\left[\frac{\partial^2 v^3}{\partial x^1 \partial x^1} + \frac{1}{x^1} \frac{\partial v^3}{\partial x^1} + \frac{1}{(x^1)^2} \frac{\partial^2 v^3}{\partial x^2 \partial x^2} + \frac{\partial^2 v^3}{\partial x^3 \partial x^3}\right] \qquad (4.3.5)$$

$$\frac{\partial v^1}{\partial x^1} + \frac{v^1}{x^1} + \frac{1}{x^1} \frac{\partial v^2}{\partial x^2} + \frac{\partial v^3}{\partial x^3} = 0 \qquad (4.3.6)$$

As a further illustration of the procedure involved in using a computer to derive the equations of fluid motion and the continuity equation, the automatic derivation of these equations in a spherical polar coordinate system will be considered. When transformed to this system of coordinates, the Navier-Stokes equations and the continuity equation are in a suitable form for oceanographic studies involving tidal motions on planetary surfaces. In physiological applications, these equations may be used to study the hydrodynamics of ocular systems. As in the previous case, it is only necessary to specify the metric tensors, derive the Christoffel symbols of the second kind, and proceed as indicated in equations (4.1.4) and (4.1.7). Equation (3.2.9) gives the metric tensors for this coordinate system. Mechanization of these operations led to the following results:

G(1,1) = 1

G(2,2) = X(1)^2

G(3,3) = SIN^2 (X(2)) X(1)^2

H(1,1) = 1

H(2,2) = 1/X(1)^2

H(3,3) = 1/SIN^2(X(2)) X(1)^2

T(1,1,1) = 0

T(1,1,2) = 0

T(1,1,3) = 0

T(1,2,1) = 0

T(1,2,2) = 1/X(1)

T(1,2,3) = 0

T(1,3,1) = 0

T(1,3,2) = 0

T(1,3,3) = 1/X(1)

T(2,1,1) = 0

T(2,1,2) = 1/X(1)

T(2,1,3) = 0

$$T(2,2,1) = -X(1)$$

$$T(2,2,2) = 0$$

$$T(2,2,3) = 0$$

$$T(2,3,1) = 0$$

$$T(2,3,2) = 0$$

$$T(2,3,3) = COS(X(2))/SIN(X(2))$$

$$T(3,1,1) = 0$$

$$T(3,1,2) = 0$$

$$T(3,1,3) = 1/X(1)$$

$$T(3,2,1) = 0$$

$$T(3,2,2) = 0$$

$$T(3,2,3) = COS\ (X(2))/SIN(X(2))$$

$$T(3,3,1) = -SIN^2(X(2))X(1)$$

$$T(3,3,2) = -COS(X(2))SIN(X(2))$$

$$T(3,3,3) = 0$$

$$A(1) = F(1)-2U(1)NU/X(1)^2-2COS(X(2))U(2)NU/SIN(X(2))X(1)^2$$

$$+U(1)^{(12)}.(X(3)),\ X(3))NU/SIN^2(X(2))X(1)^2$$

$$+U(1)^{(1)}.(X(2))COS(X(2))NU/SIN(X(2))X(1)^2$$

$$-2U(3)^{(1)}.(X(3))NU/SIN(X(2))X(1)^2$$

$$+U(1)^{(12)}.(X(2),X(2))NU/X(1)^2-2U(2)^{(1)}.(X(2))NU/X(1)^2$$

$$+2U(1)^{(1)}.(X(1))NU/X(1)+U(1)^{(12)}.(X(1),X(1))NU$$

$$-U(1)^{(1)}.(X(1))U(1)-U(1)^{(1)}.(X(2))U(2)/X(1)$$

$$-U(1)^{(1)}.(X(3))U(3)/SIN(X(2))X(1)$$

$$-U(1)^{(1)}.(X(3))U(3)/SIN(X(2))X(1)$$

$$-P^{(1)}.(X(1))/RHO+U(2)^2/X(1)+U(3)^2/X(1)$$

$$A(2) = F(2)-U(2)NU/X(1)^2-COS^2(X(2))U(2)NU/SIN^2(X(2))X(1)^2$$

$$-2U(3)^{(1)}.(X(3))COS(X(2))NU/SIN^2(X(2))X(1)^2$$

$$+U(2)^{(12)}.(X(3),X(3))NU/SIN^2(X(2))X(1)^2+U(2)^{(1)}$$

$$.(X(2))COS(X(2))NU/SIN(X(2))X(1)^2+2U(1)^{(1)}.(X(2))NU/X(1)^2$$

$$+U(2)^{(12)}.(X(2),X(2))NU/X(1)^2+2U(2)^{(1)}.(X(1))NU/X(1)$$

$$+U(2)^{(12)}.(X(1),X(1))NU-U(2)U(1)/X(1)-U(2)^{(1)}.(X(1))U(1)$$

$$-U(2)^{(1)}.(X(2))U(2)/X(1)-U(2)^{(1)}.(X(3))U(3)/SIN(X(2))X(1)$$

$$-P^{(1)}.(X(2))/X(1)RHO+COS(X(2))U(3)^2/SIN(X(2))X(1)$$

$$A(3) = F(3)-U(3)NU/X(1)^2-COS^2(X(2))U(3)NU/SIN^2(X(2))X(1)^2$$

$$+2U(2)^{(1)}.(X(3))COS(X(2))NU/SIN^2(X(2))X(1)^2$$

$$+U(3)^{(12)}.(X(3),X(3))NU/SIN^2(X(2))X(1)^2$$

$$+U(3)^{(1)}.(X(2))COS(X(2))NU/SIN(X(2))X(1)^2$$

$$+2U(1)^{(1)}.(X(3))NU/SIN(X(2))X(1)^2+U(3)^{(12)}$$

$$.(X(2),X(2))NU/X(1)^2+2U(3)^{(1)}.(X(1))NU/X(1)+U(3)^{(12)}$$

$$.(X(1),X(1))NU-U(3)U(1)/X(1)-U(3)^{(1)}$$

$$.(X(1))U(1)-COS(X(2))U(3)U(2)/SIN(X(2))X(1)-U(3)^{(1)}$$

$$.(X(2))U(2)/X(1)-U(3)^{(1)}.(X(3))U(3)/SIN(X(2))X(1)$$

$$-P^{(1)}.(X(3))/SIN(X(2))X(1)RHO$$

```
DRHO = -2U(1)RHO/X(1)-COS(X(2))U(2)RHO/SIN(X(2))X(1)
       -U(3)(1).(X(3))RHO/SIN(X(2))X(1)-U(2)(1).(X(2))RHO/X(1)
       -U(1)(1).(X(1))RHO
```

Interpretation of the output requires that the following computer notation be understood:

$$A(i) = \frac{\partial U^i}{\partial t}$$

$U(i)$ physical components of the velocity vector

$$U(i)^{(1)} \cdot (x(j)) = \frac{\partial U^i}{\partial x^j}$$

$$U(i)^{(12)} \cdot (x(j), x(k)) = \frac{\partial^2 U^i}{\partial x^j \partial x^k}$$

$$P^{(1)} \cdot (x(j)) = \frac{\partial P}{\partial x^j}$$

$$RHO = \rho$$

$$DRHO = \frac{\partial \rho}{\partial t}$$

$$NU = \nu = \frac{k}{\rho}$$

When expressed in terms of conventional mathematical symbolism these results appear as follows (ref. 1):

$$\rho\left[\frac{\partial v^1}{\partial t} + v^1 \frac{\partial v^1}{\partial x^1} + \frac{v^2}{x^1} \frac{\partial v^1}{\partial x^2} + \frac{v^3}{x^1 \sin x^2} \frac{\partial v^1}{\partial x^3} - \frac{(v^2)^2}{x^1} - \frac{(v^3)^2}{x^1}\right]$$

$$= f^1 - \frac{\partial p}{\partial x^1} + \mu\left[\frac{\partial^2 v^1}{\partial x^1 \partial x^1} + \frac{2}{x^1} \frac{\partial v^1}{\partial x^1} + \frac{1}{(x^1)^2} \frac{\partial^2 v^1}{\partial x^2 \partial x^2} + \frac{1}{(x^1)^2 \sin^2 x^2} \frac{\partial^2 v^1}{\partial x^3 \partial x^3}\right.$$

$$\left. + \frac{\cot x^2}{(x^1)^2} \frac{\partial v^1}{\partial x^2} - \frac{2}{(x^1)^2} \frac{\partial v^2}{\partial x^2} - \frac{2}{(x^1)^2 \sin x^2} \frac{\partial v^3}{\partial x^3} - \frac{2v^1}{(x^1)^2} - \frac{2\cot x^2}{(x^1)^2} v^2\right]$$

$$\rho\left[\frac{\partial v^2}{\partial t} + v^1 \frac{\partial v^2}{\partial x^1} + \frac{v^2}{x^1} \frac{\partial v^2}{\partial x^2} + \frac{v^3}{x^1 \sin x^2} \frac{\partial v^3}{\partial x^3} + \frac{v^1 v^2}{x^1} - \frac{\cot x^2}{x^1} (v^3)^2\right]$$

$$= f^2 - \frac{1}{x^1} \frac{\partial p}{\partial x^2} + \mu\left[\frac{\partial^2 v^2}{\partial x^1 \partial x^1} + \frac{2}{x^1} \frac{\partial v^2}{\partial x^2} + \frac{1}{(x^1)^2} \frac{\partial^2 v^2}{\partial x^2 \partial x^2} + \frac{1}{(x^1)^2 \sin^2 x^2} \frac{\partial^2 v^2}{\partial x^3 \partial x^3}\right.$$

$$\left. + \frac{\cot x^2}{(x^1)^2} \frac{\partial v^2}{\partial x^2} - \frac{2\cot x^2}{(x^1)^2 \sin x^2} \frac{\partial v^3}{\partial x^3} + \frac{2}{(x^1)^2} \frac{\partial v^1}{\partial x^2} - \frac{v^2}{(x^1)^2 \sin^2 x^2}\right]$$

$$\rho\left[\frac{\partial v^3}{\partial t} + v^1 \frac{\partial v^3}{\partial x^1} + \frac{v^2}{x^1} \frac{\partial v^3}{\partial x^2} + \frac{1}{x^1 \sin x^2} \frac{\partial v^3}{\partial x^3} + \frac{v^1 v^3}{x^1} + \frac{\cot x^2}{x^1} v^2 v^3\right]$$

$$= f^3 - \frac{1}{x^1 \sin x^2} \frac{\partial p}{\partial x^3} + \mu\left[\frac{\partial^2 v^3}{\partial x^1 \partial x^1} + \frac{2}{x^1} \frac{\partial v^3}{\partial x^1} + \frac{1}{(x^1)^2} \frac{\partial^2 v^3}{\partial x^2 \partial x^2}\right.$$

$$+ \frac{1}{(x^1)^2 \sin^2 x^2} \frac{\partial^2 v^3}{\partial x^3 \partial x^3} + \frac{\cot x^2}{(x^1)^2} \frac{\partial v^3}{\partial x^2} + \frac{2}{(x^1)^2 \sin x^2} \frac{\partial v^1}{\partial x^3}$$

$$\left. + \frac{2\cot x^1}{(x^1)^2 \sin x^2} \frac{\partial v^2}{\partial x^3} - \frac{1}{(x^1)^2 \sin^2 x^2} v^3\right]$$

$$\frac{\partial v^1}{\partial x^1} + \frac{\partial v^2}{x^1 \, \partial x^2} + \frac{1}{x^1 \sin x^2} \frac{\partial v^3}{\partial x^3} + \frac{2}{x^1} v^1 + \frac{\cot x^2}{x^1} v^2 = 0$$

With the exception of the section on particle dynamics, emphasis has been placed on applications occurring in cylindrical and spherical coordinate systems. This course of action has been adopted in the belief that the demonstration of principles and techniques is more effective if it is conducted in a coordinate system with which the reader is familiar. Once the technique has been thoroughly grasped, the reader can proceed to less conventional coordinate systems, if he so desires. For example, a problem of interest to physiologists is the study of fluid flow in the semicircular canals of the inner ear. These canals are the basic transducers for a vestibulo-ocular reflex which compensates the eyeball for the rotational movements of the body and of the skull which are encountered in normal life. Each inner ear has three semicircular canals in approximately mutually perpendicular planes, so that all three degrees of freedom can be sensed. These canals are small-bore circuits containing fluid which rotates relative to the skull when the skull rotates in space. The small bore of the canals ensures laminar flow of the contained fluid. The equations describing the motion of this fluid can be obtained in the manner described for the two previous cases. The curvilinear coordinates best suited to the description of fluid flow in the semicircular canals are toroidal coordinates.

A variety of other curvilinear coordinate systems is available for special situations. For example, confocal ellipsoidal coordinates have proved useful in hydrodynamics problems (ref. 5). Once a curvilinear coordinate system is chosen, the formulation can proceed by the method of equations (4.2.3) and (4.2.4) which relies on coordinate transformation equations, or by the method of equations (4.1.1) and (4.1.3) which uses metric tensor inputs. As indicated in section 3.2, the determination of the fundamental quadratic form is usually no more difficult and is frequently much simpler than finding the coordinate transformation equations. More importantly, if the functions $g_{ij}(x)$ are such that the system of equations (1.13.12) has no solution, then no admissible transformation of coordinates exists which reduces equation (3.2.1) to the Pythagorean form. In this case, the manifold is non-Euclidean and the use of coordinate transformation equations as input will fail. A non-Euclidean model is described in section 5.1.

COSMOLOGICAL APPLICATIONS

5.1 NEWTONIAN AND RELATIVISTIC FIELD EQUATIONS AND TRAJECTORY EQUATIONS

The advantages of symbolic mathematical computation are most evident in problems analogous to those described in this chapter, that is, the formulation of cosmological models and their associated trajectory equations. The field equations that govern the trajectories of bodies in space have, in general, large numbers of terms, with each term a complicated mathematical expression. The evaluation of these terms and the derivation of the equations of the geodesics that describe the trajectories of bodies in space require a substantial amount of algebraic manipulation and symbolic differentiation. For the models considered, the required operations were executed with speed and efficiency on an IBM 360/67 computer. For example, in the case of the nonhomogeneous Schwarzschild model, the computer times required to formulate the field and trajectory equations were 0.74 and 0.30 min, respectively. By mechanization of the procedure in the manner described, computation time is saved, the possibility of error is reduced, and the scope of the inquiry may be extended.

The present chapter indicates how symbolic mathematical computation can be used to formulate a variety of cosmological models. As in previous applications, all formulations evolve from a fundamental metric, and each model is determined by the metric of the Riemannian space. The only inputs required are the coefficients of the fundamental quadratic form. For illustrative purposes, only spherically symmetric static models are considered. The determination of the geodesics that describe the trajectories of bodies in space requires that the appropriate potential functions be known. The relativistic analog of Poisson's equation, which in the Newtonian theory connects a single gravitational potential function with the density of matter, is a relation between the potential functions and the components of the energy momentum tensor. In general, this relationship gives rise to 10 nonlinear partial differential equations. The solution of these equations then yields the potential functions and must precede any attempt to obtain the corresponding trajectories.

Consider the equation of motion of a particle which is moving under the influence of gravitational forces. When the equation is written in the notation of the tensor calculus, it assumes the form (ref. 1)

$$m\left(\frac{d^2x^i}{dt^2} + \left\{{i \atop jk}\right\} \frac{dx^j}{dt} \frac{dx^k}{dt}\right)\bar{a}_i = \nabla\phi \qquad (5.1.1)$$

where $i,j,k = 1,2,3$ and

$$\nabla\phi = \frac{\partial\phi}{\partial x^j} \bar{a}^j = g^{ij} \frac{\partial\phi}{\partial x^j} \bar{a}_i \qquad (5.1.2)$$

In these equations the summation convention is again assumed. That is to say, if in any term an index occurs twice, the term is to be summed with respect to that index for all admissible values of the index.

In relativistic mechanics, equation (5.1.1) is replaced by the following trajectory equation, which is the equation of a geodesic (ref. 2)

$$\left(\frac{d^2x^i}{ds^2} + \left\{{i \atop jk}\right\} \frac{dx^j}{ds} \frac{dx^k}{ds}\right) = 0 \qquad i,j,k = 1,2,3,4 \qquad (5.1.3)$$

where the line element ds satisfies the equation of the fundamental quadratic form (ref. 3)

$$ds^2 = g_{ij}dx^i dx^j \qquad (5.1.4)$$

The Newtonian theory of gravitation connects a single potential function ϕ with the density of matter. In this theory, the gravitational potential function is required to satisfy Poisson's equation (ref. 4)

$$\nabla^2\phi = -4\pi\rho \qquad (5.1.5)$$

At all points of space devoid of matter $\rho = 0$, and Poisson's equation reduces to Laplace's equation (ref. 4)

$$\nabla^2 \phi = 0 \qquad\qquad (5.1.6)$$

The relativistic analog of Poisson's equation is the following tensor equation (ref. 5)

$$R_{ij} - \frac{1}{2} R g_{ij} + \Lambda g_{ij} = -\kappa T_{ij} \qquad\qquad (5.1.7)$$

where

R_{ij} Ricci tensor (discussed in sec. 1.16 and defined in eqs. (1.16.2) and (1.16.3))

R Ricci scalar (defined in eq. (1.18.5))

g_{ij} metric tensor

Λ constant, the so-called cosmological constant

κ constant

By raising indices, the field equations can be written in the alternative form

$$R_j{}^i - \frac{1}{2} \delta_j{}^i R + \delta_j{}^i \Lambda = -\kappa T_j{}^i \qquad\qquad (5.1.8)$$

Contraction of equation (5.1.8) yields

$$R - 4\Lambda = \kappa T \qquad\qquad (5.1.9)$$

In regions of space devoid of matter, all the components of the energy momentum tensor are zero, and equation (5.1.9) simplifies accordingly. In this case

$$R = 4\Lambda \qquad\qquad (5.1.10)$$

When this result is substituted in equation (5.1.7), the field equations assume the form

$$R_{ij} = \Lambda g_{ij} \qquad\qquad (5.1.11)$$

Nevertheless, in empty space, the trajectories of bodies moving within the solar system correspond with great precision to the simpler field equations (ref. 5)

$$R_{ij} = 0 \qquad\qquad (5.1.12)$$

Equation (5.1.12) is the relativistic analog of the Laplace equation. It represents 10 nonlinear partial differential equations for the 10 unknown functions g_{ij}. Once a set of functions satisfying equation (5.1.8) or (5.1.12) is found, the corresponding trajectory equations can be formulated.

Before proceeding with the formulation of specific models, it is worthwhile to examine the equations on which the models are based. The relevant field equations are (5.1.7), (5.1.8), and (5.1.12). With the exception of the energy momentum tensor, which is unspecified, the terms of these equations consist of the Ricci tensor and the Ricci scalar. The Ricci scalar is the product of the Ricci tensor and the corresponding metric tensor. For the convenience of readers, these tensors are reproduced here. The equation for the Ricci tensor is (ref. 5)

$$R_{ij} = R^{\alpha}_{ij\alpha} = \left[\left\{ \begin{matrix} \beta \\ i\alpha \end{matrix} \right\} \left\{ \begin{matrix} \alpha \\ \beta j \end{matrix} \right\} - \left\{ \begin{matrix} \beta \\ ij \end{matrix} \right\} \left\{ \begin{matrix} \alpha \\ \beta\alpha \end{matrix} \right\} + \frac{\partial}{\partial x^j} \left\{ \begin{matrix} \alpha \\ i\alpha \end{matrix} \right\} - \frac{\partial}{\partial x^{\alpha}} \left\{ \begin{matrix} \alpha \\ ij \end{matrix} \right\} \right]$$

The equation for the Ricci scalar is simply

$$R = g^{jk} R_{jk}$$

It is seen that although the field equations are complicated, the individual terms of the equations are simply products of the Christoffel symbols of the second kind or derivatives of these symbols. Moreover, the Christoffel symbols are known when the metric tensors are known, that is, when an expression for the square of the distance between two adjacent points of the space can be formulated. The dependence of the Christoffel symbols on the fundamental metric tensor can be seen from the following equations:

$$[ij,k] = \frac{1}{2}\left(\frac{\partial g_{ik}}{\partial x^j} + \frac{\partial g_{jk}}{\partial x^i} - \frac{\partial g_{ij}}{\partial x^k} \right)$$

$$\left\{ \begin{matrix} \alpha \\ ij \end{matrix} \right\} = g^{\alpha k} [ij,k]$$

It was demonstrated in preceding chapters that, given the fundamental metric tensor, the Christoffel symbols can be formulated without difficulty. Hence, the evolution from the fundamental metric tensor to complex cosmological models is easily accomplished.

5.2 MODELING CONSIDERATIONS

For the purpose of illustrating the modeling capability of symbolic mathematical computation, a spherically symmetric static field is assumed. This assumption implies that the metric tensors g_{ij} are spherically symmetric and independent of the time. Moreover, the metric tensors must be chosen in such a way that the line element will reduce to the special relativity form for flat space time. These considerations led to the adoption of the following set of metric tensors for anisotropic space:

$$
\left.
\begin{aligned}
g_{11} &= -e^{L(x^1)} \\[6pt]
g_{22} &= -(x^1)^2 \\[6pt]
g_{33} &= -(x^1 \sin x^2)^2 \\[6pt]
g_{44} &= e^{M(x^1)}
\end{aligned}
\right\}
\tag{5.2.1}
$$

where the implicit functions $L(x^1)$ and $M(x^1)$ can be adjusted to account for the distortion of space in the presence of matter. The corresponding space-time interval is

$$
ds^2 = [-e^{L(x^1)}(dx^1)^2 - (x^1 dx^2)^2 - (x^1 \sin x^2\, dx^3)^2 + e^{M(x^1)}(dx^4)^2]
\tag{5.2.2}
$$

where

$$
dx^4 = c\, dt
$$

and

x^1 radial coordinate

x^2 polar angle coordinate

x^3 azimuthal angle coordinate

And for convenience, the velocity of light is assumed to be 1.

   If the space is assumed to be isotropic, the metric tensors are modified as follows:

$$
\left.
\begin{aligned}
g_{11} &= -e^{L(x^1)} \\[1em]
g_{22} &= -e^{L(x^1)}(x^1)^2 \\[1em]
g_{33} &= -e^{L(x^1)}(x^1 \sin x^2)^2 \\[1em]
g_{44} &= e^{M(x^1)}
\end{aligned}
\right\}
\tag{5.2.3}
$$

   The space-time interval in the isotropic case is

$$
ds^2 = \left\{ -e^{L(x^1)}[(dx^1)^2 + (x^1\,dx^2)^2 + (x^1 \sin x^2\,dx^3)^2] + e^{M(x^1)}(dx^4)^2 \right\}
$$

$$
\tag{5.2.4}
$$

   In order to demonstrate the feasibility of using symbolic mathematical computation to formulate different models of the universe, a computer program was written that required only the postulated metric tensors as inputs.

   The notation adopted in this and subsequent sections corresponds to the form of the printed output. At the time of writing, the formulated mathematical expressions could only be printed out in the form of capital letters, and the functions $L(x^1)$ and $M(x^1)$ are printed out as $L\cdot(x(1))$ and $M\cdot(x(1))$, respectively. Moreover, the ith derivative of the functions $L(x^1)$ and $M(x^1)$ with respect to x^1 are printed out as $L^{(I)}\cdot(x(1))$ and $M^{(I)}\cdot(x(1))$, respectively. To fix the ideas, the reader is referred to section (5.8) where the components of $R(I,J)$ from Section (5.3) have been interpreted in terms of conventional mathematical symbolism in equations (5.8.2) through (5.8.5).

   It should be noted that the field equations, which are denoted by $E(I,J)$ in the printed output, are related to the energy momentum tensors as indicated in equations (5.1.7) and (5.1.8). However, in empty space where all the components of the energy momentum tensor are zero, the field equations reduce to the simpler form

$$R_{ij} = 0$$

which, as already noted, is the relativistic analog of Laplace's equation.

The comments appearing between each batch of computer output are programmed statements and occur without variation during each run. This accounts for the stilted form of the language.

5.3 ANISOTROPIC MODEL

The field equations and the corresponding trajectory equations for this condition can be obtained by using the tensors defined in equations (5.2.1). With these tensors as inputs to a digital computer, which was programmed to formulate models of the universe, the following output was obtained. (The computer program and related documentation are available from Computer Software Management and Information Center (COSMIC), Barrow Hall, University of Georgia, Athens, Georgia, 30601.)

The metric coefficients determine the gravitational model being studied. In order that each run be identified with the correct inputs, the postulated metric coefficients are printed out before the main results. In the case under consideration, these have the following values:

$$G(1,1) = -E^{L \cdot (X(1))}$$

$$G(2,2) = -X(1)^2$$

$$G(3,3) = -SIN^2(X(2))X(1)^2$$

$$G(4,4) = E^{M \cdot (X(1))}$$

The program uses the metric tensor inputs to evaluate the Christoffel symbols of the first and second kinds. In order to reduce the amount of output, the Christoffel symbols of the first kind are not printed out. In terms of the system coordinates and the unknown functions L and M, the Christoffel symbols of the second kind are

$$T(1,1,1) = (1/2)L^{(1)} \cdot (X(1)) \qquad T(1,1,3) = 0$$

$$T(1,1,2) = 0 \qquad\qquad\qquad T(1,1,4) = 0$$

$T(1,2,1) = 0$

$T(1,2,2) = 1/X(1)$

$T(1,2,3) = 0$

$T(1,2,4) = 0$

$T(1,3,1) = 0$

$T(1,3,2) = 0$

$T(1,3,3) = 1/X(1)$

$T(1,3,4) = 0$

$T(1,4,1) = 0$

$T(1,4,2) = 0$

$T(1,4,3) = 0$

$T(1,4,4) = (1/2)M^{(1)} \cdot (X(1))$

$T(2,1,1) = 0$

$T(2,1,2) = 1/X(1)$

$T(2,1,3) = 0$

$T(2,1,4) = 0$

$T(2,2,1) = -E^{-L \cdot (X(1))}X(1)$

$T(2,2,2) = 0$

$T(2,2,3) = 0$

$T(2,2,4) = 0$

$T(2,3,1) = 0$

$T(2,3,2) = 0$

$T(2,3,3) = COS(X(2))/SIN(X(2))$

$T(2,3,4) = 0$

$T(2,4,1) = 0$

$T(2,4,2) = 0$

$T(2,4,3) = 0$

$T(2,4,4) = 0$

$T(3,1,1) = 0$

$T(3,1,2) = 0$

$T(3,1,3) = 1/X(1)$

$T(3,1,4) = 0$

$T(3,2,1) = 0$

$T(3,2,2) = 0$

$T(3,2,3) = COS(X(2))/SIN(X(2))$

$T(3,2,4) = 0$

$$T(3,3,1) = -E^{-L \cdot (X(1))} SIN^2(X(2))X(1)$$

$$T(3,3,2) = -COS(X(2))SIN(X(2))$$

$$T(3,3,3) = 0$$

$$T(3,3,4) = 0$$

$$T(3,4,1) = 0$$

$$T(3,4,2) = 0$$

$$T(3,4,3) = 0$$

$$T(3,4,4) = 0$$

$$T(4,1,1) = 0$$

$$T(4,1,2) = 0$$

$$T(4,1,3) = 0$$

$$T(4,1,4) = (1/2)M^{(1)} \cdot (X(1))$$

$$T(4,2,1) = 0$$

$$T(4,2,2) = 0$$

$$T(4,2,3) = 0$$

$$T(4,2,4) = 0$$

$$T(4,3,1) = 0$$

$$T(4,3,2) = 0$$

$$T(4,3,3) = 0$$

$$T(4,3,4) = 0$$

$$T(4,4,1) = (1/2)E^{-L \cdot (X(1))+M \cdot (X(1))}M^{(1)} \cdot (X(1))$$

$$T(4,4,2) = 0$$

$$T(4,4,3) = 0$$

$$T(4,4,4) = 0$$

Once the Christoffel symbols of the second kind are known, the components of the Ricci tensor can be derived. The individual components are

$$R(1,1) = -L^{(1)} \cdot (X(1))/X(1)-(1/4)M^{(1)} \cdot (X(1))L^{(1)} \cdot (X(1))$$
$$+(1/4)M^{(1)} \cdot^2 (X(1))+(1/2)M^{(2)} \cdot (X(1))$$

$$R(1,2) = 0$$

$$R(1,3) = 0$$

$$R(1,4) = 0$$

$$R(2,1) = 0$$

$$R(2,2) = -(1/2)E^{-L \cdot (X(1))}L^{(1)} \cdot (X(1))X(1)$$
$$+(1/2)E^{-L \cdot (X(1))}M^{(1)} \cdot (X(1))X(1)+E^{-L \cdot (X(1))}-1$$

$$R(2,3) = 0$$

$$R(2,4) = 0$$

$$R(3,1) = 0$$

$$R(3,2) = 0$$

$$R(3,3) = -(1/2)E^{-L \cdot (X(1))}L^{(1)} \cdot (X(1))SIN^2(X(2))X(1)$$
$$+(1/2)E^{-L \cdot (X(1))}M^{(1)} \cdot (X(1))SIN^2(X(2))X(1)$$
$$+E^{-L \cdot (X(1))}SIN^2(X(2))-SIN^2(X(2))$$

$$R(3,4) = 0$$

$R(4,1) = 0$

$R(4,2) = 0$

$R(4,3) = 0$

$$R(4,4) = -E^{-L \cdot (X(1))+M \cdot (X(1))}M^{(1)} \cdot (X(1))/X(1)$$
$$-(1/4)E^{-L \cdot (X(1))+M \cdot (X(1))}M^{(1)} \cdot {}^2(X(1))$$
$$+(1/4)E^{-L \cdot (X(1))+M \cdot (X(1))}M^{(1)} \cdot (X(1))L^{(1)} \cdot (X(1))$$
$$-(1/2)E^{-L \cdot (X(1))+M \cdot (X(1))}M^{(2)} \cdot (X(1))$$

$G(I,J)$ and $R(I,J)$ are both known at this stage of the program; therefore, the Ricci scalar can be obtained. It is given by the following equation:

$$R = -2E^{-L \cdot (X(1))}/X(1)^2+2E^{-L \cdot (X(1))}L^{(1)} \cdot (X(1))/X(1)$$
$$-2E^{-L \cdot (X(1))}M^{(1)} \cdot (X(1))/X(1)-(1/2)E^{-L \cdot (X(1))}M^{(1)} \cdot {}^2(X(1))$$
$$+(1/2)E^{-L \cdot (X(1))}M^{(1)} \cdot (X(1))L^{(1)} \cdot (X(1))$$
$$-E^{-L \cdot (X(1))}M^{(2)} \cdot (X(1))+2/X(1)^2$$

The preceding information is next used to obtain the field equations. The individual equations are

$$ET(1,1) = E^{-L \cdot (X(1))}/X(1)^2+L^{(1)} \cdot (X(1))/(E^{L \cdot (X(1))})X(1)$$
$$-E^{-L \cdot (X(1))}L^{(1)} \cdot (X(1))/X(1)+E^{-L \cdot (X(1))}M^{(1)} \cdot (X(1))/X(1)$$
$$-(1/4)M^{(1)} \cdot {}^2(X(1))/(E^{L \cdot (X(1))})$$
$$+(1/4)M^{(1)} \cdot (X(1))L^{(1)} \cdot (X(1))/(E^{L \cdot (X(1))})$$
$$-(1/2)M^{(2)} \cdot (X(1))/(E^{L \cdot (X(1))})+(1/4)E^{-L \cdot (X(1))}M^{(1)} \cdot {}^2(X(1))$$
$$-(1/4)E^{-L \cdot (X(1))}M^{(1)} \cdot (X(1))L^{(1)} \cdot (X(1))$$
$$+(1/2)E^{-L \cdot (X(1))}M^{(2)} \cdot (X(1))-1/X(1)^2$$

$$ET(1,2) = 0$$

$$ET(1,3) = 0$$

$$ET(1,4) = 0$$

$$ET(2,1) = 0$$

$$ET(2,2) = -(1/2)E^{-L \cdot (X(1))}L^{(1)} \cdot (X(1))/X(1)$$
$$+(1/2)E^{-L \cdot (X(1))}M^{(1)} \cdot (X(1))/X(1)$$
$$+(1/4)E^{-L \cdot (X(1))}M^{(1)} \cdot {}^2(X(1))$$
$$-(1/4)E^{-L \cdot (X(1))}M^{(1)} \cdot (X(1))L^{(1)} \cdot (X(1))$$
$$+(1/2)E^{-L \cdot (X(1))}M^{(2)} \cdot (X(1))$$

$$ET(2,3) = 0$$

$$ET(2,4) = 0$$

$$ET(3,1) = 0$$

$$ET(3,2) = 0$$

$$ET(3,3) = -(1/2)E^{-L \cdot (X(1))}L^{(1)} \cdot (X(1))/X(1)$$
$$+(1/2)E^{-L \cdot (X(1))}M^{(1)} \cdot (X(1))/X(1)$$
$$+(1/4)E^{-L \cdot (X(1))}M^{(1)} \cdot {}^2(X(1))$$
$$-(1/4)E^{-L \cdot (X(1))}M^{(1)} \cdot (X(1))L^{(1)} \cdot (X(1))$$
$$+(1/2)E^{-L \cdot (X(1))}M^{(2)} \cdot (X(1))$$

$$ET(3,4) = 0$$

$$ET(4,1) = 0$$

$$ET(4,2) = 0$$

$$ET(4,3) = 0$$

$$ET(4,4) = E^{-L \cdot (X(1))}/X(1)^2 - E^{-L \cdot (X(1))}L^{(1)} \cdot (X(1))/X(1) - 1/X(1)^2$$

The choice of an energy momentum tensor completes the specification of this type of model. Solution of the resulting equations gives rise to the components of the potential function. In the case under consideration, the solution yields the unknown functions $L(x^1)$ and $M(x^1)$. In terms of the postulated metric tensor inputs, the computer derives the equation of the trajectories as follows:

$$A(1) = -(1/2)L^{(1)} \cdot (X(1))V(1)^2 + E^{-L \cdot (X(1))}X(1)V(2)^2$$
$$+ E^{-L \cdot (X(1))}SIN^2(X(2))X(1)V(3)^2$$
$$- (1/2)E^{-L \cdot (X(1)) + M \cdot (X(1))}M^{(1)} \cdot (X(1))V(4)^2$$

$$A(2) = -2V(2)V(1)/X(1) - COS(X(2))SIN(X(2))V(3)^2$$

$$A(3) = -2V(3)V(1)/X(1) - 2COS(X(2))V(3)V(2)/SIN(X(2))$$

$$A(4) = -M^{(1)} \cdot (X(1))V(4)V(1)$$

5.4 ISOTROPIC MODEL

If the universe is isotropic, the line element will assume the form of equation (5.2.4). When the corresponding metric tensors (5.2.3) were used as inputs to the computer program, the following output was obtained.

The metric coefficients determine the gravitational model being studied. In order that each run be identified with the correct inputs, the postulated metric coefficients are printed out before the main results. In the case under consideration, these have the following values:

$$G(1,1) = -E^{L \cdot (X(1))}$$

$$G(2,2) = -E^{L \cdot (X(1))} X(1)^2$$

$$G(3,3) = -E^{L \cdot (X(1))} SIN^2(X(2)) X(1)^2$$

$$G(4,4) = E^{M \cdot (X(1))}$$

The program uses the metric tensor inputs to evaluate the Christoffel symbols of the first and second kinds. In order to reduce the amount of output, the Christoffel symbols of the first kind are not printed out. In terms of the system coordinates and the unknown functions L and M, the Christoffel symbols of the second kind are

$$T(1,1,1) = (1/2)L^{(1)} \cdot (X(1))$$

$$T(1,1,2) = 0$$

$$T(1,1,3) = 0$$

$$T(1,1,4) = 0$$

$$T(1,2,1) = 0$$

$$T(1,2,2) = 1/X(1) + (1/2)L^{(1)} \cdot (X(1))$$

$$T(1,2,3) = 0$$

$$T(1,2,4) = 0$$

$$T(1,3,1) = 0$$

$$T(1,3,2) = 0$$

$$T(1,3,3) = 1/X(1) + (1/2)L^{(1)} \cdot (X(1))$$

$$T(1,3,4) = 0$$

$$T(1,4,1) = 0$$

$$T(1,4,2) = 0$$

$$T(1,4,3) = 0$$

$T(1,4,4) = (1/2)M^{(1)} \cdot (X(1))$

$T(2,1,1) = 0$

$T(2,1,2) = 1/X(1)+(1/2)L^{(1)} \cdot (X(1))$

$T(2,1,3) = 0$

$T(2,1,4) = 0$

$T(2,2,1) = -X(1)-(1/2)L^{(1)} \cdot (X(1))X(1)^2$

$T(2,2,2) = 0$

$T(2,2,3) = 0$

$T(2,2,4) = 0$

$T(2,3,1) = 0$

$T(2,3,2) = 0$

$T(2,3,3) = COS(X(2))/SIN(X(2))$

$T(2,3,4) = 0$

$T(2,4,1) = 0$

$T(2,4,2) = 0$

$T(2,4,3) = 0$

$T(2,4,4) = 0$

$T(3,1,1) = 0$

$T(3,1,2) = 0$

$T(3,1,3) = 1/X(1)+(1/2)L^{(1)} \cdot (X(1))$

$T(3,1,4) = 0$

$$T(3,2,1) = 0$$

$$T(3,2,2) = 0$$

$$T(3,2,3) = COS(X(2))/SIN(X(2))$$

$$T(3,2,4) = 0$$

$$T(3,3,1) = -SIN^2(X(2))X(1)-1/2L^{(1)} \cdot (X(1))SIN^2(X(2))X(1)^2$$

$$T(3,3,2) = -COS(X(2))SIN(X(2))$$

$$T(3,3,3) = 0$$

$$T(3,3,4) = 0$$

$$T(3,4,1) = 0$$

$$T(3,4,2) = 0$$

$$T(3,4,3) = 0$$

$$T(3,4,4) = 0$$

$$T(4,1,1) = 0$$

$$T(4,1,2) = 0$$

$$T(4,1,3) = 0$$

$$T(4,1,4) = (1/2)M^{(1)} \cdot (X(1))$$

$$T(4,2,1) = 0$$

$$T(4,2,2) = 0$$

$$T(4,2,3) = 0$$

$$T(4,2,4) = 0$$

$T(4,3,1) = 0$

$T(4,3,2) = 0$

$T(4,3,3) = 0$

$T(4,3,4) = 0$

$T(4,4,1) = (1/2)E^{-L \cdot (X(1)) + M \cdot (X(1))} M^{(1)} \cdot (X(1))$

$T(4,4,2) = 0$

$T(4,4,3) = 0$

$T(4,4,4) = 0$

Once the Christoffel symbols of the second kind are known, the components of the Ricci tensor can be derived. The individual components are

$$R(1,1) = L^{(1)} \cdot (X(1))/X(1) - (1/4)M^{(1)} \cdot (X(1))L^{(1)} \cdot (X(1))$$
$$+ (1/4)M^{(1) \cdot 2}(X(1)) + L^{(2)} \cdot (X(1)) + (1/2)M^{(2)} \cdot (X(1))$$

$R(1,2) = 0$

$R(1,3) = 0$

$R(1,4) = 0$

$R(2,1) = 0$

$$R(2,2) = (3/2)L^{(1)} \cdot (X(1))X(1) + (1/2)M^{(1)} \cdot (X(1))X(1)$$
$$+ (1/4)L^{(1) \cdot 2}(X(1))X(1)^2 + (1/4)M^{(1)} \cdot (X(1))L^{(1)} \cdot (X(1))X(1)^2$$
$$+ (1/2)L^{(2)} \cdot (X(1))X(1)^2$$

$R(2,3) = 0$

$R(2,4) = 0$

$R(3,1) = 0$

$R(3,2) = 0$

$R(3,3) = (3/2)L^{(1)} \cdot (X(1))SIN^2(X(2))X(1)$

$+(1/2)M^{(1)} \cdot (X(1))SIN^2(X(2))X(1)$

$+(1/4)L^{(1)} \cdot ^2(X(1))SIN^2(X(2))X(1)^2$

$+(1/4)M^{(1)} \cdot (X(1))L^{(1)} \cdot (X(1))SIN^2(X(2))X(1)^2$

$+(1/2)L^{(2)} \cdot (X(1))SIN^2(X(2))X(1)^2$

$R(3,4) = 0$

$R(4,1) = 0$

$R(4,2) = 0$

$R(4,3) = 0$

$R(4,4) = -E^{-L \cdot (X(1))+M \cdot (X(1))}M^{(1)} \cdot (X(1))/X(1)$

$-(1/4)E^{-L \cdot (X(1))+M \cdot (X(1))}M^{(1)} \cdot ^2(X(1))$

$-(1/4)E^{-L \cdot (X(1))+M \cdot (X(1))}M^{(1)} \cdot (X(1))L^{(1)} \cdot (X(1))$

$-(1/2)E^{-L \cdot (X(1))+M \cdot (X(1))}M^{(2)} \cdot (X(1))$

$G(I,J)$ and $R(I,J)$ are both known at this stage of the program; therefore, the Ricci scalar can be obtained. It is given by the following equation:

$R = -4E^{-L \cdot (X(1))}L^{(1)} \cdot (X(1))/X(1)-2E^{-L \cdot (X(1))}M^{(1)} \cdot (X(1))/X(1)$

$-(1/2)E^{-L \cdot (X(1))}L^{(1)} \cdot ^2(X(1))-(1/2)E^{-L \cdot (X(1))}M^{(1)} \cdot ^2(X(1))$

$-(1/2)E^{-L \cdot (X(1))}M^{(1)} \cdot (X(1))L^{(1)} \cdot (X(1))-2E^{-L \cdot (X(1))}L^{(2)} \cdot (X(1))$

$-E^{-L \cdot (X(1))}M^{(2)} \cdot (X(1))$

The preceding information is next used to obtain the field equations. The individual equations are

$$ET(1,1) = -L^{(1)} \cdot (X(1))/(E^{L \cdot (X(1))})X(1) + 2E^{-L \cdot (X(1))}L^{(1)} \cdot (X(1))/X(1)$$

$$+ E^{-L \cdot (X(1))}M^{(1)} \cdot (X(1))/X(1) - (1/4)M^{(1)} \cdot {}^{2}(X(1))/(E^{L \cdot (X(1))})$$

$$+ (1/4)M^{(1)} \cdot (X(1))L^{(1)} \cdot (X(1))/(E^{L \cdot (X(1))})$$

$$- L^{(2)} \cdot (X(1))/(E^{L \cdot (X(1))}) - (1/2)M^{(2)} \cdot (X(1))/(E^{L \cdot (X(1))})$$

$$+ (1/4)E^{-L \cdot (X(1))}L^{(1)} \cdot {}^{2}(X(1)) + (1/4)E^{-L \cdot (X(1))}M^{(1)} \cdot {}^{2}(X(1))$$

$$+ (1/4)E^{-L \cdot (X(1))}M^{(1)} \cdot (X(1))L^{(1)} \cdot (X(1))$$

$$+ E^{-L \cdot (X(1))}L^{(2)} \cdot (X(1)) + (1/2)E^{-L \cdot (X(1))}M^{(2)} \cdot (X(1))$$

$$ET(1,2) = 0$$

$$ET(1,3) = 0$$

$$ET(1,4) = 0$$

$$ET(2,1) = 0$$

$$ET(2,2) = (1/2)E^{-L \cdot (X(1))}L^{(1)} \cdot (X(1))/X(1)$$

$$+ (1/2)E^{-L \cdot (X(1))}M^{(1)} \cdot (X(1))/X(1)$$

$$+ (1/4)E^{-L \cdot (X(1))}M^{(1)} \cdot {}^{2}(X(1)) + (1/2)E^{-L \cdot (X(1))}L^{(2)} \cdot (X(1))$$

$$+ (1/2)E^{-L \cdot (X(1))}M^{(2)} \cdot (X(1))$$

$$ET(2,3) = 0$$

$$ET(2,4) = 0$$

$$ET(3,1) = 0$$

$$ET(3,2) = 0$$

$$ET(3,3) = (1/2)E^{-L \cdot (X(1))}L^{(1)} \cdot (X(1))/X(1)$$

$$+ (1/2)E^{-L \cdot (X(1))}M^{(1)} \cdot (X(1))/X(1)$$

$$+(1/4)E^{-L\cdot(X(1))}M^{(1).2}(X(1))+(1/2)E^{-L\cdot(X(1))}L^{(2)}\cdot(X(1))$$
$$+(1/2)E^{-L\cdot(X(1))}M^{(2)}\cdot(X(1))$$

$ET(3,4) = 0$

$ET(4,1) = 0$

$ET(4,2) = 0$

$ET(4,3) = 0$

$$ET(4,4) = 2E^{-L\cdot(X(1))}L^{(1)}\cdot(X(1))/X(1)+(1/4)E^{-L\cdot(X(1))}L^{(1).2}(X(1))$$
$$+E^{-L\cdot(X(1))}L^{(2)}\cdot(X(1))$$

The trajectory equations for the isotropic case are

$$A(1) = -(1/2)L^{(1)}\cdot(X(1))V(1)^2+X(1)V(2)^2+(1/2)L^{(1)}\cdot(X(1))X(1)^2V(2)^2$$
$$+SIN^2(X(2))X(1)V(3)^2+(1/2)L^{(1)}\cdot(X(1))SIN^2(X(2))X(1)^2V(3)^2$$
$$-(1/2)E^{-L\cdot(X(1))+M\cdot(X(1))}M^{(1)}\cdot(X(1))V(4)^2$$

$$A(2) = -2V(2)V(1)/X(1)-L^{(1)}\cdot(X(1))V(2)V(1)+COS(X(2))SIN(X(2))V(3)^2$$

$$A(3) = -2V(3)V(1)/X(1)-L^{(1)}\cdot(X(1))V(3)V(1)$$
$$-2COS(X(2))V(3)V(2)/SIN(X(2))$$

$$A(4) = -M^{(1)}\cdot(X(1))V(4)V(1)$$

5.5 STATIC HOMOGENEOUS MODELS

In the case of a static homogeneous universe, it is evident that coordinates can be chosen so that the line element will exhibit spherical symmetry around any desired origin, since all parts of the universe are permanently alike. Hence, the line element may be taken in the spherically symmetric static form of equation (5.2.2). In obtaining this form of line element, local irregularities in the gravitational field, which would occur in the immediate neighborhood of individual stars or stellar systems, are neglected.

For the system described, it can be shown that the components of the energy momentum tensor are (ref. 5)

$$ET(1,1) = ET(2,2) = ET(3,3) = 8\pi p_O$$

$$ET(4,4) = -8\pi p_O \tag{5.5.1}$$

$$ET(I,J) = 0 \qquad \text{for} \qquad I \neq J$$

where p_O and ρ_O are the pressure and density, respectively, as measured by an observer who is at least momentarily at rest with respect to the spatial axes. The solution of these equations gives rise to the components of the potential function. In the case of the field being considered, the solution yields the unknown functions $L(x^1)$ and $M(x^1)$.

In order to satisfy the conditions of static homogeneity, it can be shown that the implicit functions $L(x^1)$ and $M(x^1)$ are subject to the following constraints: If the model is homogeneous, the pressure as measured by a local observer will be the same everywhere. Again, owing to the assumed homogeneity of the model, the density will be the same everywhere. Moreover, the line element must reduce to the special relativity form, for flat space time, owing to the known validity of the special theory in such regions. By imposing these conditions, it can be shown that there are only three possibilities for a static homogeneous model (ref. 5)

$$M = 0 \tag{5.5.2}$$

$$L + M = 0 \tag{5.5.3}$$

$$L = M = 0 \tag{5.5.4}$$

These conditions lead, respectively, to the Einstein, the De Sitter, and the special relativity line elements.

5.6 THE EINSTEIN MODEL UNIVERSE

Substitution from equation (5.5.2) in equation (5.2.2) yields the following metric for a homogeneous model which is not isotropic

$$ds^2 = [-e^{L(x^1)}(dx^1)^2 - (x^1 dx^2)^2 - (x^1 \sin x^2 \, dx^3)^2 + (dx^4)^2] \tag{5.6.1}$$

If the model were assumed to be homogeneous and isotropic, it would be necessary to use equation (5.2.4) subject to the constraint equation (5.5.2).

Cosmological considerations led Einstein to consider a universe defined by the metric (5.6.1). When the metric coefficients were supplied as input to the computer program, the following output was obtained.

Again the metric coefficients determine the gravitational model being studied. In order that each run be identified with the correct inputs, the postulated metric coefficients are printed out before the main results. In the case under consideration, these have the following values:

$$G(1,1) = -E^{L \cdot (X(1))}$$

$$G(2,2) = -X(1)^2$$

$$G(3,3) = -SIN^2(X(2))X(1)^2$$

$$G(4,4) = 1$$

The program uses the metric tensor inputs to evaluate the Christoffel symbols of the first and second kinds. In order to reduce the amount of output, the Christoffel symbols of the first kind are not printed out. In terms of the system coordinates and the unknown functions L and M, the Christoffel symbols of the second kind are

$T(1,1,1) = (1/2)L^{(1)} \cdot (X(1))$	$T(1,3,1) = 0$
$T(1,1,2) = 0$	$T(1,3,2) = 0$
$T(1,1,3) = 0$	$T(1,3,3) = 1/X(1)$
$T(1,1,4) = 0$	$T(1,3,4) = 0$
$T(1,2,1) = 0$	$T(1,4,1) = 0$
$T(1,2,2) = 1/X(1)$	$T(1,4,2) = 0$
$T(1,2,3) = 0$	$T(1,4,3) = 0$
$T(1,2,4) = 0$	$T(1,4,4) = 0$

$$T(2,1,1) = 0$$

$$T(2,1,2) = 1/X(1)$$

$$T(2,1,3) = 0$$

$$T(2,1,4) = 0$$

$$T(2,2,1) = -E^{-L \cdot (X(1))}X(1)$$

$$T(2,2,2) = 0$$

$$T(2,2,3) = 0$$

$$T(2,2,4) = 0$$

$$T(2,3,1) = 0$$

$$T(2,3,2) = 0$$

$$T(2,3,3) = COS(X(2))/SIN(X(2))$$

$$T(2,3,4) = 0$$

$$T(2,4,1) = 0$$

$$T(2,4,2) = 0$$

$$T(2,4,3) = 0$$

$$T(2,4,4) = 0$$

$$T(3,1,1) = 0$$

$$T(3,1,2) = 0$$

$$T(3,1,3) = 1/X(1)$$

$$T(3,1,4) = 0$$

$$T(3,2,1) = 0$$

$$T(3,2,2) = 0$$

$$T(3,2,3) = COS(X(2))/SIN(X(2))$$

$$T(3,2,4) = 0$$

$$T(3,3,1) = -E^{-L \cdot (X(1))} SIN^2(X(2))X(1)$$

$$T(3,3,2) = -COS(X(2))SIN(X(2))$$

$$T(3,3,3) = 0$$

$$T(3,3,4) = 0$$

$$T(3,4,1) = 0$$

$$T(3,4,2) = 0$$

$$T(3,4,3) = 0$$

$$T(3,4,4) = 0$$

$$T(4,1,1) = 0$$

$$T(4,1,2) = 0$$

$$T(4,1,3) = 0$$

$$T(4,1,4) = 0$$

$$T(4,2,1) = 0$$

$$T(4,2,2) = 0$$

$$T(4,2,3) = 0$$

$$T(4,2,4) = 0$$

$$T(4,3,1) = 0$$

$$T(4,3,2) = 0$$

$$T(4,3,3) = 0$$

$$T(4,3,4) = 0$$

$$T(4,4,1) = 0$$

$$T(4,4,2) = 0$$

$$T(4,4,3) = 0$$

$$T(4,4,4) = 0$$

Once the Christoffel symbols of the second kind are known, the components of the Ricci tensor can be derived. The individual components are

$$R(1,1) = -L^{(1)} \cdot (X(1))/X(1)$$

$$R(1,2) = 0$$

$$R(1,3) = 0$$

$$R(1,4) = 0$$

$$R(2,1) = 0$$

$$R(2,2) = -(1/2)E^{-L \cdot (X(1))}L^{(1)} \cdot (X(1))X(1) + E^{-L \cdot (X(1))} - 1$$

$$R(2,3) = 0$$

$$R(2,4) = 0$$

$$R(3,1) = 0$$

$$R(3,2) = 0$$

$$R(3,3) = -(1/2)E^{-L \cdot (X(1))}L^{(1)} \cdot (X(1))SIN^2(X(2))X(1)$$
$$+E^{-L \cdot (X(1))}SIN^2(X(2)) - SIN^2(X(2))$$

R(3,4) = 0

R(4,1) = 0

R(4,2) = 0

R(4,3) = 0

R(4,4) = 0

$G(I,J)$ and $R(I,J)$ are both known at this stage of the program; therefore, the Ricci scalar can be obtained. It is given by the following equation:

$$R = -2E^{-L \cdot (X(1))}/X(1)^2 + 2E^{-L \cdot (X(1))}L^{(1)} \cdot (X(1))/X(1) + 2/X(1)^2$$

The preceding information is next used to obtain the field equations. The individual equations are

$$ET(1,1) = E^{-L \cdot (X(1))}/X(1)^2 + L^{(1)} \cdot (X(1))/(E^{L \cdot (X(1))})X(1)$$
$$-E^{-L \cdot (X(1))}L^{(1)} \cdot (X(1))/X(1) - 1/X(1)^2$$

ET(1,2) = 0

ET(1,3) = 0

ET(1,4) = 0

ET(2,1) = 0

$$ET(2,2) = -(1/2)E^{-L \cdot (X(1))}L^{(1)} \cdot (X(1))/X(1)$$

ET(2,3) = 0

ET(2,4) = 0

$ET(3,1) = 0$

$ET(3,2) = 0$

$ET(3,3) = -(1/2)E^{-L \cdot (X(1))}L^{(1)} \cdot (X(1))/X(1)$

$ET(3,4) = 0$

$ET(4,1) = 0$

$ET(4,2) = 0$

$ET(4,3) = 0$

$ET(4,4) = E^{-L \cdot (X(1))}/X(1)^2 - E^{-L \cdot (X(1))}L^{(1)} \cdot (X(1))/X(1) - 1/X(1)^2$

The equations of the corresponding trajectories are

$$A(1) = E^{-L \cdot (X(1))}SIN^2(X(2))X(1)V(3)^2 - (1/2)L^{(1)} \cdot (X(1))V(1)^2$$
$$+ E^{-L \cdot (X(1))}X(1)V(2)^2$$

$$A(2) = -2V(2)V(1)/X(1) + COS(X(2))SIN(X(2))V(3)^2$$

$$A(3) = -2V(1)V(3)/X(1) - 2COS(X(2))V(2)V(3)/SIN(X(2))$$

$$A(4) = 0$$

5.7 THE De SITTER MODEL

As already indicated, the only other general relativistic model that is static and homogeneous is the De Sitter universe. In the next section the Schwarzschild model will be considered. It will be found to have the same form, although not the same content, as the De Sitter model. Although the Schwarzschild universe is inhomogeneous, the implicit functions $L(x^1)$ and $M(x^1)$ that satisfy its field equations also satisfy equation (5.5.3). In view of these considerations the De Sitter model will not be formulated.

5.8 A NONHOMOGENEOUS CASE

The Schwarzschild model represents a specially important application of relativity theory, since it provides a treatment of the gravitational field surrounding the sun. This problem was first studied by Schwarzschild in 1916, and the results obtained were used to distinguish between the predictions of the Newtonian theory of gravitation and the more exact predictions of relativity theory. Since the space surrounding the sun is assumed to be devoid of matter, all the components of the energy momentum tensor are zero. In this case, the field equations have been shown to satisfy equation (5.1.12), that is

$$R_{ij} = 0 \qquad (5.8.1)$$

Therefore, the components of the Ricci tensor obtained for the anisotropic model and satisfying equation (5.8.1) yield the components of the potential function for the field surrounding a single attracting mass, which is spherically symmetric.

In terms of conventional mathematical symbolism, the Schwarzschild field equations assume the following form:

$$R_{11} = \left[\frac{1}{2} \frac{d^2 M}{dx^1 dx^1} - \frac{1}{4} \frac{dL}{dx^1} \frac{dM}{dx^1} + \frac{1}{4} \left(\frac{dM}{dx^1} \right)^2 - \frac{1}{x^1} \frac{dL}{dx^1} \right] = 0 \qquad (5.8.2)$$

$$R_{22} = \left\{ e^{-L} \left[1 + \frac{1}{2} x^1 \left(\frac{dM}{dx^1} - \frac{dL}{dx^1} \right) \right] - 1 \right\} = 0 \qquad (5.8.3)$$

$$R_{33} = \sin^2 x^2 \left\{ e^{-L} \left[1 + \frac{x^1}{2} \left(\frac{dM}{dx^1} - \frac{dL}{dx^1} \right) \right] - 1 \right\} = 0 \qquad (5.8.4)$$

$$R_{44} = e^{M-L} \left[\frac{1}{4} \frac{dL}{dx^1} \frac{dM}{dx^1} - \frac{1}{2} \frac{d^2 M}{dx^1 dx^1} - \frac{1}{x^1} \frac{dM}{dx^1} - \frac{1}{4} \left(\frac{dM}{dx^1} \right)^2 \right] = 0 \qquad (5.8.5)$$

$$R_{ij} = 0 \quad \text{for} \quad i \neq j$$

The corresponding trajectory equations are

$$\frac{d^2x^1}{ds^2} = \left[-\frac{1}{2} \frac{dL}{dx^1} \left(\frac{dx^1}{ds}\right)^2 + e^{-L} x^1 \left(\frac{dx^2}{ds}\right)^2 + x^1 \left(\frac{dx^3}{ds}\right)^2 e^{-L} \sin^2 x^2 \right.$$

$$\left. -\frac{1}{2} e^{M-L} \frac{dM}{dx^1} \left(\frac{dx^4}{ds}\right)^2 \right] \qquad (5.8.6)$$

$$\frac{d^2x^2}{ds^2} = \left[-\frac{2}{x^1} \frac{dx^1}{ds} \frac{dx^2}{ds} + \sin x^2 \cos x^2 \left(\frac{dx^3}{ds}\right)^2 \right] \qquad (5.8.7)$$

$$\frac{d^2x^3}{ds^2} = \left(-\frac{2}{x^1} \frac{dx^1}{ds} \frac{dx^3}{ds} - 2 \cot x^2 \frac{dx^2}{ds} \frac{dx^3}{ds} \right) \qquad (5.8.8)$$

$$\frac{d^2x^4}{ds^2} = \left(-\frac{dM}{dx^1} \frac{dx^1}{ds} \frac{dx^4}{ds} \right) \qquad (5.8.9)$$

It is seen that

$$R_{33} = \sin^2 x^2 R_{22}$$

and there are, therefore, only three equations in L and M. In this connection, it should be noted that the 10 equations given by equation (5.1.8) or (5.1.12) are not all independent since, theoretically at least, they would then determine completely the metric tensor and would restrict the choice of reference system. Therefore, there can be no more than six independent conditions between the components of R_{ij} to permit a free choice of coordinate system in four-dimensional space.

The system of 10 nonlinear partial differential equations

$$R_{ij} = 0$$

for the 10 unknown functions g_{ij} is very complicated. The general solution of this system is not known. However, for the case considered in this section, it is possible to obtain a closed-form solution. It can easily be deduced that

$$L = -M$$

and

$$e^M = 1 + \frac{a}{x^1} = e^{-L}$$

Hence,

$$g_{11} = -\frac{1}{[1 + (a/x^1)]}$$

$$g_{22} = -(x^1)^2$$

$$g_{33} = -(x^1 \sin x^2)^2 \qquad\qquad (5.8.10)$$

$$g_{44} = 1 + \frac{a}{x^1}$$

If $a = -2m$, the metric (5.8.10) is consistent with the existence of one gravitating mass m situated at the origin and surrounded by empty space.

If the metric tensor inputs (5.2.1) consisting of unknown functions of x^1 are now replaced by the known functions (5.8.10), and the program rerun, the trajectory equations are obtained in the following form:

$$\left\{ \frac{d^2 x^1}{ds^2} + \frac{a}{2(x^1)^2[1 + (a/x^1)]} \left(\frac{dx^1}{ds}\right)^2 - x^1 \left(1 + \frac{a}{x^1}\right)\left(\frac{dx^2}{ds}\right)^2 \right.$$

$$\left. - x^1 \sin^2 x^2 \left(1 + \frac{a}{x^1}\right)\left(\frac{dx^3}{ds}\right)^2 - \frac{a[1 + (a/x^1)]}{2(x^1)^2}\left(\frac{dx^4}{ds}\right)^2 \right\} = 0 \qquad (5.8.11)$$

$$\left[\frac{d^2 x^2}{ds^2} + \frac{2}{x^1} \frac{dx^1}{ds} \frac{dx^2}{ds} - \sin x^2 \cos x^2 \left(\frac{dx^3}{ds}\right)^2 \right] = 0 \qquad (5.8.12)$$

$$\left(\frac{d^2 x^3}{ds^2} + \frac{2}{x^1} \frac{dx^1}{ds} \frac{dx^3}{ds} + 2 \cot x^2 \frac{dx^2}{ds} \frac{dx^3}{ds} \right) = 0 \qquad (5.8.13)$$

$$\left\{ \frac{d^2 x^4}{ds^2} - \frac{a}{(x^1)^2[1 + (a/x^1)]} \frac{dx^1}{ds} \frac{dx^4}{ds} \right\} = 0 \qquad (5.8.14)$$

5.9 CONCLUDING COMMENTS

Symbolic mathematical computation can facilitate the formulation of mathematical models. This has been demonstrated by using the method to formulate several cosmological models and their associated trajectory equations. It has been shown that such models can be derived with speed and efficiency on present generation computers, provided they are equipped with formula manipulation compilers. For example, in the case of the Einstein and De Sitter models, the computer times required to formulate the field and the trajectory equations were 0.66 and 0.32 min, respectively. For the nonhomogeneous Schwarzschild model, the corresponding times were 0.74 and 0.30 min, respectively. In addition to saving time and eliminating the errors to which humans are prone, the method facilitates the study of a greater variety of models.

5.10 REFERENCES

1. Sokolnikoff, Ivan S.: Tensor Analysis; Theory and Applications. John Wiley & Sons, Inc., 1960.

2. McConnell, Albert J.: Applications of Tensor Analysis. Dover Publications, Inc., 1957.

3. Spain, Barry: Tensor Calculus. Third ed., revised. Oliver and Boyd, Edinburgh, 1960.

4. Goldstein, Herbert: Classical Mechanics. Addison-Wesley Publishing Company, Inc., 1959.

5. Tolman, Richard C.: Relativity, Thermodynamics and Cosmology. Clarendon Press, Oxford, 1958.

ALTERNATIVE TECHNIQUES

6.1 REDUCE AND MACSYMA

A variety of other languages is available for carrying out symbolic manipulations. The choice of language depends on accessibility, personal preference, the type and magnitude of the models to be formulated, and the computer facilities available to the user. At the time of writing, the two most important contenders in the symbolic manipulation field appeared to the author to be REDUCE and MACSYMA (ref. 1). REDUCE is a language which was developed by Professor Anthony Hearn of the University of Utah. It is designed for general algebraic computations of interest to mathematicians, physicists, and engineers. In addition to the usual algebraic manipulations, it has the capability of performing calculations of special interest to high energy physicists.

REDUCE, in one form or another, has been available for over 10 years. Originally it began as a system for solving special problems that arise in high energy physics, where much tedious repetitive calculation is involved. However, it was quickly recognized that the simplification processes being used were quite general, and in 1967 REDUCE was announced as a system for general purpose algebraic simplification and released for distribution.

Although REDUCE can operate in a batch processing mode, it is intended primarily for interactive calculations in a time-shared environment. Hence, it is command-oriented rather than program-oriented, since the result of a given command may be required before proceeding to the next step. Since REDUCE is well known to computer users and is available for use on most IBM 360 or 370 series computers, the DEC PDP-10 and the CDC 6400, 6500, 6600, and 7600 machines, it will not be discussed further but, instead, the MACSYMA system, which is less well known, will be examined.

At the time of writing, the MACSYMA system was available only at MIT through the ADVANCED RESEARCH and PROJECT AGENCY (ARPA) network. Since

MACSYMA appears to the author to be a very promising system, which is flexible and continuously evolving to meet the needs of users, it will be demonstrated by using it to reformulate some of the problems considered in previous chapters. It is a large computer programming system which can be used to perform symbolic as well as numerical mathematical computations. It was developed by the MATHLAB group of project MAC's Automatic Programming Division specifically for interactive use, and has capabilities for manipulating algebraic expressions involving numbers, variables, and functions. It can differentiate, integrate, take limits, solve systems of linear or polynomial equations, factor polynomials, expand functions, plot curves, and manipulate matrices. Since, however, the tensor operations contemplated here only require differentiation and summation, attention will be confined to these two operations.

6.2 USE OF MACSYMA TO TRANSFORM AERODYNAMIC STABILITY DERIVATIVES

In chapter 2, section 2.7, it was deduced that the aerodynamic stability derivatives transform like the components of a mixed tensor, having one index of contravariance and one index of covariance. Moreover, due to the equivalence of covariant and contravariant transformations in orthogonal Cartesian systems of coordinates, the transformations can be treated as doubly covariant or doubly contravariant, if this simplifies the formulation (see sec. 2.8.2). When the doubly contravariant form is used, the transformation law assumes the following form (ref. 2):

$$Y^{ij} = \frac{\partial y^i}{\partial x^\alpha} \frac{\partial y^j}{\partial x^\beta} X^{\alpha\beta} \tag{6.2.1}$$

where the first index denotes the component of the aerodynamic force or moment, and the second index denotes the component of the motion vector with respect to which the derivatives are obtained.

When $y = y(x)$ is specified, equation (6.2.1) can be evaluated. Let it be assumed that a transformation from wind axes to body axes is required (see sec. 2.2). In this case the transformation equations are

$$\left.\begin{aligned}
y^1 &= x^1 \cos A \cos B - x^2 \cos A \sin B - x^3 \sin A \\
y^2 &= x^1 \sin B + x^2 \cos B \\
y^3 &= x^1 \sin A \cos B - x^2 \sin A \sin B + x^3 \cos A
\end{aligned}\right\} \tag{6.2.2}$$

As noted previously, the angles A and B correspond to the angle of attack and the angle of sideslip, respectively.

Implementation of the technique of symbolic mathematical computation requires that equation (6.2.2) be used as input to a computer program to permit expansion of equation (6.2.1). To illustrate the technique of using MACSYMA interactively, let us carry out the steps involved in expanding equation (6.2.1).

When a user has established communication with the system, MACSYMA responds by typing the label $C(1)$, which means that the system is ready to accept the first command from the user. The user than types a command or statement in the MACSYMA input language. The first three input statements are the three equations (6.2.2). The MACSYMA system requires that these be given in the following modified form, where asterisks again denote multiplication:

```
(C1)  Y[1]:X[1]*COS(A)*COS(B)-X[2]*COS(A)*SIN(B)-X[3]*SIN(A)$

(C2)  Y[2]:X[1]*SIN(B)+X[2]*COS(B)$

(C3)  Y[3]:X[1]*SIN(A)*COS(B)-X[2]*SIN(A)*SIN(B)+X[3]*COS(A)$
```

Note that the corresponding FORMAC input statements employed the conventional equality sign and enclosed the identifying indices in parentheses, whereas MACSYMA replaces the equal sign with a colon and encloses the indices in brackets. Again, it will be seen that although the FORMAC indices retain the parentheses when printed out, MACSYMA prints the output statements in conventional subscripted form.

When the user has finished typing the three transformation equations in the MACSYMA input language, the system prompts the user by typing another command label. In this case MACSYMA types the label $C(4)$. The same notation for the aerodynamic parameters is used in this program as was used previously; that is, $C(I,J)$ denotes an aerodynamic stability derivative and $TC(I,J)$ denotes a transformed derivative. The simplicity of the program is evident from the fact that only three additional program steps are required: (1) an initializing command, (2) a DO loop, and (3) a DISPLAY command which replaces the printout statement in the FORMAC program.

The following program and the displayed output are taken from actual computer printout:

```
(C4) TC[I,J]:=0$

(C5) FOR I:1 THRU 3 DO FOR J:1 THRU 3 DO
 FOR M:1 THRU 3 DO FOR N:1 THRU 3 DO
 TC[I,J]:TC[I,J]+DIFF(Y[I],X[M])*DIFF(Y[J],X[N])*C[M,N]$

(C6) FOR I:1 THRU 3 DO FOR J:1 THRU 3 DO
 DISPLAY(TC[I,J])$
```

$$TC_{1,1} = C_{2,2} \cos^2(A) \sin^2(B) - C_{2,1} \cos^2(A) \cos(B) \sin(B)$$

$$- C_{1,2} \cos^2(A) \cos(B) \sin(B) + C_{3,2} \cos(A) \sin(A) \sin(B)$$

$$+ C_{2,3} \cos(A) \sin(A) \sin(B) + C_{1,1} \cos^2(A) \cos^2(B)$$

$$- C_{3,1} \cos(A) \sin(A) \cos(B) - C_{1,3} \cos(A) \sin(A) \cos(B)$$

$$+ C_{3,3} \sin^2(A)$$

$$TC_{1,2} = - C_{2,1} \cos(A) \sin^2(B) - C_{2,2} \cos(A) \cos(B) \sin(B)$$

$$+ C_{1,1} \cos(A) \cos(B) \sin(B) - C_{3,1} \sin(A) \sin(B)$$

$$+ C_{1,2} \cos(A) \cos^2(B) - C_{3,2} \sin(A) \cos(B)$$

$$TC_{1,3} = C_{2,2} \cos(A) \sin(A) \sin^2(B)$$

$$- C_{2,1} \cos(A) \sin(A) \cos(B) \sin(B)$$

$$- C_{1,2} \cos(A) \sin(A) \cos(B) \sin(B) + C_{3,2} \sin^2(A) \sin(B)$$

$$- C_{2,3} \cos^2(A) \sin(B) + C_{1,1} \cos(A) \sin(A) \cos^2(B)$$

$$- C_{3,\ 1}\ SIN^2(A)\ COS(B) + C_{1,\ 3}\ COS^2(A)\ COS(B)$$

$$- C_{3,\ 3}\ COS(A)\ SIN(A)$$

$$TC_{2,\ 1} = - C_{1,\ 2}\ COS(A)\ SIN^2(B) - C_{2,\ 2}\ COS(A)\ COS(B)\ SIN(B)$$

$$+ C_{1,\ 1}\ COS(A)\ COS(B)\ SIN(B) - C_{1,\ 3}\ SIN(A)\ SIN(B)$$

$$+ C_{2,\ 1}\ COS(A)\ COS^2(B) - C_{2,\ 3}\ SIN(A)\ COS(B)$$

$$TC_{2,\ 2} = C_{1,\ 1}\ SIN^2(B) + C_{2,\ 1}\ COS(B)\ SIN(B)$$

$$+ C_{1,\ 2}\ COS(B)\ SIN(B) + C_{2,\ 2}\ COS^2(B)$$

$$TC_{2,\ 3} = - C_{1,\ 2}\ SIN(A)\ SIN^2(B) - C_{2,\ 2}\ SIN(A)\ COS(B)\ SIN(B)$$

$$+ C_{1,\ 1}\ SIN(A)\ COS(B)\ SIN(B) + C_{1,\ 3}\ COS(A)\ SIN(B)$$

$$+ C_{2,\ 1}\ SIN(A)\ COS^2(B) + C_{2,\ 3}\ COS(A)\ COS(B)$$

$$TC_{3,\ 1} = C_{2,\ 2}\ COS(A)\ SIN(A)\ SIN^2(B)$$

$$- C_{2,\ 1}\ COS(A)\ SIN(A)\ COS(B)\ SIN(B)$$

$$- C_{1,\ 2}\ COS(A)\ SIN(A)\ COS(B)\ SIN(B) + C_{2,\ 3}\ SIN^2(A)\ SIN(B)$$

$$- C_{3,\ 2}\ COS^2(A)\ SIN(B) + C_{1,\ 1}\ COS(A)\ SIN(A)\ COS^2(B)$$

$$- C_{1,\ 3}\ SIN^2(A)\ COS(B) + C_{3,\ 1}\ COS^2(A)\ COS(B)$$

$$- C_{3,\ 3}\ COS(A)\ SIN(A)$$

.3

$$TC_{3,\,2} = -\,C_{2,\,1}\ SIN(A)\ SIN^2(B) - C_{2,\,2}\ SIN(A)\ COS(B)\ SIN(B)$$

$$+\ C_{1,\,1}\ SIN(A)\ COS(B)\ SIN(B) + C_{3,\,1}\ COS(A)\ SIN(B)$$

$$+\ C_{1,\,2}\ SIN(A)\ COS^2(B) + C_{3,\,2}\ COS(A)\ COS(B)$$

$$TC_{3,\,3} = C_{2,\,2}\ SIN^2(A)\ SIN^2(B) - C_{2,\,1}\ SIN^2(A)\ COS(B)\ SIN(B)$$

$$-\ C_{1,\,2}\ SIN^2(A)\ COS(B)\ SIN(B) - C_{3,\,2}\ COS(A)\ SIN(A)\ SIN(B)$$

$$-\ C_{2,\,3}\ COS(A)\ SIN(A)\ SIN(B) + C_{1,\,1}\ SIN^2(A)\ COS^2(B)$$

$$+\ C_{3,\,1}\ COS(A)\ SIN(A)\ COS(B) + C_{1,\,3}\ COS(A)\ SIN(A)\ COS(B)$$

$$+\ C_{3,\,3}\ COS^2(A)$$

6.3 PRINTOUT OF INPUT COMMANDS

If the user wishes to have any input command printed out, this can be accomplished by terminating each command with a semicolon (;) instead of a dollar sign ($). For example, when the input commands to the preceding program for the transformation of aerodynamic stability derivatives are terminated in this manner, the results appear as follows:

(C1) Y[1]:X[1]*COS(A)*COS(B)-X[2]*COS(A)*SIN(B)-X[3]*SIN(A);

(D1) $-\ X_2\ COS(A)\ SIN(B) + X_1\ COS(A)\ COS(B) - X_3\ SIN(A)$

(C2) Y[2]:X[1]*SIN(B)+X[2]*COS(B);

(D2) $X_1\ SIN(B) + X_2\ COS(B)$

(C3) Y[3]:X[1]*SIN(A)*COS(B)-X[2]*SIN(A)*SIN(B)+X[3]*COS(A);

(D3) $-\ X_2\ SIN(A)\ SIN(B) + X_1\ SIN(A)\ COS(B) + X_3\ COS(A)$

6.4 CENTRAL PROCESSING UNIT TIMES

MACSYMA makes it easy for the user to obtain the central processing unit (CPU) time required to perform each operation. This is a convenient facility that enables the programmer to keep track of costs.

By typing the command

$$SHOWTIME:TRUE\$$$

the CPU time is printed out after each operation. When this command is used in the preceding program, the time required for each operation is printed out as follows:

```
(C1) SHOWTIME:TRUE$

time= 1 msec.

(C2) Y[1]:X[1]*COS(A)*COS(B)-X[2]*COS(A)*SIN(B)-X[3]*SIN(A)$

time=22 msec.

(C3) Y[2]:X[1]*SIN(B)+X[2]*COS(B)$

time= 8 msec.

(C4) Y[3]:X[1]*SIN(A)*COS(B)-X[2]*SIN(A)*SIN(B)+X[3]*COS(A)$

time= 17 msec.

(C5) TC[I,J]:=0$

time= 2 msec.

(C6) FOR I:1 THRU 3 DO FOR J:1 THRU 3 DO
 FOR M:1 THRU 3 DO FOR N:1 THRU 3 DO
 TC[I,J]:TC[I,J]+DIFF(Y[I],X[M])*DIFF(Y[J],X[N])*C[M,N]$

time=2247 msec.

(C7) FOR I:1 THRU 3 DO FOR J:1 THRU 3 DO DISPLAY(TC[I,J])$
```

$$TC_{1,\,1} = C_{2,\,2}\ \cos^2(A)\ \sin^2(B) - C_{2,\,1}\ \cos^2(A)\ \cos(B)\ \sin(B)$$

$$- C_{1,\,2}\ \cos^2(A)\ \cos(B)\ \sin(B) + C_{3,\,2}\ \cos(A)\ \sin(A)\ \sin(B)$$

$$+ C_{2,\,3}\ \cos(A)\ \sin(A)\ \sin(B) + C_{1,\,1}\ \cos^2(A)\ \cos^2(B)$$

$$- C_{3,\,1}\ \cos(A)\ \sin(A)\ \cos(B) - C_{1,\,3}\ \cos(A)\ \sin(A)\ \cos(B)$$

$$+ C_{3,\,3}\ \sin^2(A)$$

$$TC_{1,\,2} = - C_{2,\,1}\ \cos(A)\ \sin^2(B) - C_{2,\,2}\ \cos(A)\ \cos(B)\ \sin(B)$$

$$+ C_{1,\,1}\ \cos(A)\ \cos(B)\ \sin(B) - C_{3,\,1}\ \sin(A)\ \sin(B)$$

$$+ C_{1,\,2}\ \cos(A)\ \cos^2(B) - C_{3,\,2}\ \sin(A)\ \cos(B)$$

$$TC_{1,\,3} = C_{2,\,2}\ \cos(A)\ \sin(A)\ \sin^2(B)$$

$$- C_{2,\,1}\ \cos(A)\ \sin(A)\ \cos(B)\ \sin(B)$$

$$- C_{1,\,2}\ \cos(A)\ \sin(A)\ \cos(B)\ \sin(B) + C_{3,\,2}\ \sin^2(A)\ \sin(B)$$

$$- C_{2,\,3}\ \cos^2(A)\ \sin(B) + C_{1,\,1}\ \cos(A)\ \sin(A)\ \cos^2(B)$$

$$- C_{3,\,1}\ \sin^2(A)\ \cos(B) + C_{1,\,3}\ \cos^2(A)\ \cos(B)$$

$$- C_{3,\,3}\ \cos(A)\ \sin(A)$$

$$TC_{2,\,1} = - C_{1,\,2}\ \cos(A)\ \sin^2(B) - C_{2,\,2}\ \cos(A)\ \cos(B)\ \sin(B)$$

$$+ C_{1,\,1}\ \cos(A)\ \cos(B)\ \sin(B) - C_{1,\,3}\ \sin(A)\ \sin(B)$$

$$+ C_{2,\,1}\ \cos(A)\ \cos^2(B) - C_{2,\,3}\ \sin(A)\ \cos(B)$$

$$TC_{2,\,2} = C_{1,\,1}\,\text{SIN}^2(B) + C_{2,\,1}\,\text{COS}(B)\,\text{SIN}(B) + C_{1,\,2}\,\text{COS}(B)\,\text{SIN}(B)$$
$$+\ C_{2,\,2}\,\text{COS}^2(B)$$

$$TC_{2,\,3} = -\ C_{1,\,2}\,\text{SIN}(A)\,\text{SIN}^2(B) - C_{2,\,2}\,\text{SIN}(A)\,\text{COS}(B)\,\text{SIN}(B)$$
$$+\ C_{1,\,1}\,\text{SIN}(A)\,\text{COS}(B)\,\text{SIN}(B) + C_{1,\,3}\,\text{COS}(A)\,\text{SIN}(B)$$
$$+\ C_{2,\,1}\,\text{SIN}(A)\,\text{COS}^2(B) + C_{2,\,3}\,\text{COS}(A)\,\text{COS}(B)$$

$$TC_{3,\,1} = C_{2,\,2}\,\text{COS}(A)\,\text{SIN}(A)\,\text{SIN}^2(B)$$
$$-\ C_{2,\,1}\,\text{COS}(A)\,\text{SIN}(A)\,\text{COS}(B)\,\text{SIN}(B)$$
$$-\ C_{1,\,2}\,\text{COS}(A)\,\text{SIN}(A)\,\text{COS}(B)\,\text{SIN}(B) + C_{2,\,3}\,\text{SIN}^2(A)\,\text{SIN}(B)$$
$$-\ C_{3,\,2}\,\text{COS}^2(A)\,\text{SIN}(B) + C_{1,\,1}\,\text{COS}(A)\,\text{SIN}(A)\,\text{COS}^2(B)$$
$$-\ C_{1,\,3}\,\text{SIN}^2(A)\,\text{COS}(B) + C_{3,\,1}\,\text{COS}^2(A)\,\text{COS}(B)$$
$$-\ C_{3,\,3}\,\text{COS}(A)\,\text{SIN}(A)$$

$$TC_{3,\,2} = -\ C_{2,\,1}\,\text{SIN}(A)\,\text{SIN}^2(B) - C_{2,\,2}\,\text{SIN}(A)\,\text{COS}(B)\,\text{SIN}(B)$$
$$+\ C_{1,\,1}\,\text{SIN}(A)\,\text{COS}(B)\,\text{SIN}(B) + C_{3,\,1}\,\text{COS}(A)\,\text{SIN}(B)$$
$$+\ C_{1,\,2}\,\text{SIN}(A)\,\text{COS}^2(B) + C_{3,\,2}\,\text{COS}(A)\,\text{COS}(B)$$

$$TC_{3,\,3} = C_{2,\,2}\, SIN^2(A)\, SIN^2(B) - C_{2,\,1}\, SIN^2(A)\, COS(B)\, SIN(B)$$

$$- C_{1,\,2}\, SIN^2(A)\, COS(B)\, SIN(B) \doteq C_{3,\,2}\, COS(A)\, SIN(A)\, SIN(B)$$

$$- C_{2,\,3}\, COS(A)\, SIN(A)\, SIN(B) + C_{1,\,1}\, SIN^2(A)\, COS^2(B)$$

$$+ C_{3,\,1}\, COS(A)\, SIN(A)\, COS(B) + C_{1,\,3}\, COS(A)\, SIN(A)\, COS(B)$$

$$+ C_{3,\,3}\, COS^2(A)$$

time= 34 msec.

It should be noted that all times are given in milliseconds.

6.5 FORMULATION OF CHRISTOFFEL'S SYMBOLS

The Christoffel symbols of the first and second kinds are related to the fundamental metric tensors as shown in equations (6.5.1) and (6.5.2), respectively (ref. 3).

$$[ij,k] = \frac{1}{2}\left(\frac{\partial g_{ik}}{\partial x^j} + \frac{\partial g_{jk}}{\partial x^i} - \frac{\partial g_{ij}}{\partial x^k}\right) \tag{6.5.1}$$

$$\left\{\begin{matrix}k\\ij\end{matrix}\right\} = g^{kl}[ij,l] \tag{6.5.2}$$

Given the metric tensors for the coordinate system being considered, it is a simple matter to program the formulation of these symbols.

As in the previous example, only three program commands are required, that is, an initializing command, a DO command, and a DISPLAY command.

Since the Christoffel symbols of the second kind are required in the formulation of mathematical models, and will be used in later applications, a MACSYMA program will be written to derive them.

The first step in this program, as in the corresponding FORMAC program, is to input the metric tensors. In this application both the covariant and the contravariant form of the metric tensors are required. Since the applications being considered are confined to orthogonal coordinate systems, it is possible to use the following simplifications:

$$g^{ij} = g_{ij} = 0 \quad \text{for} \quad i \neq j$$

$$g^{ij} = \frac{1}{g_{ij}} \quad \text{for} \quad i = j$$

that is

$$g^{(ii)} = \frac{1}{g_{(ii)}}$$

where a subscript in parentheses denotes suspension of the summation convention.

MACSYMA will accept the metric tensor inputs if the following substitutions are made:

$$g_{ii} = G[I,I]$$

$$g^{ii} = H[I,I]$$

$$\left\{ \begin{matrix} k \\ ij \end{matrix} \right\} = T[K,I,J]$$

The following notation will be used by MACSYMA to display the formulated Christoffel symbols:

$$T_{K,I,J} = \left\{ \begin{matrix} k \\ ij \end{matrix} \right\}$$

The computer program and the formulated Christoffel symbols for a cylindrical polar coordinate system assume the following form:

```
(C1) SHOWTIME:TRUE$
time= 1 msec.
```

```
(C2) G[1,1]:1$
time= 3 msec.

(C3) G[2,2]:X[1]**2$
time= 4 msec.

(C4) G[3,3]:1$
time= 1 msec.

(C5) FOR I: 1 THRU 3 DO H[I,I]:1/G[I,I]$
time= 17 msec.

(C6) T[I,J,K]:=0$
time= 2 msec.

(C7) FOR I:1 THRU 3 DO FOR J:1 THRU 3 DO FOR K:1 THRU 3 DO
 T[I,J,K]:T[I,J,K]+H[I,I]*(DIFF(G[J,I],X[K])+DIFF(G[K,I],X[J])
-DIFF(G[J,K],X[I]))/2$
time= 794 msec.

(C8) FOR I:1 THRU 3 DO FOR J:1 THRU 3 DO
 FOR K:1 THRU 3 DO DISPLAY(T[I,J,K])$
```

$$T_{1, 1, 1} = 0 \qquad\qquad T_{1, 3, 1} = 0$$

$$T_{1, 1, 2} = 0 \qquad\qquad T_{1, 3, 2} = 0$$

$$T_{1, 1, 3} = 0 \qquad\qquad T_{1, 3, 3} = 0$$

$$T_{1, 2, 1} = 0 \qquad\qquad T_{2, 1, 1} = 0$$

$$T_{1, 2, 2} = - X_1 \qquad\qquad T_{2, 1, 2} = \frac{1}{X_1}$$

$$T_{1, 2, 3} = 0$$

$$T_{2, 1, 3} = 0$$

$$T_{2, 2, 1} = \frac{1}{x_1}$$

$$T_{3, 1, 3} = 0$$

$$T_{2, 2, 2} = 0$$

$$T_{3, 2, 1} = 0$$

$$T_{2, 2, 3} = 0$$

$$T_{3, 2, 2} = 0$$

$$T_{2, 3, 1} = 0$$

$$T_{3, 2, 3} = 0$$

$$T_{2, 3, 2} = 0$$

$$T_{3, 3, 1} = 0$$

$$T_{2, 3, 3} = 0$$

$$T_{3, 3, 2} = 0$$

$$T_{3, 1, 1} = 0$$

$$T_{3, 3, 3} = 0$$

$$T_{3, 1, 2} = 0$$

time= 104 msec.

The same program may be used to formulate the Christoffel symbols of the second kind for a spherical polar coordinate system. The only difference is that the metric tensor inputs must be changed to correctly describe an element of arc in this system, that is

$$ds^2 = (dx^1)^2 + (x^1 dx^2)^2 + (x^1 \sin x^2 dx^3)^2$$

where x^1 is the radial distance, x^2 the polar angle, and x^3 the azimuth. Therefore

$$G[1,1] = 1$$

$$G[2,2] = (x^1)^2$$

$$G[3,3] = (x^1 \sin x^2)^2$$

With these inputs, the program and the corresponding Christoffel symbols assume the following form:

```
(C1) SHOWTIME:TRUE$
time= 1 msec.

(C2) G[1,1]:1$
time= 3 msec.

(C3) G[2,2]:X[1]**2$
time= 4 msec.

(C4) G[3,3]:(X[1]*SIN(X[2]))**2$
time= 9 msec.

(C5) FOR I:1 THRU 3 DO H[I,I]:1/G[I,I]$
time= 20 msec.

(C6) T[I,J,K]:=0$
time= 2 msec.

(C7) FOR I:1 THRU 3 DO FOR J:1 THRU 3 DO FOR K:1 THRU 3 DO
 T[I,J,K]:T[I,J,K]+H[I,I]*(DIFF(G[J,I],X[K])+DIFF(G[K,I],X[J])
-DIFF(G[J,K],X[I]))/2$
time= 843 msec.

(C8) FOR I:1 THRU 3 DO FOR J:1 THRU 3 DO
 FOR K:1 THRU 3 DO DISPLAY(T[I,J,K])$
```

$$T_{1,1,1} = 0 \qquad\qquad T_{1,2,2} = -X_1$$

$$T_{1,1,2} = 0 \qquad\qquad T_{1,2,3} = 0$$

$$T_{1,1,3} = 0 \qquad\qquad T_{1,3,1} = 0$$

$$T_{1,2,1} = 0 \qquad\qquad T_{1,3,2} = 0$$

$$T_{1,\,3,\,3} = -X_1 \, SIN^2(X_2)$$

$$T_{2,\,1,\,1} = 0$$

$$T_{2,\,1,\,2} = \frac{1}{X_1}$$

$$T_{2,\,1,\,3} = 0$$

$$T_{2,\,2,\,1} = \frac{1}{X_1}$$

$$T_{2,\,2,\,2} = 0$$

$$T_{2,\,2,\,3} = 0$$

$$T_{2,\,3,\,1} = 0$$

$$T_{2,\,3,\,2} = 0$$

$$T_{2,\,3,\,3} = -COS(X_2) \, SIN(X_2)$$

$$T_{3,\,1,\,1} = 0$$

$$T_{3,\,1,\,2} = 0$$

$$T_{3,\,1,\,3} = \frac{1}{X_1}$$

$$T_{3,\,2,\,1} = 0$$

$$T_{3,\,2,\,2} = 0$$

$$T_{3,\,2,\,3} = \frac{COS(X_2)}{SIN(X_2)}$$

$$T_{3, 3, 1} = \frac{1}{X_1}$$

$$T_{3, 3, 2} = \frac{COS(X_2)}{SIN(X_2)}$$

$$T_{3, 3, 3} = 0$$

```
time= 104 msec.
```

A Christoffel symbol which occurs in cosmological studies is derived from metric tensors that are exponential functions of the coordinates. The formulation of Christoffel symbols of this type will be used to demonstrate the method employed by MACSYMA to differentiate a function of a variable. In MACSYMA, functional dependence can be declared by using a DEPENDENCIES function. For example, since MACSYMA knows the chain rule for symbolic differentiation, the statement

```
DEPENDENCIES(F(X,Y),X(T),Y(T));
```

followed by

```
DIFF(F,T);
```

will yield

$$\frac{DF}{DY}\frac{DY}{DT} + \frac{DF}{DX}\frac{DX}{DT}$$

Moreover, it will be seen that executing DEPENDENCIES(Y(X)) will cause the differential of Y with respect to X to be displayed as DY/DX.

Apart from the use of the DEPENDENCIES function, the program required to formulate Christoffel symbols of this type is the same as before, except that the extent of each dimension is increased from 3 to 4. Hence, the range of the DO command and the DISPLAY command must be increased to account for this change.

Since the processing time increases with the complexity of the formulation, the CPU time required for each step of the operation will again be requested by typing the additional command

SHOWTIME:TRUE$

The DEPENDENCIES functions together with the components of the fundamental metric tensor and the initializing statement are

```
(C1) SHOWTIME:TRUE$
time= 1 msec.

(C2) DEPENDENCIES(L(X[1]))$
time= 1 msec.

(C3) DEPENDENCIES(M(X[1]))$
time= 1 msec.

(C4) G[1,1]:-%E**L(X[1])$
time= 12 msec.

(C5) G[2,2]:-X[1]**2$
time= 4 msec.

(C6) G[3,3]:-(X[1]*SIN(X[2]))**2$
time= 10 msec.

(C7) G[4,4]:%E**M(X[1])$
time= 5 msec.

(C8) FOR I:1 THRU 4 DO H[I,I]:1/G[I,I]$
time= 27 msec.

(C9) T[I,J,K]:=0$
time= 2 msec.
```

Note that the base of natural logarithms e is written as %E, hence the statement

$$\%E**L(x[1])$$

is equivalent to

$$e^{L(x^1)}$$

and similarly

$$\%E**M(x[1])$$

is equivalent to

$$_eM(x^1)$$

The following program statements are the same as before, except that the range of the DO command and the DISPLAY command has been increased to account for the fact that the space is now four dimensional. With these inputs, the MACSYMA program and the formulated Christoffel symbols are

```
(C10) FOR I:1 THRU 4 DO FOR J:1 THRU 4 DO
 FOR K:1 THRU 4 DO T[I,J,K]:T[I,J,K]
 +H[I,I]*(DIFF(G[J,I],X[K])+DIFF(G[K,I],X[J])-DIFF(G[J,K],X[I]))/2$
time= 2155 msec.

(C11) FOR I:1 THRU 4 DO FOR J:1 THRU 4 DO
 FOR K:1 THRU 4 DO DISPLAY(T[I,J,K]);
```

$$T_{1,1,1} = \frac{\frac{d}{dX_1}L(X_1)}{2}$$

$$T_{1,1,2} = 0$$

$$T_{1,1,3} = 0$$

$$T_{1,1,4} = 0$$

$$T_{1,2,1} = 0$$

$$T_{1,2,2} = -X_1 \%E^{-L(X_1)}$$

$$T_{1,2,3} = 0$$

$$T_{1,2,4} = 0$$

$$T_{1,3,1} = 0$$

$$T_{1,3,2} = 0$$

$$T_{1,3,3} = -X_1 \%E^{-L(X_1)} SIN^2(X_2)$$

$$T_{1,3,4} = 0$$

$$T_{1,4,1} = 0$$

$$T_{1,4,2} = 0$$

$$T_{1,4,3} = 0$$

$$T_{1,4,4} = \frac{\%E^{M(X_1) - L(X_1)} (-\frac{d}{dX_1} M(X_1))}{2}$$

$$T_{2,1,1} = 0$$

$$T_{2,1,2} = \frac{1}{X_1}$$

$$T_{2,1,3} = 0$$

$$T_{2, 1, 4} = 0$$

$$T_{2, 2, 1} = \frac{1}{X_1}$$

$$T_{2, 2, 2} = 0$$

$$T_{2, 2, 3} = 0$$

$$T_{2, 2, 4} = 0$$

$$T_{2, 3, 1} = 0$$

$$T_{2, 3, 2} = 0$$

$$T_{2, 3, 3} = - COS(X_2) \; SIN(X_2)$$

$$T_{2, 3, 4} = 0$$

$$T_{2, 4, 1} = 0$$

$$T_{2, 4, 2} = 0$$

$$T_{2, 4, 3} = 0$$

$$T_{2, 4, 4} = 0$$

$$T_{3, 1, 1} = 0$$

$$T_{3, 1, 2} = 0$$

$$T_{3, 1, 3} = \frac{1}{X_1}$$

$$T_{3, 1, 4} = 0$$

$$T_{3, 2, 1} = 0$$

$$T_{3, 2, 2} = 0$$

$$T_{3, 2, 3} = \frac{COS(X_2)}{SIN(X_2)}$$

$$T_{3, 2, 4} = 0$$

$$T_{3, 3, 1} = \frac{1}{X_1}$$

$$T_{3, 3, 2} = \frac{COS(X_2)}{SIN(X_2)}$$

$$T_{3, 3, 3} = 0$$

$$T_{3, 3, 4} = 0$$

$$T_{3, 4, 1} = 0$$

$$T_{3, 4, 2} = 0$$

$$T_{3, 4, 3} = 0$$

$$T_{3, 4, 4} = 0$$

$$T_{4, 1, 1} = 0$$

$$T_{4, 1, 2} = 0$$

$$T_{4, 1, 3} = 0$$

$$T_{4, 1, 4} = \frac{\dfrac{d}{dX_1} M(X_1)}{2}$$

$$T_{4, 2, 1} = 0$$

$$T_{4, 2, 2} = 0$$

$$T_{4, 2, 3} = 0$$

$$T_{4, 2, 4} = 0$$

$$T_{4, 3, 1} = 0$$

$$T_{4, 3, 2} = 0$$

$$T_{4, 3, 3} = 0$$

$$T_{4, 3, 4} = 0$$

$$T_{4,\,4,\,1} = \frac{\dfrac{d}{dX_1} M(X_1)}{2}$$

$$T_{4,\,4,\,2} = 0$$

$$T_{4,\,4,\,3} = 0$$

$$T_{4,\,4,\,4} = 0$$

```
time=519 msec.
```

6.6 EQUATIONS OF MOTION OF A PARTICLE

A form of the equations of motion of a particle which is valid in all orthogonal curvilinear coordinate systems is given in section 3.5, equation (3.5.9). For the convenience of readers this equation is reproduced here (ref. 4)

$$M\left(\frac{d^2x^i}{dt^2} + \left\{\begin{matrix} i \\ jk \end{matrix}\right\} \frac{dx^j}{dt}\frac{dx^k}{dt}\right) = \frac{1}{g_{(ii)}}\frac{\partial \phi}{\partial x^i} + \frac{\tau^i}{\sqrt{g_{(ii)}}} \qquad (6.6.1)$$

If the components d^2x^i/dt^2 be denoted by $A[I]$ and the component

$$\left\{\begin{matrix} i \\ jk \end{matrix}\right\} \frac{dx^j}{dt}\frac{dx^k}{dt} = \left\{\begin{matrix} i \\ jk \end{matrix}\right\} v^j v^k$$

be denoted by $R[I]$, then a program to formulate the equations of motion of a particle in a given coordinate system would proceed as follows.

The first and most important step in the formulation of equation (6.6.1) is the determination of the components of the Christoffel symbols, given the components of the fundamental metric tensor. The facility with which MACSYMA can formulate the Christoffel symbols was demonstrated in the preceding section, where the notation

$$T_{i,j,k} = \left\{\begin{matrix} i \\ jk \end{matrix}\right\}$$

was used to display the formulated symbols.

A program to formulate the equations of motion of a particle in a cylindrical polar coordinate system would require, as input, the metric tensors for this system of coordinates. These are

$$(C1) \quad G[1,1]:1\$$$

$$(C2) \quad G[2,2]:X[1]**2\$$$

$$(C3) \quad G[3,3]:1\$$$

Since, for the cases being considered, the contravariant metric tensors are simply the reciprocals of the corresponding covariant metric tensors, they can be formulated as follows:

$$(C4) \quad FOR \ I:1 \ THRU \ 3 \ DO \ H[I,I]:1/G[I,I]\$$$

As already indicated, the notation $T[I,J,K]$ will be treated by MACSYMA as a Christoffel symbol of the second kind. Hence, the initializing statement and the DO command required to formulate these symbols, can be taken from any one of the three preceding programs. These two programming steps assume the following form:

$$(C5) \quad T[I,J,K]:=0\$$$

```
(C6) FOR I:1 THRU 3 DO FOR J:1 THRU 3 DO FOR K:1 THRU 3 DO
 T[I,J,K]:T[I,J,K]+H[I,I]*(DIFF(G[J,I],X[K])+DIFF(G[K,I],X[J])
 -DIFF(G[J,K],X[I]))/2$
```

The equation for $R[I]$ has already been defined, and it will be formulated by instructing MACSYMA to execute the following DO command:

$$(C7) \quad R[I]:=0\$$$

```
 (C8) FOR I:1 THRU 3 DO FOR J:1 THRU 3 DO FOR K:1 THRU 3 DO
 R(I):R(I)+T[I,J,K]*V[J]*V[K]$
```

The gravitational potential function denoted by ϕ in equation (6.6.1) is a function of the coordinates. To facilitate the printout of results it will be denoted by P in the present program. As in previous applications, functional dependence can be declared by using the **DEPENDENCIES** function. The *i*th component of the gravitational force is given by the partial differential coefficient of P with respect to x^i. Denoting by F^i the *i*th component of the thrust vector, and declaring the dependence of P on x^i, the programmed version of equation (6.6.1) and the formulated equations are obtained as follows:

```
(C9) DEPENDENCIES(P(X[I]))$

(C10) FOR I:1 THRU 3 DO
 A[I]:(H[I,I]*DIFF(P(X[I]),X[I])+F[I]*SQRT(H[I,I]))/M-R[I]$

(C11) FOR I:1 THRU 3 DO DISPLAY(A[I])$
```

$$A_1 = \frac{\frac{d}{dX_1}P(X_1) + F_1}{M} + X_1 V_2^2$$

$$A_2 = \frac{\frac{\frac{d}{dX_2}P(X_2)}{X_1^2} + \frac{F_2}{X_1}}{M} - \frac{2\,V_1\,V_2}{X_1}$$

$$A_3 = \frac{\frac{d}{dX_3}P(X_3) + F_3}{M}$$

Exactly the same procedure may be used to formulate the equations of motion of a particle in a spherical polar coordinate system. When the metric tensors which characterize an element of arc in this coordinate system are used as input, the MACSYMA program and the formulated equations assume the form

```
(C1) G[1,1]:1$

(C2) G[2,2]:X[1]**2$

(C3) G[3,3]:(X[1]*SIN(X[2]))**2$

(C4) FOR I:1 THRU 3 DO H[I,I]:1/G[I,I]$

(C5) T[I,J,K]:=0$

(C6) FOR I:1 THRU 3 DO FOR J:1 THRU 3 DO FOR K:1 THRU 3 DO
 T[I,J,K]:T[I,J,K]+H[I,I]*(DIFF(G[J,I],X[K])+DIFF(G[K,I],X[J])
-DIFF(G[J,K],X[I]))/2$

(C7) R[I]:=0$

(C8) FOR I:1 THRU 3 DO FOR J:1 THRU 3 DO FOR K:1 THRU 3 DO
 R[I]:R[I]+T[I,J,K]*V[J]*V[K]$

(C9) DEPENDENCIES(P(X[I]))$

(C10) FOR I:1 THRU 3 DO
 A[I]:(H[I,I]*DIFF(P(X[I]),X[I])+F[I]*SQRT(H[I,I]))/M-R[I]$

(C11) FOR I:1 THRU 3 DO DISPLAY(A[I])$
```

$$A_1 = \frac{\frac{d}{dX_1} P(X_1) + F_1}{M} + X_1 V_3^2 \operatorname{SIN}^2(X_2) + X_1 V_2^2$$

$$A_2 = \frac{\frac{\frac{d}{dX_2} P(X_2)}{X_1^2} + \frac{F_2}{X_1}}{M} + V_3^2 \operatorname{COS}(X_2) \operatorname{SIN}(X_2) - \frac{2 V_1 V_2}{X_1}$$

$$A_3 = \frac{\dfrac{d}{dX_3} P(X_3)}{X_1^2 \, SIN^2(X_2)} M + \frac{F_3}{X_1 \, SIN(X_2)} - \frac{2 \, V_2 \, V_3 \, COS(X_2)}{SIN(X_2)} - \frac{2 \, V_1 \, V_3}{X_1}$$

6.7 FORMULATING MODELS OF AERONAUTICAL SYSTEMS

The formulation of models of aeronautical systems for simulation and other purposes involves at least 12 equations: 3 force equations; 3 moment equations; 3 Euler angle equations, or 9 direction cosine equations to determine the spatial orientation of the body; and 3 equations to determine the location of the body in inertial space. In view of this complexity, it is important to mechanize as much of the formulation as possible. An important aspect of the formulation of models of aeronautical systems is the specification of the system of forces and moments. In aeronautical applications, the thrust and gravity forces can be formulated without difficulty, but the aerodynamic forces and moments require more detailed consideration. These are represented by the static forces and moments and the aerodynamic stability derivatives. These forces and moments have to be transformed from wind or wind-tunnel stability axes to body axes before the formulation can proceed. Although the aerodynamic transformations and formulations are not complicated, they are complex and unwieldy and are likely to contain errors when formulated manually. In summary, formula manipulation as implemented by the MACSYMA system can be used to facilitate the formulation of complex mathematical models and reduce the errors to which human operators are prone. The interactive capability, versatility, and simplicity of the system make it attractive to programmers and nonprogrammers alike. In order to illustrate these aspects of the system, a mathematical model of an aeronautical system has been formulated.

6.8 AERONAUTICAL REFERENCE SYSTEMS

There are many coordinate systems in use in aeronautical research. Aerodynamic data obtained from wind-tunnel experiments may be referred to wind axes or to wind-tunnel stability axes. When the wind axes are used, the X_1 axis is aligned with the relative wind at all times. Most wind-tunnel data are referred to the wind-tunnel stability axes system. For this system, the X_1 axis is in the same horizontal plane as the relative wind at all times. In addition to the wind axes and the wind-tunnel stability axes, there are other systems of axes fixed in the body and moving with the body. These are referred to as body axes. In aerospace applications, a body axis system has the Y_1 axis fixed along the longitudinal centerline of the body, the Y_2 axis normal to the plane of symmetry, and the Y_3 axis in the plane of symmetry. The equations of motion of aerospace vehicles are formulated with respect to body axes. The main advantage of these axes in motion calculations is that vehicle moments and products of inertia about the axes are constants. When the body axes are chosen so that the products of inertia vanish, they are known as principal axes. A system of axes, which is frequently used to study the stability of aircraft in the presence of disturbing forces that produce small perturbations, is the flight stability axes. This is an orthogonal system fixed to the vehicle, the Y_1 axis of which is aligned with the relative wind vector when the vehicle is in a steady-state condition, but then rotates with the vehicle after a disturbance as the vehicle changes angle of attack and sideslip (ref. 5). Some of these axes are shown in figure 6.8.1.

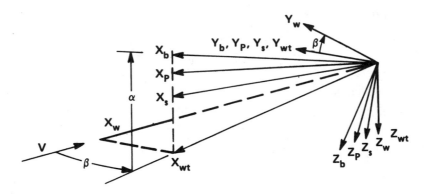

Figure 6.8.1.— Systems of reference axes, including body, principal, wind, flight stability, and wind-tunnel stability.

6.9 TRANSFORMATION EQUATIONS

As indicated in section 6.7 the first step in the formulation is the transformation of relevant data. The elements of the matrices defining a transformation from wind or wind-tunnel stability axes to body axes are functions of the angle of attack (A) and the angle of sideslip (B). Moreover, coordinates in wind-tunnel axes are denoted by a column vector of coordinates X_i, and the body axes coordinates by a column vector Y_i. To bring a reference frame from the wind axes into coincidence with the body axes involves a negative rotation (B) about the Y_3 axis, followed by a positive rotation (A) about the Y_2 axis. These matrices may be entered and multiplied when communication has been established and the system prints (C1). When this occurs, the user types ENTERMATRIX(m,n) which allows one to enter a matrix, element by element, with MACSYMA requesting values for each of the (m,n) entries as follows:

```
(C1) ENTERMATRIX(3,3);

ROW 1 COLUMN 1  COS(A);

ROW 1 COLUMN 2  0;

ROW 1 COLUMN 3 -SIN(A);

ROW 2 COLUMN 1  0;

ROW 2 COLUMN 2  1;

ROW 2 COLUMN 3  0;

ROW 3 COLUMN 1  SIN(A);

ROW 3 COLUMN 2  0;

ROW 3 COLUMN 3  COS(A);

MATRIX-ENTERED
```

$$(D1) \qquad \begin{bmatrix} COS(A) & 0 & -SIN(A) \\ 0 & 1 & 0 \\ SIN(A) & 0 & COS(A) \end{bmatrix}$$

```
(C2) ENTERMATRIX(3,3);


ROW 1 COLUMN 1  COS(B);

ROW 1 COLUMN 2  -SIN(B);
```

ROW 1 COLUMN 3 0;

ROW 2 COLUMN 1 SIN(B);

ROW 2 COLUMN 2 COS(B);

ROW 2 COLUMN 3 0;

ROW 3 COLUMN 1 0;

ROW 3 COLUMN 2 0;

ROW 3 COLUMN 3 1;

MATRIX-ENTERED

(D2)
$$\begin{bmatrix} COS(B) & -SIN(B) & 0 \\ SIN(B) & COS(B) & 0 \\ 0 & 0 & 1 \end{bmatrix}$$

(C3) ENTERMATRIX(3,1);

ROW 1 COLUMN 1 X[1];

ROW 2 COLUMN 1 X[2];

ROW 3 COLUMN 1 X[3];

MATRIX-ENTERED

(D3)
$$\begin{bmatrix} X_1 \\ X_2 \\ X_3 \end{bmatrix}$$

(C4) (D1).(D2).(D3);

$$(D4) \quad \begin{bmatrix} \cos(A) \ (X_1 \ \cos(B) - X_2 \ \sin(B)) - X_3 \ \sin(A) \\ \\ X_1 \ \sin(B) + X_2 \ \cos(B) \\ \\ \sin(A) \ (X_1 \ \cos(B) - X_2 \ \sin(B)) + X_3 \ \cos(A) \end{bmatrix}$$

(C5) FOR I:1 THRU 3 DO ROW[1]:FIRST(ROW((D4),I))$

(C6) FOR I:1 THRU 3 DO (Y[I]:ROW[I][1],DISPLAY(Y[I]));

$$Y_1 = \cos(A) \ (X_1 \ \cos(B) - X_2 \ \sin(B)) - X_3 \ \sin(A)$$

$$(D6) \qquad Y_2 = X_1 \ \sin(B) + X_2 \ \cos(B)$$

$$Y_3 = \sin(A) \ (X_1 \ \cos(B) - X_2 \ \sin(B)) + X_3 \ \cos(A)$$

In order to more fully appreciate the results obtained so far, the reader should note that MACSYMA requests the ith row and the jth column of the matrix being entered by typing ROWICOLUMNJ. The user merely provides the corresponding element. When all $m \times n$ elements have been entered, the system types MATRIX-ENTERED, formulates the matrix and assigns an identifying number (DI). When the user types the command (C4), that is, (D1).(D2).(D3), the three matrices are multiplied in the order requested and the product matrix is displayed in (D4).

The two programming steps shown in (C5) and (C6) lead to the functional form (D6), which represents the required transformation from wind axes X_I to body axes Y_J.

6.10 TRANSFORMATION LAW FOR STATIC FORCES

The static aerodynamic forces transform like the components of a contravariant vector; that is, if S_{F_n} denotes a static aerodynamic force in the X frame of reference, and SF_i denotes the corresponding transformed force in the Y reference frame, then (ref. 6)

$$SF_i = \frac{\partial Y^i}{\partial X^n} \, S_{F_n} \tag{6.10.1}$$

where $Y = Y(X)$ is obtained from the displayed output (D6).

Given the transformation equations (D6), the transformed aerodynamic static forces are obtained by expanding equation (6.10.1). The three programming steps used in previous applications may again be employed to formulate the required values. The simple program and the displayed results are

```
(C7) SF[I]:=0$
```

```
(C8) FOR I:1 THRU 3 DO FOR M:1 THRU 3 DO
SF[I]:SF[I]+DIFF(Y[I],X[M])*S[F[M]]$
```

```
(C9) FOR I:1 THRU 3 DO DISPLAY(SF[I]);
```

$$SF_1 = -S_{F_2} \cos(A)\sin(B) + S_{F_1}\cos(A)\cos(B) - S_{F_3}\sin(A)$$

$$SF_2 = S_{F_1}\sin(B) + S_{F_2}\cos(B)$$

$$SF_3 = -S_{F_2}\sin(A)\sin(B) + S_{F_1}\sin(A)\cos(B) + S_{F_3}\cos(A)$$

(D9)   DONE

6.11 TRANSFORMATION LAW FOR CONTROL FORCE DERIVATIVES

The control force derivatives obey the same transformation law as the static forces; that is, if D_{F_n,C_K} denotes the nth control force derivative with respect to the Kth control surface as measured in the X reference frame, and $TD_{I,C}$ denotes the corresponding transformed derivative in the Y frame, then

$$TD_{i,c} = \frac{\partial Y^i}{\partial X^n} \, D_{F_n,C_k} \tag{6.11.1}$$

where $Y = Y(X)$ is again obtained from the displayed output (D6).

As in the preceding section, the transformed control derivatives are obtained by expanding the transformation law for derivatives given the transformation equations (D6). The transformed derivatives are obtained by executing the following simple program, which has exactly the same form as the program used to transform the static forces in section 6.10. These are:

```
(C10) TD[I,C]:=0$
```

```
(C11) FOR I:1 THRU 3 DO FOR M:1 THRU 3 DO
 TD[I,C]:TD[I,C]+DIFF(Y[I],X[M])*D[F[M],C[K]]$
```

```
(C12) FOR I:1 THRU 3 DO DISPLAY(TD[I,C]);
```

$$TD_{1,C} = -D_{F_2,C_K}\cos(A)\sin(B) + D_{F_1,C_K}\cos(A)\cos(B) - D_{F_3,C_K}\sin(A)$$

$$TD_{2,C} = D_{F_1,C_K}\sin(B) + D_{F_2,C_K}\cos(B)$$

$$TD_{3,C} = -D_{F_2,C_K}\sin(A)\sin(B) + D_{F_1,C_K}\sin(A)\cos(B) + D_{F_3,C_K}\cos(A)$$

(D12)   DONE

The corresponding control forces are obtained by multiplying the control derivatives by the appropriate control increments ΔC_K. The following two programming steps are sufficient to ensure evaluation of the required forces. These are denoted by CF_i in the displayed output.

```
(C13) FOR I:1 THRU 3 DO CF[I]:TD[I,C]*DEL(C[K])$
```

```
(C14) FOR I:1 THRU 3 DO DISPLAY(CF[I]);
```

$$CF_1 = (-D_{F_2,C_K}\cos(A)\sin(B) + D_{F_1,C_K}\cos(A)\cos(B) - D_{F_3,C_K}\sin(A))$$

$$DEL(C_K)$$

$$CF_2 = (D_{F_1,C_K} SIN(B) + D_{F_2,C_K} COS(B))DEL(C_K)$$

$$CF_3 = (-D_{F_2,C_K} SIN(A)SIN(B) + D_{F_1,C_K} SIN(A)COS(B) + D_{F_3,C_K} COS(A))$$

$$DEL(C_K)$$

(D14)   DONE

6.12 FORCES PRODUCED BY LINEAR VELOCITY PERTURBATIONS

The next step in the formulation involves the determination of the aerodynamic forces produced when an aircraft is subjected to linear velocity perturbations ΔU_j. Before these forces can be determined, the aerodynamic stability derivatives, with respect to linear velocity components, must be transformed from wind or wind-tunnel stability axes to aircraft axes. For a detailed discussion of the transformation of these derivatives, the reader is referred to section 2.7. The program used for the transformation in section 6.2 can be used in this case also. In this application, the aerodynamic stability derivatives of the ith force with respect to the jth velocity components are denoted by D_{F_i,U_j}. The corresponding transformed derivatives are denoted by TD_{F_i,U_j}. When the program of section 6.2 is rewritten to accommodate the notational changes required for this application, it assumes the form

(C15) TDU[I,J]:=0$

(C16) FOR I:1 THRU 3 DO FOR J:1 THRU 3 DO
FOR M:1 THRU 3 DO FOR N:1 THRU 3 DO
 TDU[I,J]:TDU[I,J]+DIFF(Y[I],X[M])*DIFF(Y[J],X[N])*D[F[M],U[N]]$

It only remains to multiply the transformed derivatives by the appropriate velocity increments to obtain the required forces, which are denoted by FDU_i. The next three programming steps instruct MACSYMA to evaluate and display the forces produced by linear velocity perturbations. These are

```
(C17) FDU[I]:=0$

(C18) FOR I:1 THRU 3 DO FOR J:1 THRU 3 DO
 FDU[I]:FDU[I]+TDU[I,J]*DEL(U[J])$

(C19) FOR I:1 THRU 3 DO DISPLAY(FDU[I]);
```

$$FDU_1 = (D_{F_2,U_2} \, COS(A)SIN(A)SIN^2(B)$$

$$- D_{F_2,U_1} \, COS(A)SIN(A)COS(B)SIN(B)$$

$$- D_{F_1,U_2} \, COS(A)SIN(A)COS(B)SIN(B) + D_{F_3,U_2} \, SIN^2(A)SIN(B)$$

$$- D_{F_2,U_3} \, COS^2(A)SIN(B) + D_{F_1,U_1} \, COS(A)SIN(A)COS^2(B)$$

$$- D_{F_3,U_1} \, SIN^2(A)COS(B) + D_{F_1,U_3} \, COS^2(A)COS(B)$$

$$- D_{F_3,U_3} \, COS(A)SIN(A))DEL(U_3)$$

$$+ (-D_{F_2,U_1} \, COS(A)SIN^2(B) - D_{F_2,U_2} \, COS(A)COS(B)SIN(B)$$

$$+ D_{F_1,U_1} \, COS(A)COS(B)SIN(B) - D_{F_3,U_1} \, SIN(A)SIN(B)$$

$$+ D_{F_1,U_2} \, COS(A)COS^2(B) - D_{F_3,U_2} \, SIN(A)COS(B))DEL(U_2)$$

$$+ (D_{F_2,U_2} \, COS^2(A)SIN^2(B) - D_{F_2,U_1} \, COS^2(A)COS(B)SIN(B)$$

$$- D_{F_1,U_2} \, COS^2(A)COS(B)SIN(B) + D_{F_3,U_2} \, COS(A)SIN(A)SIN(B)$$

$$+ D_{F_2,U_3} \cos(A)\sin(A)\sin(B) + D_{F_1,U_1} \cos^2(A)\cos^2(B)$$

$$- D_{F_3,U_1} \cos(A)\sin(A)\cos(B) - D_{F_1,U_3} \cos(A)\sin(A)\cos(B)$$

$$+ D_{F_3,U_3} \sin^2(A))\text{DEL}(U_1)$$

$$FDU_2 = (-D_{F_1,U_2} \sin(A)\sin^2(B) - D_{F_2,U_2} \sin(A)\cos(B)\sin(B)$$

$$+ D_{F_1,U_1} \sin(A)\cos(B)\sin(B) + D_{F_1,U_3} \cos(A)\sin(B)$$

$$+ D_{F_2,U_1} \sin(A)\cos^2(B) + D_{F_2,U_3} \cos(A)\cos(B))\text{DEL}(U_3)$$

$$+ (D_{F_1,U_1} \sin^2(B) + D_{F_2,U_1} \cos(B)\sin(B) + D_{F_1,U_2} \cos(B)\sin(B)$$

$$+ D_{F_2,U_2} \cos^2(B))\text{DEL}(U_2) + (-D_{F_1,U_2} \cos(A)\sin^2(B)$$

$$- D_{F_2,U_2} \cos(A)\cos(B)\sin(B) + D_{F_1,U_1} \cos(A)\cos(B)\sin(B)$$

$$- D_{F_1,U_3} \sin(A)\sin(B) + D_{F_2,U_1} \cos(A)\cos^2(B)$$

$$- D_{F_2,U_3} \sin(A)\cos(B))\text{DEL}(U_1)$$

$$FDU_3 = (D_{F_2,U_2} \sin^2(A)\sin^2(B) - D_{F_2,U_1} \sin^2(A)\cos(B)\sin(B)$$

$$- D_{F_1,U_2} \sin^2(A)\cos(B)\sin(B) - D_{F_3,U_2} \cos(A)\sin(A)\sin(B)$$

$$- D_{F_2,U_3} \cos(A)\sin(A)\sin(B) + D_{F_1,U_1} \sin^2(A)\cos^2(B)$$

$$+ D_{F_3,U_1} \cos(A)\sin(A)\cos(B) + D_{F_1,U_3} \cos(A)\sin(A)\cos(B)$$

$$+ D_{F_3,U_3} \cos^2(A))\,\mathrm{DEL}(U_3) + (-D_{F_2,U_1} \sin(A)\sin^2(B)$$

$$- D_{F_2,U_2} \sin(A)\cos(B)\sin(B) + D_{F_1,U_1} \sin(A)\cos(B)\sin(B)$$

$$+ D_{F_3,U_1} \cos(A)\sin(B) + D_{F_1,U_2} \sin(A)\cos^2(B)$$

$$+ D_{F_3,U_2} \cos(A)\cos(B))\,\mathrm{DEL}(U_2)$$

$$+ (D_{F_2,U_2} \cos(A)\sin(A)\sin^2(B)$$

$$- D_{F_2,U_1} \cos(A)\sin(A)\cos(B)\sin(B)$$

$$- D_{F_1,U_2} \cos(A)\sin(A)\cos(B)\sin(B) + D_{F_2,U_3} \sin^2(A)\sin(B)$$

$$- D_{F_3,U_2} \cos^2(A)\sin(B) + D_{F_1,U_1} \cos(A)\sin(A)\cos^2(B)$$

$$- D_{F_1,U_3} \sin^2(A)\cos(B) + D_{F_3,U_1} \cos^2(A)\cos(B)$$

$$- D_{F_3,U_3} \cos(A)\sin(A))\,\mathrm{DEL}(U_1)$$

6.13 FORCES PRODUCED BY ANGULAR VELOCITY PERTURBATIONS

The program used in section 6.12 can, with suitable notational changes, be used to formulate the forces produced by angular velocity perturbations. However,

6.13

whereas in the preceding application the required forces were obtained by multiplying the transformed aerodynamic stability derivatives by linear velocity increments, in the present case the transformed derivatives must be multiplied by angular velocity increments. In view of these similarities, the following program and displayed forces will be presented without further comment, except to point out that the aerodynamic stability derivatives of the *i*th force with respect to the *j*th angular velocity component are denoted by D_{F_i,P_j}. The corresponding transformed derivatives are denoted by TD_{F_i,P_j}, and the resulting forces by FDP_i.

(C20) TDP[I,J]:=0$

(C21) FOR I:1 THRU 3 DO FOR J:1 THRU 3 DO
 FOR M:1 THRU 3 DO FOR N:1 THRU 3 DO
 TDP[I,J]:TDP[I,J]+DIFF(Y[I],X[M])*DIFF(Y[J],X[N])*D[F[M],P[N]]$

(C22) FDP[I]:=0$

(C23) FOR I:1 THRU 3 DO FOR J:1 THRU 3 DO
 FDP[I]:FDP[I]+TDP[I,J]*DEL(P[J])$

(C24) FOR I:1 THRU 3 DO DISPLAY(FDP[I]);

$$FDP_1 = (D_{F_2,P_2} \cos(A)\sin(A)\sin^2(B)$$

$$- D_{F_2,P_1} \cos(A)\sin(A)\cos(B)\sin(B)$$

$$- D_{F_1,P_2} \cos(A)\sin(A)\cos(B)\sin(B) + D_{F_3,P_2} \sin^2(A)\sin(B)$$

$$- D_{F_2,P_3} \cos^2(A)\sin(B) + D_{F_1,P_1} \cos(A)\sin(A)\cos^2(B)$$

$$- D_{F_3,P_1} \sin^2(A)\cos(B) + D_{F_1,P_3} \cos^2(A)\cos(B)$$

$$- D_{F_3,P_3} \cos(A)\sin(A))DEL(P_3)$$

328

$$+ \ (-D_{F_2,P_1} \ \cos(A)\sin^2(B) \ - \ D_{F_2,P_2} \ \cos(A)\cos(B)\sin(B)$$

$$+ \ D_{F_1,P_1} \ \cos(A)\cos(B)\sin(B) \ - \ D_{F_3,P_1} \ \sin(A)\sin(B)$$

$$+ \ D_{F_1,P_2} \ \cos(A)\cos^2(B) \ - \ D_{F_3,P_2} \ \sin(A)\cos(B))\text{DEL}(P_2)$$

$$+ \ (D_{F_2,P_2} \ \cos^2(A)\sin^2(B) \ - \ D_{F_2,P_1} \ \cos^2(A)\cos(B)\sin(B)$$

$$- \ D_{F_1,P_2} \ \cos^2(A)\cos(B)\sin(B) \ + \ D_{F_3,P_2} \ \cos(A)\sin(A)\sin(B)$$

$$+ \ D_{F_2,P_3} \ \cos(A)\sin(A)\sin(B) \ + \ D_{F_1,P_1} \ \cos^2(A)\cos^2(B)$$

$$- \ D_{F_3,P_1} \ \cos(A)\sin(A)\cos(B) \ - \ D_{F_1,P_3} \ \cos(A)\sin(A)\cos(B)$$

$$+ \ D_{F_3,P_3} \ \sin^2(A))\text{DEL}(P_1)$$

$$\text{FDP}_2 = (-D_{F_1,P_2} \ \sin(A)\sin^2(B) \ - \ D_{F_2,P_2} \ \sin(A)\cos(B)\sin(B)$$

$$+ \ D_{F_1,P_1} \ \sin(A)\cos(B)\sin(B) \ + \ D_{F_1,P_3} \ \cos(A)\sin(B)$$

$$+ \ D_{F_2,P_1} \ \sin(A)\cos^2(B) \ + \ D_{F_2,P_3} \ \cos(A)\cos(B))\text{DEL}(P_3)$$

$$+ \ D_{F_1,P_1} \ \sin^2(B) \ + \ D_{F_2,P_1} \ \cos(B)\sin(B) \ + \ D_{F_1,P_2} \ \cos(B)\sin(B)$$

$$+ \ D_{F_2,P_2} \ \cos^2(B))\text{DEL}(P_2) \ + \ (-D_{F_1,P_2} \ \cos(A)\sin^2(B)$$

$$- \ D_{F_2,P_2} \ \cos(A)\cos(B)\sin(B) \ + \ D_{F_1,P_1} \ \cos(A)\cos(B)\sin(B)$$

$$- D_{F_1,P_3} \, SIN(A)SIN(B) + D_{F_2,P_1} \, COS(A)COS^2(B)$$

$$- D_{F_2,P_3} \, SIN(A)COS(B))DEL(P_1)$$

$$FDP_3 = (D_{F_2,P_2} \, SIN^2(A)SIN^2(B) - D_{F_2,P_1} \, SIN^2(A)COS(B)SIN(B)$$

$$- D_{F_1,P_2} \, SIN^2(A)COS(B)SIN(B) - D_{F_3,P_2} \, COS(A)SIN(A)SIN(B)$$

$$- D_{F_2,P_3} \, COS(A)SIN(A)SIN(B) + D_{F_1,P_1} \, SIN^2(A)COS^2(B)$$

$$+ D_{F_3,P_1} \, COS(A)SIN(A)COS(B) + D_{F_1,P_3} \, COS(A)SIN(A)COS(B)$$

$$+ D_{F_3,P_3} \, COS^2(A))DEL(P_3) + (-D_{F_2,P_1} \, SIN(A)SIN^2(B)$$

$$- D_{F_2,P_2} \, SIN(A)COS(B)SIN(B) + D_{F_1,P_1} \, SIN(A)COS(B)SIN(B)$$

$$+ D_{F_3,P_1} \, COS(A)SIN(B) + D_{F_1,P_2} \, SIN(A)COS^2(B)$$

$$+ D_{F_3,P_2} \, COS(A)COS(B))DEL(P_2)$$

$$+ (D_{F_2,P_2} \, COS(A)SIN(A)SIN^2(B)$$

$$- D_{F_2,P_1} \, COS(A)SIN(A)COS(B)SIN(B)$$

$$- D_{F_1,P_2} \, COS(A)SIN(A)COS(B)SIN(B)$$

$$+ D_{F_2,P_3} \, SIN^2(A)SIN(B) - D_{F_3,P_2} \, COS^2(A)SIN(B)$$

$$+ D_{F_1, P_1} \cos(A)\sin(A)\cos^2(B) - D_{F_1, P_3} \sin^2(A)\cos(B)$$

$$+ D_{F_3, P_1} \cos^2(A)\cos(B) - D_{F_3, P_3} \cos(A)\sin(A))DEL(P_1)$$

6.14 FORCES PRODUCED BY LINEAR ACCELERATION PERTURBATIONS

The procedure used in the preceding two sections may, with equal facility, be used to formulate the aerodynamic forces produced by linear acceleration perturbations. However, in this case the required forces are obtained by multiplying the transformed aerodynamic stability derivatives, with respect to acceleration components, by linear acceleration increments. The aerodynamic stability derivatives of the ith force component F_i with respect to the jth linear acceleration component A_j are denoted by D_{F_j, A_j}, and the transformed derivatives by TD_{F_i, A_j}. The corresponding force components in body axes are denoted by FDA_i.

Due to the fact that lift responds in a transient manner when, for example, the angle of attack A or the linear velocity component U_3 is suddenly changed, the acceleration derivatives are very different from the velocity derivatives, which can be determined on the basis of steady-state aerodynamics. This is a consequence of the fact that the pressure distribution on a wing or tail surface does not adjust itself instantaneously to its equilibrium value when the angle of attack or the velocity components are suddenly changed. Hence, in order to get a sufficiently accurate description of these derivatives during the indicial response phase, it may be necessary to use function generation or look-up tables (ref. 7).

When the program of the preceding section has been modified to incorporate the necessary notational changes, it assumes the following form:

```
(C25) TDA[I,J]:=0$

(C26) FOR I:1 THRU 3 DO FOR J:1 THRU 3 DO
 FOR M:1 THRU 3 DO FOR N:1 THRU 3 DO
 TDA[I,J]:TDA[I,J]+DIFF(Y[I],X[M])*DIFF(Y[J],X[N])*D[F[M],A[N]]$

(C27) FDA[I]:=0$

(C28) FOR I:1 THRU 3 DO FOR J:1 THRU 3 DO
 FDA[I]:FDA[I]+TDA[I,J]*DEL(A[J])$
```

```
(C29) FOR I:1 THRU 3 DO DISPLAY(FDA[I])$
```

Execution of this program yields the aerodynamic forces produced by linear acceleration perturbations. These are

$$FDA_1 = (D_{F_2,A_2} \, COS(A)SIN(A)SIN^2(B)$$

$$- \, D_{F_2,A_1} \, COS(A)SIN(A)COS(B)SIN(B)$$

$$- \, D_{F_1,A_2} \, COS(A)SIN(A)COS(B)SIN(B) + D_{F_3,A_2} \, SIN^2(A)SIN(B)$$

$$- \, D_{F_2,A_3} \, COS^2(A)SIN(B) + D_{F_1,A_1} \, COS(A)SIN(A)COS^2(B)$$

$$- \, D_{F_3,A_1} \, SIN^2(A)COS(B) + D_{F_1,A_3} \, COS^2(A)COS(B)$$

$$- \, D_{F_3,A_3} \, COS(A)SIN(A)DEL(A_3)$$

$$+ \, (-D_{F_2,A_1} \, COS(A)SIN^2(B) - D_{F_2,A_2} \, COS(A)COS(B)SIN(B)$$

$$+ \, D_{F_1,A_1} \, COS(A)COS(B)SIN(B) - D_{F_3,A_1} \, SIN(A)SIN(B)$$

$$+ \, D_{F_1,A_2} \, COS(A)COS^2(B) - D_{F_3,A_2} \, SIN(A)COS(B))DEL(A_2)$$

$$+ \, (D_{F_2,A_2} \, COS^2(A)SIN^2(B) - D_{F_2,A_1} \, COS^2(A)COS(B)SIN(B)$$

$$- \, D_{F_1,A_2} \, COS^2(A)COS(B)SIN(B) + D_{F_3,A_2} \, COS(A)SIN(A)SIN(B)$$

$$+ \, D_{F_2,A_3} \, COS(A)SIN(A)SIN(B) + D_{F_1,A_1} \, COS^2(A)COS^2(B)$$

$$- D_{F_3,A_1} \cos(A)\sin(A)\cos(B) - D_{F_1,A_3} \cos(A)\sin(A)\cos(B)$$

$$+ D_{F_3,A_3} \sin^2(A))\mathrm{DEL}(A_1)$$

$$FDA_2 = (-D_{F_1,A_2} \sin(A)\sin^2(B) - D_{F_2,A_2} \sin(A)\cos(B)\sin(B)$$

$$+ D_{F_1,A_1} \sin(A)\cos(B)\sin(B) + D_{F_1,A_3} \cos(A)\sin(B)$$

$$+ D_{F_2,A_1} \sin(A)\cos^2(B) + D_{F_2,A_3} \cos(A)\cos(B))\mathrm{DEL}(A_3)$$

$$+ (D_{F_1,A_1} \sin^2(B) + D_{F_2,A_1} \cos(B)\sin(B) + D_{F_1,A_2} \cos(B)\sin(B)$$

$$+ D_{F_2,A_2} \cos^2(B))\mathrm{DEL}(A_2) + (-D_{F_1,A_2} \cos(A)\sin^2(B)$$

$$- D_{F_2,A_2} \cos(A)\cos(B)\sin(B) + D_{F_1,A_1} \cos(A)\cos(B)\sin(B)$$

$$- D_{F_1,A_3} \sin(A)\sin(B) + D_{F_2,A_1} \cos(A)\cos^2(B)$$

$$- D_{F_2,A_3} \sin(A)\cos(B)\mathrm{DEL}(A_1)$$

$$FDA_3 = (D_{F_2,A_2} \sin^2(A)\sin^2(B) - D_{F_2,A_1} \sin^2(A)\cos(B)\sin(B)$$

$$- D_{F_1,A_2} \sin^2(A)\cos(B)\sin(B) - D_{F_3,A_2} \cos(A)\sin(A)\sin(B)$$

$$- D_{F_2,A_3} \cos(A)\sin(A)\sin(B) + D_{F_1,A_1} \sin^2(A)\cos^2(B)$$

$$+ D_{F_3,A_1} \cos(A)\sin(A)\cos(B) + D_{F_1,A_3} \cos(A)\sin(A)\cos(B)$$

$$+ D_{F_3,A_3} \cos^2(A))DEL(A_3) + (-D_{F_2,A_1} \sin(A)\sin^2(B)$$

$$- D_{F_2,A_2} \sin(A)\cos(B)\sin(B) + D_{F_1,A_1} \sin(A)\cos(B)\sin(B)$$

$$+ D_{F_3,A_1} \cos(A)\sin(B) + D_{F_1,A_2} \sin(A)\cos^2(B)$$

$$+ D_{F_3,A_2} \cos(A)\cos(B))DEL(A_2)$$

$$+ (D_{F_2,A_2} \cos(A)\sin(A)\sin^2(B)$$

$$- D_{F_2,A_1} \cos(A)\sin(A)\cos(B)\sin(B)$$

$$- D_{F_1,A_2} \cos(A)\sin(A)\cos(B)\sin(B)$$

$$+ D_{F_2,A_3} \sin^2(A)\sin(B) - D_{F_3,A_2} \cos^2(A)\sin(B)$$

$$+ D_{F_1,A_1} \cos(A)\sin(A)\cos^2(B) - D_{F_1,A_3} \sin^2(A)\cos(B)$$

$$+ D_{F_3,A_1} \cos^2(A)\cos(B) - D_{F_3,A_3} \cos(A)\sin(A))DEL(A_1)$$

The components of the resultant aerodynamic force are

(C27) FOR I:1 THRU 3 DO FA[I]:FDU[I]+FDP[I]+FDA[I]+CF[I]+SF[I]$

6.15 GRAVITY FORCES

The gravitational force vector acting on an aircraft has the value $M\bar{g}$, where M is the mass of the aircraft and $\bar{g}$ is the gravitational acceleration vector. The magnitude of $\bar{g}$ is assumed constant, which is tantamount to the assumption of a flat earth. The gravity vector is specified in an earth-fixed reference frame; and it is required to find the components of this vector in aircraft body axes. In accordance with aeronautical convention, a transformation from earth-fixed axes to aircraft body axes involves a rotation R_3 about the Y_3 body axis, followed by a rotation R_2 about the Y_2 body axis, and a rotation R_1 about the Y_1 body axis. Hence, if it is assumed that the body axes and the earth-fixed axes are initially coincident, the components of the gravitational force FG_i in body axes are given by an equation of the form

$$[FG] = [R_1][R_2][R_3][Mg]$$

where $[FG]$ is a column vector of body axes components, $[R_1]$, $[R_2]$, and $[R_3]$ are rotation matrices, and $[Mg]$ is a column vector of earth-fixed axes components. These matrix operations can be performed by MACSYMA to yield the required force components in body axes as follows:

```
(C31) ENTERMATRIX(3,3);

ROW 1 COLUMN 1  COS(R[3]);

ROW 1 COLUMN 2  SIN(R[3]);

ROW 1 COLUMN 3  0;

ROW 2 COLUMN 1  -SIN([3]);

ROW 2 COLUMN 2  COS(R[3]);

ROW 2 COLUMN 3  0;

ROW 3 COLUMN 1  0;

ROW 3 COLUMN 2  0;

ROW 3 COLUMN 3  1;

MATRIX-ENTERED
```

$$(D31) \quad \begin{bmatrix} \cos(R_3) & \sin(R_3) & 0 \\ -\sin(R_3) & \cos(R_3) & 0 \\ 0 & 0 & 1 \end{bmatrix}$$

```
(C32) ENTERMATRIX(3,3);

ROW 1 COLUMN 1  COS(R[2]);

ROW 1 COLUMN 2  0;

ROW 1 COLUMN 3  -SIN(R[2]);

ROW 2 COLUMN 1  0;

ROW 2 COLUMN 2  1;

ROW 2 COLUMN 3  0;

ROW 3 COLUMN 1  SIN(R[2]);

ROW 3 COLUMN 2  0;

ROW 3 COLUMN 3  COS(R[2]);
```

MATRIX-ENTERED

(D32)
$$\begin{bmatrix} \cos(R_2) & 0 & -\sin(R_2) \\ 0 & 1 & 0 \\ \sin(R_2) & 0 & \cos(R_2) \end{bmatrix}$$

```
(C33) ENTERMATRIX(3,3);

ROW 1 COLUMN 1  1;

ROW 1 COLUMN 2  0;

ROW 1 COLUMN 3  0;

ROW 2 COLUMN 1  0;

ROW 2 COLUMN 2  COS(R[1]);
```

ROW 2 COLUMN 3 SIN(R[1]);

ROW 3 COLUMN 1 0;

ROW 3 COLUMN 2 -SIN(R[1]);

ROW 3 COLUMN 3 COS(R[1]);

MATRIX-ENTERED

(D33)
$$\begin{bmatrix} 1 & 0 & 0 \\ 0 & \cos(R_1) & \sin(R_1) \\ 0 & -\sin(R_1) & \cos(R_1) \end{bmatrix}$$

(C34) ENTERMATRIX(3,1);

ROW 1 COLUMN 1 0;

ROW 2 COLUMN 1 0;

ROW 3 COLUMN 1 M*G;

MATRIX-ENTERED

(D34)
$$\begin{bmatrix} 0 \\ 0 \\ G\,M \end{bmatrix}$$

The product of these four matrices gives the following column vector of gravitational force components relative to aircraft body axes:

(C35) (D33).(D32).(D31).(D34);

(D35)

$$
\begin{bmatrix}
-\text{SIN}(R_2) \ G \ M \\
\text{SIN}(R_1) \ \text{COS}(R_2) \ G \ M \\
\text{COS}(R_1) \ \text{COS}(R_2) \ G \ M
\end{bmatrix}
$$

These vector components may be expressed in conventional form by executing the following two programming steps, which yield:

(C36) FOR I:1 THRU 3 DO ROW[I]:FIRST(ROW((D35),I))$

(C37) FOR I:1 THRU 3 DO (FG[I]:ROW[I][1],DISPLAY(FG[I]))$

$$
FG_1 = -\text{SIN}(R_2) \ G \ M
$$
$$
FG_2 = \text{SIN}(R_1) \ \text{COS}(R_2) \ G \ M
$$
$$
FG_3 = \text{COS}(R_1) \ \text{COS}(R_2) \ G \ M
$$

where $R_i = (R_i^O + \delta R_i)$, R_i^O are equilibrium values, and δR_i are angular perturbations.

6.16 INERTIA FORCES

The formulation of the inertia forces involves the determination of the product of an angular velocity matrix and a column vector of linear velocity components. This product is the matrix equivalent of the familiar vector product $\bar{\omega} \times \bar{V}$. By adding to the components of this vector, the components of linear acceleration relative to aircraft body axes, the components of inertial acceleration relative to these axes are obtained. The required matrices may be entered and multiplied as follows:

(C38) ENTERMATRIX(3,3);

ROW 1 COLUMN 1 0;

ROW 1 COLUMN 2 -P[3];

ROW 1 COLUMN 3 P[2];

ROW 2 COLUMN 1 P[3];

ROW 2 COLUMN 2 0;

ROW 2 COLUMN 3 -P[1];

ROW 3 COLUMN 1 -P[2];

ROW 3 COLUMN 2 P[1];

ROW 3 COLUMN 3 0;

MATRIX-ENTERED

(D38)
$$\begin{bmatrix} 0 & -P_3 & P_2 \\ P_3 & 0 & -P_1 \\ -P_2 & P_1 & 0 \end{bmatrix}$$

(C39) ENTERMATRIX(3,1);

ROW 1 COLUMN 1 U[1];

ROW 2 COLUMN 1 U[2];

ROW 3 COLUMN 1 U[3];

MATRIX-ENTERED

$$(D39) \qquad \begin{bmatrix} U_1 \\ U_2 \\ U_3 \end{bmatrix}$$

(C40) (D38).(D39);

$$(D40) \qquad \begin{bmatrix} P_2\, U_3 - U_2\, P_3 \\ U_1\, P_3 - P_1\, U_3 \\ P_1\, U_2 - U_1\, P_2 \end{bmatrix}$$

(C41) FOR I:1 THRU 3 DO ROW[I]:FIRST(ROW((D40),I))$

(C42) FOR I:1 THRU 3 DO (C[I]:ROW[I][1],DISPLAY(C[I]))$

$$C_1 = P_2\, U_3 - U_2\, P_3$$

$$C_2 = U_1\, P_3 - P_1\, U_3$$

$$C_3 = P_1\, U_2 - U_1\, P_2$$

A statement of the fact that the ith component of the linear velocity vector is a function of time, requires the use of the DEPENDENCIES function. The use of this function permits the system to differentiate the components U_i with respect to time. The remaining two programming statements request the system to add the components, multiply the individual sums by the mass M of the vehicle, and display the resulting inertial force components FR_i as follows:

(C43) DEPENDENCIES(U(I,T))$

(C44) FOR I:1 THRU 3 DO FR[I]:M*(C[I]+DIFF(U[I],T))$

(C45) FOR I:1 THRU 3 DO DISPLAY(FR[I])$

$$FR_1 = \left(\frac{d}{dT} U_1 + P_2 U_3 - U_2 P_3\right) M$$

$$FR_2 = \left(\frac{d}{dT} U_2 - P_1 U_3 + U_1 P_3\right) M$$

$$FR_3 = M \left(-U_1 P_2 + \frac{d}{dT} U_3 + P_1 U_2\right)$$

6.17 RESULTANT FORCES

It only remains to request MACSYMA to combine the aerodynamic, gravitational, and inertia forces that were formulated in preceding sections and display the results. The ith component of the resultant force will be denoted by FT_i where T_i is the ith component of thrust. The two programming steps and the formulated equations follow.

(C46) FOR I:1 THRU 3 DO FT[I]:FR[I]-FG[I]-FA[I]$

(C47) FOR I:1 THRU 3 DO DISPLAY(FT[I])$

$FT_1 = -(-D_{F_2,C_k} \cos(A)\sin(B) + D_{F_1,C_K} \cos(A)\cos(B)$

$- D_{F_3,C_K} \sin(A))DEL(C_K) - (D_{F_2,U_2} \cos(A)\sin(A)\sin^2(B)$

$- D_{F_2,U_1} \cos(A)\sin(A)\cos(B)\sin(B)$

$- D_{F_1,U_2} \cos(A)\sin(A)\cos(B)\sin(B) + D_{F_3,U_2} \sin^2(A)\sin(B)$

$- D_{F_2,U_3} \cos^2(A)\sin(B) + D_{F_1,U_1} \cos(A)\sin(A)\cos^2(B)$

$- D_{F_3,U_1} \sin^2(A)\cos(B) + D_{F_1,U_3} \cos^2(A)\cos(B)$

$- D_{F_3,U_3} \cos(A)\sin(A))DEL(U_3) - (D_{F_2,P_2} \cos(A)\sin(A)\sin^2(B)$

$$- D_{F_2,P_1} \cos(A)\sin(A)\cos(B)\sin(B)$$

$$- D_{F_1,P_2} \cos(A)\sin(A)\cos(B)\sin(B) + D_{F_3,P_2} \sin^2(A)\sin(B)$$

$$- D_{F_2,P_3} \cos^2(A)\sin(B) + D_{F_1,P_1} \cos(A)\sin(A)\cos^2(B)$$

$$- D_{F_3,P_1} \sin^2(A)\cos(B) + D_{F_1,P_3} \cos^2(A)\cos(B)$$

$$- D_{F_3,P_3} \cos(A)\sin(A))\mathrm{DEL}(P_3) - (D_{F_2,A_2} \cos(A)\sin(A)\sin^2(B)$$

$$- D_{F_2,A_1} \cos(A)\sin(A)\cos(B)\sin(B)$$

$$- D_{F_1,A_2} \cos(A)\sin(A)\cos(B)\sin(B) + D_{F_3,A_2} \sin^2(A)\sin(B)$$

$$- D_{F_2,A_3} \cos^2(A)\sin(B) + D_{F_1,A_1} \cos(A)\sin(A)\cos^2(B)$$

$$- D_{F_3,A_1} \sin^2(A)\cos(B) + D_{F_1,A_3} \cos^2(A)\cos(B)$$

$$- D_{F_3,A_3} \cos(A)\sin(A))\mathrm{DEL}(A_3) - (-D_{F_2,U_1} \cos(A)\sin^2(B)$$

$$- D_{F_2,U_2} \cos(A)\cos(B)\sin(B) + D_{F_1,U_1} \cos(A)\cos(B)\sin(B)$$

$$- D_{F_3,U_1} \sin(A)\sin(B) + D_{F_1,U_2} \cos(A)\cos^2(B)$$

$$- D_{F_3,U_2} \sin(A)\cos(B))\mathrm{DEL}(U_2) - (-D_{F_2,P_1} \cos(A)\sin^2(B)$$

$$- D_{F_2,P_2} \cos(A)\cos(B)\sin(B) + D_{F_1,P_1} \cos(A)\cos(B)\sin(B)$$

$$- D_{F_3,P_1} \, SIN(A)SIN(B) - D_{F_1,P_2} \, COS(A)COS^2(B)$$

$$- D_{F_3,P_2} \, SIN(A)COS(B))DEL(P_2) - (-D_{F_2,A_1} \, COS(A)SIN^2(B)$$

$$- D_{F_2,A_2} \, COS(A)COS(B)SIN(B) - D_{F_1,A_1} \, COS(A)COS(B)SIN(B)$$

$$- D_{F_3,A_1} \, SIN(A)SIN(B) + D_{F_1,A_2} \, COS(A)COS^2(B)$$

$$- D_{F_3,A_2} \, SIN(A)COS(B))DEL(A_2) - (D_{F_2,U_2} \, COS^2(A)SIN^2(B)$$

$$- D_{F_2,U_1} \, COS^2(A)COS(B)SIN(B) - D_{F_1,U_2} \, COS^2(A)COS(B)SIN(B)$$

$$+ D_{F_3,U_2} \, COS(A)SIN(A)SIN(B) + D_{F_2,U_3} \, COS(A)SIN(A)SIN(B)$$

$$+ D_{F_1,U_1} \, COS^2(A)COS^2(B) - D_{F_3,U_1} \, COS(A)SIN(A)COS(B)$$

$$- D_{F_1,U_3} \, COS(A)SIN(A)COS(B) + D_{F_3,U_3} \, SIN^2(A))DEL(U_1)$$

$$- (D_{F_2,P_2} \, COS^2(A)SIN^2(B) - D_{F_2,P_1} \, COS^2(A)COS(B)SIN(B)$$

$$- D_{F_1,P_2} \, COS^2(A)COS(B)SIN(B) + D_{F_3,P_2} \, COS(A)SIN(A)SIN(B)$$

$$+ D_{F_2,P_3} \, COS(A)SIN(A)SIN(B) + D_{F_1,P_1} \, COS^2(A)COS^2(B)$$

$$- D_{F_3,P_1} \, COS(A)SIN(A)COS(B) - D_{F_1,P_3} \, COS(A)SIN(A)COS(B)$$

$$+ D_{F_3,P_3} \, SIN^2(A))DEL(P_1) - (D_{F_2,A_2} \, COS^2(A)SIN^2(B)$$

$$- D_{F_2,A_1} \cos^2(A)\cos(B)\sin(B) - D_{F_1,A_2} \cos^2(A)\cos(B)\sin(B)$$

$$+ D_{F_3,A_2} \cos(A)\sin(A)\sin(B) + D_{F_2,A_3} \cos(A)\sin(A)\sin(B)$$

$$+ D_{F_1,A_1} \cos^2(A)\cos^2(B) - D_{F_3,A_1} \cos(A)\sin(A)\cos(B)$$

$$- D_{F_1,A_3} \cos(A)\sin(A)\cos(B) + D_{F_3,A_3} \sin^2(A))DEL(A_1)$$

$$+ \sin(R_2)GM + (\frac{d}{dT} U_1 + P_2U_3 - U_2P_3)M$$

$$+ S_{F_2} \cos(A)\sin(B) - S_{F_1} \cos(A)\cos(B) + S_{F_3} \sin(A)$$

$$FT_2 = -(D_{F_1,C_K} \sin(B) + D_{F_2,C_K} \cos(B))DEL(C_K)$$

$$- (-D_{F_1,U_2} \sin(A)\sin^2(B) - D_{F_2,U_2} \sin(A)\cos(B)\sin(B)$$

$$+ D_{F_1,U_1} \sin(A)\cos(B)\sin(B) + D_{F_1,U_3} \cos(A)\sin(B)$$

$$+ D_{F_2,U_1} \sin(A)\cos^2(B) + D_{F_2,U_3} \cos(A)\cos(B))DEL(U_3)$$

$$- (-D_{F_1,P_2} \sin(A)\sin^2(B) - D_{F_2,P_2} \sin(A)\cos(B)\sin(B)$$

$$+ D_{F_1,P_1} \sin(A)\cos(B)\sin(B) + D_{F_1,P_3} \cos(A)\sin(B)$$

$$+ D_{F_2,P_1} \sin(A)\cos^2(B) + D_{F_2,P_3} \cos(A)\cos(B))DEL(P_3)$$

$$- (-D_{F_1,A_2} \sin(A)\sin^2(B) - D_{F_2,A_2} \sin(A)\cos(B)\sin(B)$$

$$+ D_{F_1,A_1} SIN(A)COS(B)SIN(B) + D_{F_1,A_3} COS(A)SIN(B)$$

$$+ D_{F_2,A_1} SIN(A)COS^2(B) + D_{F_2,A_3} COS(A)COS(B))DEL(A_3)$$

$$- (D_{F_1,U_1} SIN^2(B) + D_{F_2,U_1} COS(B)SIN(B) + D_{F_1,U_2} COS(B)SIN(B)$$

$$+ D_{F_2,U_2} COS^2(B))DEL(U_2) - (D_{F_1,P_1} SIN^2(B) + D_{F_2,P_1} COS(B)SIN(B)$$

$$+ D_{F_1,P_2} COS(B)SIN(B) + D_{F_2,P_2} COS^2(B))DEL(P_2) - (D_{F_1,A_1} SIN^2(B)$$

$$+ D_{F_2,A_1} COS(B)SIN(B) + D_{F_1,A_2} COS(B)SIN(B)$$

$$+ D_{F_2,A_2} COS^2(B))DEL(A_2) - (-D_{F_1,U_2} COS(A)SIN^2(B)$$

$$- D_{F_2,U_2} COS(A)COS(B)SIN(B) + D_{F_1,U_1} COS(A)COS(B)SIN(B)$$

$$- D_{F_1,U_3} SIN(A)SIN(B) + D_{F_2,U_1} COS(A)COS^2(B)$$

$$- D_{F_2,U_3} SIN(A)COS(B))DEL(U_1) - (-D_{F_1,P_2} COS(A)SIN^2(B)$$

$$- D_{F_2,P_2} COS(A)COS(B)SIN(B) + D_{F_1,P_1} COS(A)COS(B)SIN(B)$$

$$- D_{F_1,P_3} SIN(A)SIN(B) + D_{F_2,P_1} COS(A)COS^2(B)$$

$$- D_{F_2,P_3} SIN(A)COS(B))DEL(P_1) - (-D_{F_1,A_2} COS(A)SIN^2(B)$$

$$- D_{F_2,A_2} COS(A)COS(B)SIN(B) + D_{F_1,A_1} COS(A)COS(B)SIN(B)$$

$$- D_{F_1,A_3} SIN(A)SIN(B) + D_{F_2,A_1} COS(A)COS^2(B)$$

$$- D_{F_2,A_3} \, SIN(A)COS(B))DEL(A_1) - SIN(R_1)COS(R_2)GM$$

$$+ (\frac{d}{dT} U_2 - P_1 U_3 + U_1 P_3)M - S_{F_1} \, SIN(B) - S_{F_2} \, COS(B)$$

$$FT_3 = -(-D_{F_2,C_K} \, SIN(A)SIN(B) + D_{F_1,C_K} \, SIN(A)COS(B)$$

$$+ D_{F_3,C_K} \, COS(A))DEL(C_K) - (D_{F_2,U_2} \, SIN^2(A)SIN^2(B)$$

$$- D_{F_2,U_1} \, SIN^2(A)COS(B)SIN(B) - D_{F_1,U_2} \, SIN^2(A)COS(B)SIN(B)$$

$$- D_{F_3,U_2} \, COS(A)SIN(A)SIN(B) - D_{F_2,U_3} \, COS(A)SIN(A)SIN(B)$$

$$+ D_{F_1,U_1} \, SIN^2(A)COS^2(B) + D_{F_3,U_1} \, COS(A)SIN(A)COS(B)$$

$$+ D_{F_1,U_3} \, COS(A)SIN(A)COS(B) + D_{F_3,U_3} \, COS^2(A))DEL(U_3)$$

$$- (D_{F_2,P_2} \, SIN^2(A)SIN^2(B) - D_{F_2,P_1} \, SIN^2(A)COS(B)SIN(B)$$

$$- D_{F_1,P_2} \, SIN^2(A)COS(B)SIN(B) - D_{F_3,P_2} \, COS(A)SIN(A)SIN(B)$$

$$- D_{F_2,P_3} \, COS(A)SIN(A)SIN(B) + D_{F_1,P_1} \, SIN^2(A)COS^2(B)$$

$$+ D_{F_3,P_1} \, COS(A)SIN(A)COS(B) + D_{F_1,P_1} \, COS(A)SIN(A)COS^2(B)$$

$$- D_{F_1,P_3} \, SIN^2(A)COS(B) + D_{F_3,P_1} \, COS^2(A)COS(B)$$

$$- D_{F_3,P_3} \, COS(A)SIN(A))DEL(P_1) - (D_{F_2,A_2} \, COS(A)SIN(A)SIN^2(B)$$

$- D_{F_2,A_1} \cos(A)\sin(A)\cos(B)\sin(B)$

$- D_{F_1,A_2} \cos(A)\sin(A)\cos(B)\sin(B) + D_{F_2,A_3} \sin^2(A)\sin(B)$

$- D_{F_3,A_2} \cos^2(A)\sin(B) + D_{F_1,A_1} \cos(A)\sin(A)\cos^2(B)$

$- D_{F_1,A_3} \sin^2(A)\cos(B) + D_{F_3,A_1} \cos^2(A)\cos(B)$

$- D_{F_3,A_3} \cos(A)\sin(A))DEL(A_1) + M(-U_1 P_2 + \frac{d}{dT} U_3$

$+ P_1 U_2) - \cos(R_1)\cos(R_2)GM + S_{F_2} \sin(A)\sin(B)$

$- S_{F_1} \sin(A)\cos(B) - S_{F_3} \cos(A)$

$+ D_{F_1,P_2} \sin(A)\cos^2(B) + D_{F_3,P_2} \cos(A)\cos(B)) DEL(P_2)$

$- (-D_{F_2,A_1} \sin(A)\sin^2(B) - D_{F_2,A_2} \sin(A)\cos(B)\sin(B)$

$+ D_{F_1,A_1} \sin(A)\cos(B)\sin(B) + D_{F_3,A_1} \cos(A)\sin(B)$

$+ D_{F_1,A_2} \sin(A)\cos^2(B) + D_{F_3,A_2} \cos(A)\cos(B)) DEL(A_2)$

$- (D_{F_2,U_2} \cos(A)\sin(A)\sin^2(B) - D_{F_2,U_1} \cos(A)\sin(A)\cos(B)\sin(B)$

$- D_{F_1,U_2} \cos(A)\sin(A)\cos(B)\sin(B) + D_{F_2,U_3} \sin^2(A)\sin(B)$

$- D_{F_3,U_2} \cos^2(A)\sin(B) + D_{F_1,U_1} \cos(A)\sin(A)\cos^2(B)$

$- D_{F_1,U_3} \sin^2(A)\cos(B) + D_{F_3,U_1} \cos^2(A)\cos(B)$

$- D_{F_3,U_3} \cos(A) \sin(A)) \, DEL(U_1) - (D_{F_2,P_2} \cos(A) \sin(A) \sin^2(B)$

$- D_{F_2,P_1} \cos(A) \sin(A) \cos(B) \sin(B)$

$- D_{F_1,P_2} \cos(A) \sin(A) \cos(B) \sin(B)$

$+ D_{F_2,P_3} \sin^2(A) \sin(B) - D_{F_3,P_2} \cos^2(A) \sin(B)$

$+ D_{F_1,P_3} \cos(A)\sin(A)\cos(B)$

$+ D_{F_3,P_3} \cos^2(A)) \, DEL(P_3) - (D_{F_2,A_2} \sin^2(A)\sin^2(B)$

$- D_{F_2,A_1} \sin^2(A)\cos(B)\sin(B) - D_{F_1,A_2} \sin^2(A)\cos(B)\sin(B)$

$- D_{F_3,A_2} \cos(A)\sin(A)\sin(B) - D_{F_2,A_3} \cos(A)\sin(A)\sin(B)$

$+ D_{F_1,A_1} \sin^2(A)\cos^2(B) + D_{F_3,A_1} \cos(A)\sin(A)\cos(B)$

$+ D_{F_1,A_3} \cos(A)\sin(A)\cos(B) + D_{F_3,A_3} \cos^2(A)) \, DEL(A_3)$

$- (-D_{F_2,U_1} \sin(A)\sin^2(B) - D_{F_2,U_2} \sin(A)\cos(B)\sin(B)$

$+ D_{F_1,U_1} \sin(A)\cos(B)\sin(B) + D_{F_3,U_1} \cos(A)\sin(B)$

$+ D_{F_1,U_2} \sin(A)\cos^2(B) + D_{F_3,U_2} \cos(A)\cos(B)) \, DEL(U_2)$

$- (-D_{F_2,P_1} \sin(A)\sin^2(B) - D_{F_2,P_2} \sin(A)\cos(B)\sin(B)$

$+ D_{F_1,P_1} \sin(A)\cos(B)\sin(B) + D_{F_3,P_1} \cos(A)\sin(B)$

6.18 SPECIAL FORMS OF THE EQUATIONS OF MOTION

In aeronautical studies involving small perturbations about the equilibrium or trim condition, the investigator sometimes wants to know how the vehicle will respond if the motion is restricted in some way. For example, he might wish to determine vehicle response in the absence of sideslip (ref. 8). MACSYMA is well equipped to implement assumptions of this type. By using a substitution command, MACSYMA goes through the equations, makes the required substitutions, and displays the modified results. For the case of zero sideslip the program requests MACSYMA to make the substitutions: $SIN(B) = 0$ and $COS(B) = 1$ in each force equation. The required substitution and display commands and the modified equations assume the following form:

```
(C48) FOR I:1 THRU 3 DO FT[I]:SUBST([SIN(B)=0,COS(B)=1],FT[I])$
```

```
(C49) FOR I:1 THRU 3 DO DISPLAY(FT[I])$
```

$$FT_1 = -(D_{F_1,C_K} COS(A) - D_{F_3,C_K} SIN(A))DEL(C_K) - (-D_{F_3,U_1} SIN^2(A)$$

$$- D_{F_3,U_3} COS(A)SIN(A) + D_{F_1,U_1} COS(A)SIN(A)$$

$$+ D_{F_1,U_3} COS^2(A))DEL(U_3) - (-D_{F_3,P_1} SIN^2(A) - D_{F_3,P_3} COS(A)SIN(A)$$

$$+ D_{F_1,P_1} COS(A)SIN(A) + D_{F_1,P_3} COS^2(A))DEL(P_3) - (-D_{F_3,A_1} SIN^2(A)$$

$$- D_{F_3,A_3} COS(A)SIN(A) + D_{F_1,A_1} COS(A)SIN(A)$$

$$+ D_{F_1,A_3} COS^2(A))DEL(A_3) - (D_{F_1,U_2} COS(A) - D_{F_3,U_2} SIN(A))DEL(U_2)$$

$$- (D_{F_1,P_2} COS(A) - D_{F_3,P_2} SIN(A))DEL(P_2) - (D_{F_1,A_2} COS(A)$$

$$- D_{F_3,A_2} \text{ SIN(A))DEL}(A_2) - (D_{F_3,U_3} \text{ SIN}^2\text{(A)} - D_{F_3,U_1} \text{ COS(A)SIN(A)}$$

$$- D_{F_1,U_3} \text{ COS(A)SIN(A)} + D_{F_1,U_1} \text{ COS}^2\text{(A))DEL}(U_1) - (D_{F_3,P_3} \text{ SIN}^2\text{(A)}$$

$$- D_{F_3,P_1} \text{ COS(A)SIN(A)} - D_{F_1,P_3} \text{ COS(A)SIN(A)}$$

$$+ D_{F_1,P_1} \text{ COS}^2\text{(A))DEL}(P_1) - (D_{F_3,A_3} \text{ SIN}^2\text{(A)} - D_{F_3,A_1} \text{ COS(A)SIN(A)}$$

$$- D_{F_1,A_3} \text{ COS(A)SIN(A)} + D_{F_1,A_1} \text{ COS}^2\text{(A))DEL}(A_1) + \text{SIN}(R_2)\text{GM}$$

$$+ (\frac{d}{dT} U_1 + P_2 U_3 - U_2 P_3)M + S_{F_3} \text{ SIN(A)} - S_{F_1} \text{ COS(A)}$$

$$FT_2 = -D_{F_2,C_K} \text{ DEL}(C_K) - (D_{F_2,U_1} \text{ SIN(A)} + D_{F_2,U_3} \text{ COS(A))DEL}(U_3)$$

$$- (D_{F_2,P_1} \text{ SIN(A)} + D_{F_2,P_3} \text{ COS(A))DEL}(P_3) - (D_{F_2,A_1} \text{ SIN(A)}$$

$$+ D_{F_2,A_3} \text{ COS(A))DEL}(A_3) - D_{F_2 U_2} \text{ DEL}(U_2) - D_{F_2,P_2} \text{ DEL}(P_2)$$

$$- D_{F_2,A_2} \text{ DEL}(A_2) - (D_{F_2,U_1} \text{ COS(A)} - D_{F_2,U_3} \text{ SIN(A))DEL}(U_1)$$

$$- (D_{F_2,P_1} \text{ COS(A)} - D_{F_2,P} \text{ SIN(A)) DEL}(P_1) - (D_{F_2,A_1} \text{ COS(A)}$$

$$- D_{F_2,A_3} \text{ SIN(A))DEL}(A_1) - \text{SIN}(R_1)\text{COS}(R_2)\text{GM}$$

$$+ (\frac{d}{dT} U_2 - P_1 U_3 + U_1 P_3)M - S_{F_2}$$

$$FT_3 = -(D_{F_1,C_K} SIN(A) + D_{F_3,C_K} COS(A))DEL(C_K)$$

$$- (D_{F_1,U_1} SIN^2(A) + D_{F_3,U_1} COS(A)SIN(A) + D_{F_1,U_3} COS(A)SIN(A)$$

$$+ D_{F_3,U_3} COS^2(A))DEL(U_3) - (D_{F_1,P_1} SIN^2(A) + D_{F_3,P_1} COS(A)SIN(A)$$

$$+ D_{F_1,P_3} COS(A)SIN(A) + D_{F_3,P_3} COS^2(A))DEL(P_3) - (D_{F_1,A_1} SIN^2(A)$$

$$+ D_{F_3,A_1} COS(A)SIN(A) + D_{F_1,A_3} COS(A)SIN(A)$$

$$+ D_{F_3,A_3} COS^2(A))DEL(A_3) - (D_{F_1,U_2} SIN(A) + D_{F_3,U_2} COS(A))DEL(U_2)$$

$$- (D_{F_1,P_2} SIN(A) + D_{\Gamma_3,P_2} COS(A))DEL(P_2) - (D_{F_1,A_2} SIN(A)$$

$$+ D_{F_3,A_2} COS(A))DEL(A_2) - (-D_{F_1,U_3} SIN^2(A) - D_{F_3,U_3} COS(A)SIN(A)$$

$$+ D_{F_1,U_1} COS(A)SIN(A) + D_{F_3,U_1} COS^2(A))DEL(U_1)$$

$$- (-D_{F_1,P_3} SIN^2(A) - D_{F_3,P_3} COS(A)SIN(A) + D_{F_1,P_1} COS(A)SIN(A)$$

$$+ D_{F_3,P_1} COS^2(A))DEL(P_1) - (-D_{F_1,A_3} SIN^2(A)$$

$$- D_{F_3,A_3} COS(A)SIN(A) + D_{F_1,A_1} COS(A)SIN(A)$$

$$+ D_{F_3,A_1} COS^2(A))DEL(A_1) + M(-U_1P_2 + \frac{d}{dT} U_3 + P_1U_2)$$

$$- COS(R_1)COS(R_2)GM - S_{F_1} SIN(A) - S_{F_3} COS(A)$$

In addition to the zero sideslip condition, the investigator might wish to determine vehicle response when the angle of attack *A* is limited to small values. For this condition the program would request MACSYMA to make the substitution SIN(A)=A. Moreover, if the angle of attack were sufficiently small, the program would request MACSYMA to make the additional substitution COS(A)=1.

In this case, the required substitutions and display commands give rise to the following modified equations:

```
(C50) FOR I:1 THRU 3 DO FT[I]:SUBST([SIN(A)=A,COS(A)=1],FT[I])$

(C51) FOR I:1 THRU 3 DO DISPLAY(FT[I])$
```

$$FT_1 = -(D_{F_1,C_K} - D_{F_3,C_K} A)DEL(C_K) - (-D_{F_3,U_1} A^2 - D_{F_3,U_3} A$$

$$+ D_{F_1,U_1} A + D_{F_1,U_3})DEL(U_3) - (-D_{F_3,P_1} A^2 - D_{F_3,P_3} A + D_{F_1,P_1} A$$

$$+ D_{F_1,P_3})DEL(P_3) - (-D_{F_3,A_1} A^2 - D_{F_3,A_3} A + D_{F_1,A_1} A$$

$$+ D_{F_1,A_3})DEL(A_3) - (D_{F_1,U_2} - D_{F_3,U_2} A)DEL(U_2) - (D_{F_1,P_2}$$

$$- D_{F_3,P_2} A)DEL(P_2) - (D_{F_1,A_2} - D_{F_3,A_2} A)DEL(A_2) - (D_{F_3,U_3} A^2$$

$$- D_{F_3,U_1} A - D_{F_1,U_3} A + D_{F_1,U_1})DEL(U_1) - (D_{F_3,P_3} A^2 - D_{F_3,P_1} A$$

$$- D_{F_1,P_3} A + D_{F_1,P_1})DEL(P_1) - (D_{F_3,A_3} A^2 - D_{F_3,A_1} A - D_{F_1,A_3} A$$

$$+ D_{F_1,A_1})DEL(A_1) + SIN(R_2)GM + (\frac{d}{dT} U_1 + P_2U_3 - U_2P_3)M + S_{F_3} A - S_{F_1}$$

$$FT_2 = -D_{F_2,C_K} DEL(C_K) - (D_{F_2,U_1} A + D_{F_2,U_3}) DEL(U_3) - (D_{F_2,P_1} A$$

$$+ D_{F_2,P_3}) DEL(P_3) - (D_{F_2,A_1} A + D_{F_2,A_3}) DEL(A_3) - D_{F_2,U_2} DEL(U_2)$$

$$- D_{F_2,P_2} DEL(P_2) - D_{F_2,A_2} DEL(A_2) - (D_{F_2,U_1} - D_{F_2,U_3} A) DEL(U_1)$$

$$- (D_{F_2,P_1} - D_{F_2,P_3} A) DEL(P_1) - (D_{F_2,A_1} - D_{F_2,A_3} A) DEL(A_1)$$

$$- SIN(R_1) COS(R_2) GM + (\frac{d}{dT} U_2 - P_1 U_3 + U_1 P_3) M - S_{F_2}$$

$$FT_3 = -(D_{F_1,C_K} A + D_{F_3,C_K}) DEL(C_K) - (D_{F_1,U_1} A^2 + D_{F_3,U_1} A$$

$$+ D_{F_1,U_3} A + D_{F_3,U_3}) DEL(U_3) - (D_{F_1,P_1} A^2 + D_{F_3,P_1} A + D_{F_1,P_3} A$$

$$+ D_{F_3,P_3}) DEL(P_3) - (D_{F_1,A_1} A^2 + D_{F_3,A_1} A + D_{F_1,A_3} A$$

$$+ D_{F_3,A_3}) DEL(A_3) - (D_{F_1,U_2} A + D_{F_3,U_2}) DEL(U_2) - (D_{F_1,P_2} A$$

$$+ D_{F_3,P_2}) DEL(P_2) - (D_{F_1,A_2} A + D_{F_3,A_2}) DEL(A_2) - (-D_{F_1,U_3} A^2$$

$$- D_{F_3,U_3} A + D_{F_1,U_1} A + D_{F_3,U_1}) DEL(U_1) - (-D_{F_1,P_3} A^2 - D_{F_3,P_3} A$$

$$+ D_{F_1,P_1} A + D_{F_3,P_1}) DEL(P_1) - (-D_{F_1,A_3} A^2 - D_{F_3,A_3} A + D_{F_1,A_1} A$$

$$+ D_{F_3,A_1}) DEL(A_1) + M(-U_1 P_2 + \frac{d}{dT} U_3 + P_1 U_2) - COS(R_1) COS(R_2) GM$$

$$- S_{F_1} A - S_{F_3}$$

Examination of these equations reveals the existence of terms such as A^2. If it is assumed that second-order terms in A are negligible, a program statement instructing MACSYMA to make the substitution $A^2 = 0$ would simplify the equations as follows:

```
(C52) FOR I:1 THRU 3 DO FT[I]:SUBST([A**2=0],FT[I])$

(C53) FOR I:1 THRU 3 DO DISPLAY(FT[I])$
```

$$FT_1 = -(D_{F_1,C_K} - D_{F_3,C_K} A)DEL(C_K) - (-D_{F_3,U_3} A + D_{F_1,U_1} A$$

$$+ D_{F_1,U_3})DEL(U_3) - (-D_{F_3,P_3} A + D_{F_1,P_1} A + D_{F_1,P_3})DEL(P_3)$$

$$- (-D_{F_3,A_3} A + D_{F_1,A_1} A + D_{F_1,A_3})DEL(A_3) - (D_{F_1,U_2}$$

$$- D_{F_3,U_2} A)DEL(U_2) - (D_{F_1,P_2} - D_{F_3,P_2} A)DEL(P_2) - (D_{F_1,A_2}$$

$$- D_{F_3,A_2} A)DEL(A_2) - (-D_{F_3,U_1} A - D_{F_1,U_3} A + D_{F_1,U_1})DEL(U_1)$$

$$- (-D_{F_3,P_1} A - D_{F_1,P_3} A + D_{F_1,P_1})DEL(P_1) - (-D_{F_3,A_1} A - D_{F_1,A_3} A$$

$$+ D_{F_1,A_1})DEL(A_1) + SIN(R_2)GM + (\frac{d}{dT} U_1 + P_2 U_3 - U_2 P_3)M + S_{F_3} A - S_{F_1}$$

$$FT_2 = -D_{F_2,C_K} DEL(C_K) - (D_{F_2,U_1} A + D_{F_2,U_3})DEL(U_3) - (D_{F_2,P_1} A$$

$$+ D_{F_2,P_3})DEL(P_3) - (D_{F_2,A_1} A + D_{F_2,A_3})DEL(A_3) - D_{F_2,U_2} DEL(U_2)$$

$$- D_{F_2,P_2} DEL(P_2) - D_{F_2,A_2} DEL(A_2) - (D_{F_2,U_1} - D_{F_2,U_3} A)DEL(U_1)$$

$$- (D_{F_2,P_1} - D_{F_2,P_3} A)\text{DEL}(P_1) - (D_{F_2,A_1} - D_{F_2,A_3} A)\text{DEL}(A_1)$$

$$- \text{SIN}(R_1)\text{COS}(R_2)\text{GM} + (\frac{d}{dT} U_2 - P_1 U_3 + U_1 P_3)M - S_{F_2}$$

$$\text{FT}_3 = -(D_{F_1,C_K} A + D_{F_3,C_K})\text{DEL}(C_K) - (D_{F_3,U_1} A + D_{F_1,U_3} A + D_{F_3,U_3})$$

$$\text{DEL}(U_3) - (D_{F_3,P_1} A + D_{F_1,P_3} A + D_{F_3,P_3})\text{DEL}(P_3) - (D_{F_3,A_1} A$$

$$+ D_{F_1,A_3} A + D_{F_3,A_3})\text{DEL}(A_3) - (D_{F_1,U_2} A + D_{F_3,U_2})\text{DEL}(U_2)$$

$$- (D_{F_1,P_2} A + D_{F_3,P_2})\text{DEL}(P_2) - (D_{F_1,A_2} A + D_{F_3,A_2})\text{DEL}(A_2)$$

$$- (-D_{F_3,U_3} A + D_{F_1,U_1} A + D_{F_3,U_1})\text{DEL}(U_1) - (-D_{F_3,P_3} A + D_{F_1,P_1} A$$

$$+ D_{F_3,P_1})\text{DEL}(P_1) - (-D_{F_3,A_3} A + D_{F_1,A_1} A + D_{F_3,A_1})\text{DEL}(A_1)$$

$$+ M(-U_1 P_2 + \frac{d}{dT} U_3 + P_1 U_2) - \text{COS}(R_1)\text{COS}(R_2)\text{GM} - S_{F_1} A - S_{F_3}$$

Additional simplifications are possible if it is assumed that angular velocity perturbations are negligible. This assumption can be implemented by again using the substitution command, which yields the following greatly simplified equations:

```
(C54) FOR I:1 THRU 3 DO FOR J:1 THRU 3 DO
  FT[I]:SUBST([DEL(P[J])=0],FT[I])$

(C56) FOR I:1 THRU 3 DO DISPLAY(FT[I])$
```

$$FT_1 = -(D_{F_1,C_K} - D_{F_3,C_K} A)DEL(C_K) - (-D_{F_3,U_3} A + D_{F_1,U_1} A$$

$$+ D_{F_1,U_3})DEL(U_3) - (-D_{F_3,A_3} A + D_{F_1,A_1} A + D_{F_1,A_3})DEL(A_3)$$

$$- (D_{F_1,U_2} - D_{F_3,U_2} A)DEL(U_2) - (D_{F_1,A_2} - D_{F_3,A_2} A)DEL(A_2)$$

$$- (-D_{F_3,U_1} A - D_{F_1,U_3} A + D_{F_1,U_1})DEL(U_1) - (-D_{F_3,A_1} A - D_{F_1,A_3} A$$

$$+ D_{F_1,A_1})DEL(A_1) + SIN(R_2)GM + (\tfrac{d}{dT} U_1 + P_2 U_3 - U_2 P_3)M + S_{F_3} A - S_{F_1}$$

$$FT_2 = -D_{F_2,C_K} DEL(C_K) - (D_{F_2,U_1} A + D_{F_2,U_3})DEL(U_3) - (D_{F_2,A_1} A$$

$$+ D_{F_2,A_3})DEL(A_3) - D_{F_2,U_2} DEL(U_2) - D_{F_2,A_2} DEL(A_2)$$

$$- (D_{F_2,U_1} - D_{F_2,U_3} A)DEL(U_1) - (D_{F_2,A_1} - D_{F_2,A_3} A)DEL(A_1)$$

$$- SIN(R_1)COS(R_2)GM + (\tfrac{d}{dT} U_2 - P_1 U_3 + U_1 P_3)M - S_{F_2}$$

$$FT_3 = -(D_{F_1,C_K} A + D_{F_3,C_K})DEL(C_K) - (D_{F_3,U_1} A + D_{F_1,U_3} A + D_{F_3,U_3})$$

$$DEL(U_3) - (D_{F_3,A_1} A + D_{F_1,A_3} A + D_{F_3,A_3})DEL(A_3) - (D_{F_1,U_2} A$$

$$+ D_{F_3,U_2})DEL(U_2) - (D_{F_1,A_2} A + D_{F_3,A_2})DEL(A_2) - (-D_{F_3,U_3} A$$

$$+ D_{F_1,U_1} A + D_{F_3,U_1})DEL(U_1) - (-D_{F_3,A_3} A + D_{F_1,A_1} A$$

$$+ \, D_{F_3,A_1} \,)DEL(A_1) + M(-U_1 P_2 + \frac{d}{dT} U_3 + P_1 U_2)$$

$$- \, COS(R_1)COS(R_2)GM - S_{F_1} A - S_{F_3}$$

Finally, it may be of interest to consider the effect of omitting the linear acceleration terms. By comparing the response of the system with and without acceleration perturbations, the influence of these perturbations can be determined. Again, a simple substitution command is all that is required to implement the assumption that $DEL(A_i)=0$. Execution of this command yields the modified equations as follows:

```
(C57) FOR I:1 THRU 3 DO FOR J:1 THRU 3 DO
 FT[I]:SUBST([DEL(A[J])=0],FT[I])$
```

```
(C58) FOR I:1 THRU 3 DO DISPLAY(FT[I])$
```

$$FT_1 = -(D_{F_1,C_K} - D_{F_3,C_K} A)DEL(C_K) - (-D_{F_3,U_3} A + D_{F_1,U_1} A$$

$$+ \, D_{F_1,U_3})DEL(U_3) - (D_{F_1,U_2} - D_{F_3,U_2} A)DEL(U_2) - (-D_{F_3,U_1} A$$

$$- \, D_{F_1,U_3} A + D_{F_1,U_1})DEL(U_1) + SIN(R_2)GM + (\frac{d}{dT} U_1 + P_2 U_3 - U_2 P_3)M$$

$$+ \, S_{F_3} A - S_{F_1}$$

$$FT_2 = -D_{F_2,C_K} DEL(C_K) - (D_{F_2,U_1} A + D_{F_2,U_3})DEL(U_3) - D_{F_2,U_2} DEL(U_2)$$

$$- \, (D_{F_2,U_1} - D_{F_2,U_3} A)DEL(U_1) - SIN(R_1)COS(R_2)GM$$

$$+ \, (\frac{d}{dT} U_2 - P_1 U_3 + U_1 P_3)M - S_{F_2}$$

$$FT_3 = -(D_{F_1,C_K} A + D_{F_3,C_K})DEL(C_K) - (D_{F_3,U_1} A + D_{F_1,U_3} A + D_{F_3,U_3})$$

$$DEL(U_3) - (D_{F_1,U_2} A + D_{F_3,U_2})DEL(U_2) - (-D_{F_3,U_3} A + D_{F_1,U_1} A$$

$$+ D_{F_3,U_1})DEL(U_1) + M(-U_1P_2 + \frac{d}{dT} U_3 + P_1U_2)$$

$$- COS(R_1)COS(R_2)GM - S_{F_1} A - S_{F_3}$$

6.19 THRUST FORCES

It should be noted that the thrust forces T_i appearing on the left-hand side of these equations are the resultant of a number of thrust generating systems, each contributing a thrust vector T_n. Each thrust vector is referred to a thrust axes system $X_n{}^i$ with origin at the point of application of the thrust vector. The axes are chosen such that each thrust vector coincides with the $X_n{}^1$ axis of the system. Moreover, each thrust vector is then transformed to a coordinate system $Y_n{}^i$ which has the same origin as the thrust axes, but is parallel to the body axes system. Finally, the components of thrust in the $Y_n{}^i$ system of axes are transformed to the body axes system, which has its origin at the center of gravity of the aircraft. Each thrust axis $X_n{}^1$ is related to the $Y_n{}^i$ system by the following transformation equations (see sketch):

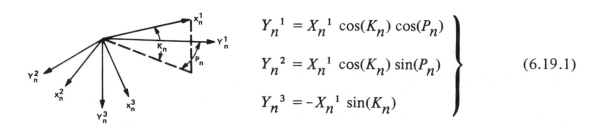

$$Y_n{}^1 = X_n{}^1 \cos(K_n) \cos(P_n)$$

$$Y_n{}^2 = X_n{}^1 \cos(K_n) \sin(P_n) \quad\Bigg\} \quad (6.19.1)$$

$$Y_n{}^3 = -X_n{}^1 \sin(K_n)$$

Hence, the components of the thrust vector T_n in the $Y_n{}^i$ system of coordinates are

$$\frac{\partial Y_n{}^1}{\partial X_n{}^1} T_n \; ; \qquad \frac{\partial Y_n{}^2}{\partial X_n{}^1} T_n \; ; \qquad \frac{\partial Y_n{}^3}{\partial X_n{}^1} T_n \qquad\qquad (6.19.2)$$

These are also the components of thrust in the y^i system of coordinates, which has its origin at the center of gravity of the aircraft. The thrust components due to all thrust generating systems are obtained by summing the right-hand side of the following equation

$$T^i = \frac{\partial Y_n{}^i}{\partial X_n{}^1} T_n \qquad\qquad (6.19.3)$$

The expanded form of equation (6.19.3), when summed over n will yield the resultant thrust components. When the number of thrust generating systems is known, the components T^i can be formulated and displayed by using equation (6.19.3), and executing the following two commands, which yield the components contributed by the nth thrust generating system. These are

```
(C1) Y[1,N]:X[1,N]*COS(K[N])*COS(P[N])$

(C2) Y[2,N]:X[1,N]*COS(K[N])*SIN(P[N])$

(C3) Y[3,N]:-X[1,N]*SIN(K[N])$

(C4) FOR I THRU 3 DO T[I]:DIFF(Y[I,N],X[1,N],1)*T[N]$

(C5) FOR I:1 THRU 3 DO DISPLAY (T[I])$
```

$$T_1 = T_N \cos(K_N) \cos(P_N)$$

$$T_2 = T_N \cos(K_N) \sin(P_N)$$

$$T_3 = -T_N \sin(K_N)$$

6.20 DETERMINATION OF THE GEOGRAPHICAL LOCATION
OF AIRCRAFT

In order to determine the geographical location of an aircraft relative to some initial location, it is necessary to transform the components of the aircraft's velocity vector from aircraft body axes to a system of Earth-fixed axes. The transformed components can then be integrated to find the location of the aircraft as a function of time. The product of the three rotation matrices (D33), (D32), and (D31), which were used to transform the gravity vector from an Earth-fixed axes system to aircraft body axes, may be transposed and used to transform the aircraft velocity components to an Earth-fixed system. If the column vector (D39) of aircraft velocity components is premultiplied by the transposed matrix, the velocity components relative to the Earth-fixed system are obtained as follows:

(C56) TRANSPOSE((D33).(D32).(D31)).(D39);

(D56) MATRIX([U_3(SIN(R_1)SIN(R_3) + COS(R_1)SIN(R_2)COS(R_3))

  + U_2(SIN(R_1)SIN(R_2)COS(R_3) - COS(R_1)SIN(R_3)) + U_1COS(R_2)COS(R_3)],

[U_2(SIN(R_1)SIN(R_2)SIN(R_3) + COS(R_1)COS(R_3))

  + U_3(COS(R_1)SIN(R_2)SIN(R_3) - SIN(R_1)COS(R_3)) + U_1COS(R_2)SIN(R_3)],

[-U_1SIN(R_2) + U_2SIN(R_1)COS(R_2) + U_3COS(R_1)COS(R_2)])

If the components $\dot{X}_i$ relative to the Earth-fixed system be denoted by DX_i, execution of the following programming steps will ensure that the required velocity components are displayed in conventional form.

(C57) FOR I:1 THRU 3 DO ROW[I]:FIRST(ROW((D56),I))$

(C58) FOR I:1 THRU 3 DO (DX[I]:ROW[I][1],DISPLAY(DX[I]));

DX_1 = U_3(SIN(R_1)SIN(R_3) + COS(R_1)SIN(R_2)COS(R_3))

  + U_2(SIN(R_1)SIN(R_2)COS(R_3) - COS(R_1)SIN(R_3)) + U_1COS(R_2)COS(R_3)

$$DX_2 = U_2(SIN(R_1)SIN(R_2)SIN(R_3) + COS(R_1)COS(R_3))$$

$$+ U_3(COS(R_1)SIN(R_2)SIN(R_3) - SIN(R_1)COS(R_3)) + U_1COS(R_2)SIN(R_3)$$

$$DX_3 = -U_1SIN(R_2) + U_2SIN(R_1)COS(R_2) + U_3COS(R_1)COS(R_2)$$

Integration of these velocity components will yield the required coordinates of the aircraft relative to a set of Earth-fixed reference axes. These are

$$X_E{}^i = X_{EO}^i + \int DX[I]\,dt$$

where X_{EO}^i are the initial values of the coordinates in the Earth-fixed reference frame.

6.21 TRANSFORMATION LAW FOR STATIC MOMENTS

The static aerodynamic moments obey the same transformation law as the static aerodynamic forces; that is, if S_{M_n} denotes a static moment in the X frame of reference, and SM_i denotes the corresponding transformed moment in the Y reference frame, then

$$SM_i = \frac{\partial Y^i}{\partial X^n} S_{M_n} \tag{6.21.1}$$

where $Y = Y(X)$ is obtained from the displayed output (D6) and reentered here to facilitate the formulation of the moment equations. Given the transformation equations (D6), the transformed aerodynamic static moments are obtained by expanding equation (6.21.1). The three programming steps used to transform the static forces may again be employed to transform the static moments. The simple program and the displayed results are

```
(C1) Y[1]:X[1]*COS(A)*COS(B)-X[2]*COS(A)*SIN(B)-X[3]*SIN(A)$

(C2) Y[2]:X[1]*SIN(B)+X[2]*COS(B)$
```

```
(C3) Y[3]:X[1]*SIN(A)*COS(B)-X[2]*SIN(A)*SIN(B)+X[3]*COS(A)$

(C4) SM[I]:=0$

(C5) FOR I:1 THRU 3 DO FOR N:1 THRU 3 DO
 SM[I]:SM[I]+DIFF(Y[I],X[N])*S[M[N]]$

(C6) FOR I:1 THRU 3 DO DISPLAY (SM[I])$
```

$$SM_1 = -S_{M_2} \cos(A)\sin(B) + S_{M_1} \cos(A)\cos(B) - S_{M_3} \sin(A)$$

$$SM_2 = S_{M_1} \sin(B) + S_{M_2} \cos(B)$$

$$SM_3 = -S_{M_2} \sin(A)\sin(B) + S_{M_1} \sin(A)\cos(B) + S_{M_3} \cos(A)$$

6.22 TRANSFORMATION LAW FOR CONTROL MOMENT DERIVATIVES

The control moment derivatives obey the same transformation law as the static moments; that is, if D_{M_n,C_K} denotes the nth control moment derivative with respect to the Kth control surface as measured in the X reference frame, and $TD_{I,C}$ denotes the corresponding transformed derivative in the Y frame, then

$$TD_{I,C} = \frac{\partial Y^i}{\partial X^n} D_{M_n,C_k} \tag{6.22.1}$$

where $Y = Y(X)$ is again obtained from the displayed output (D6).

As in the preceding section, the transformed control derivatives are obtained by expanding the transformation law (6.22.1) given the transformation equations (D6). The transformed derivatives are obtained by executing the following simple program, which has exactly the same form as the program used to transform the static moments in section 6.21. These are

```
(C7) TD[I,C]:=0$
```

```
(C8) FOR I:1 THRU 3 DO FOR N:1 THRU 3 DO
 TD[I,C]:TD[I,C]+DIFF(Y[I],X[N])*D[M[N],C[K]]$
```

```
(C9) FOR I:1 THRU 3 DO DISPLAY(TD[I,C])$
```

$$TD_{1,C} = -D_{M_2,C_K} \cos(A)\sin(B) + D_{M_1,C_K} \cos(A)\cos(B) - D_{M_3,C_K} \sin(A)$$

$$TD_{2,C} = D_{M_1,C_K} \sin(B) + D_{M_2,C_K} \cos(B)$$

$$TD_{3,C} = -D_{M_2,C_K} \sin(A)\sin(B) + D_{M_1,C_K} \sin(A)\cos(B) + D_{M_3,C_K} \cos(A)$$

The corresponding control moments are obtained by multiplying the control derivatives by the appropriate control increments $DEL(C_K)$. The following two programming steps are sufficient to formulate the required moments. These are denoted by CM_i in the displayed output.

```
(C10) FOR I:1 THRU 3 DO CM[I]:TD[I,C]*DEL(C[K])$
```

```
(C11) FOR I:1 THRU 3 DO DISPLAY(CM[I])$
```

$$CM_1 = (-D_{M_2,C_K} \cos(A)\sin(B) + D_{M_1,C_K} \cos(A)\cos(B)$$
$$- D_{M_3,C_K} \sin(A))DEL(C_K)$$

$$CM_2 = (D_{M_1,C_K} \sin(B) + D_{M_2,C_K} \cos(B))DEL(C_K)$$

$$CM_3 = (-D_{M_2,C_K} \sin(A)\sin(B) + D_{M_1,C_K} \sin(A)\cos(B)$$
$$+ D_{M_3,C_K} \cos(A))DEL(C_K)$$

6.23 MOMENTS PRODUCED BY LINEAR VELOCITY PERTURBATIONS

The next step in the formulation involves the determination of the aerodynamic moments produced when an aircraft is subjected to linear velocity perturbations $DEL(U_J)$. Before these moments can be determined, the aerodynamic stability derivatives with respect to linear velocity components must be transformed from wind or wind-tunnel stability axes to body axes. For a detailed discussion of the transformation of these derivatives, the reader is referred to section 6.2. The program used for that transformation can be used in this case also. In this application, the aerodynamic stability derivative of the *i*th moment with respect to the *j*th velocity component will be denoted by D_{M_i, U_j}. The corresponding transformed derivatives are denoted by TD_{M_i, U_j}. When the program of section 6.2 is rewritten to accommodate the notational changes required for this application, it assumes the following form:

```
(C12) TDU[I,J]:=0$
```

```
(C13) FOR I:1 THRU 3 DO FOR J:1 THRU 3 DO
 FOR R:1 THRU 3 DO FOR N:1 THRU 3 DO
 TDU[I,J]:TDU[I,J]+DIFF(Y[I],X[R])*DIFF(Y[J],X[N])*D[M[R],U[N]]$
```

It only remains to multiply the transformed derivatives by the appropriate velocity increments to obtain the required moments, which are denoted by MDU_i. The next three programming steps instruct MACSYMA to evaluate and display the moments produced by linear velocity perturbations. These are

```
(C14) MDU[I]:=0$
```

```
(C15) FOR I:1 THRU 3 DO FOR J:1 THRU 3 DO
 MDU[I]:MDU[I]+TDU[I,J]*DEL(U[J])$
```

```
(C16) FOR I:1 THRU 3 DO DISPLAY(MDU[I])$
```

$$MDU_1 = (D_{M_2, U_2} \cos(A)\sin(A)\sin^2(B)$$

$$- D_{M_2, U_1} \cos(A)\sin(A)\cos(B)\sin(B)$$

$$- D_{M_1,U_2} \cos(A)\sin(A)\cos(B)\sin(B) + D_{M_3,U_2} \sin^2(A)\sin(B)$$

$$- D_{M_2,U_3} \cos^2(A)\sin(B) + D_{M_1,U_1} \cos(A)\sin(A)\cos^2(B)$$

$$- D_{M_3,U_1} \sin^2(A)\cos(B) + D_{M_1,U_3} \cos^2(A)\cos(B)$$

$$- D_{M_3,U_3} \cos(A)\sin(A))DEL(U_3) + (-D_{M_2,U_1} \cos(A)\sin^2(B)$$

$$- D_{M_2,U_2} \cos(A)\cos(B)\sin(B) + D_{M_1,U_1} \cos(A)\cos(B)\sin(B)$$

$$- D_{M_3,U_1} \sin(A)\sin(B) + D_{M_1,U_2} \cos(A)\cos^2(B)$$

$$- D_{M_3,U_2} \sin(A)\cos(B))DEL(U_2) + (D_{M_2,U_2} \cos^2(A)\sin^2(B)$$

$$- D_{M_2,U_1} \cos^2(A)\cos(B)\sin(B) - D_{M_1,U_2} \cos^2(A)\cos(B)\sin(B)$$

$$+ D_{M_3,U_2} \cos(A)\sin(A)\sin(B) + D_{M_2,U_3} \cos(A)\sin(A)\sin(B)$$

$$+ D_{M_1,U_1} \cos^2(A)\cos^2(B) - D_{M_3,U_1} \cos(A)\sin(A)\cos(B)$$

$$- D_{M_1,U_3} \cos(A)\sin(A)\cos(B) + D_{M_3,U_3} \sin^2(A))DEL(U_1)$$

$$MDU_2 = (-D_{M_1,U_2} \sin(A)\sin^2(B) - D_{M_2,U_2} \sin(A)\cos(B)\sin(B)$$

$$+ D_{M_1,U_1} \sin(A)\cos(B)\sin(B) + D_{M_1,U_3} \cos(A)\sin(B)$$

$$+ D_{M_2,U_1} \sin(A)\cos^2(B) + D_{M_2,U_3} \cos(A)\cos(B))DEL(U_3)$$

$$+ (D_{M_1,U_1} \, \text{SIN}^2(B) + D_{M_2,U_1} \, \text{COS(B)SIN(B)} + D_{M_1,U_2} \, \text{COS(B)SIN(B)}$$

$$+ D_{M_2,U_2} \, \text{COS}^2(B))\text{DEL}(U_2) + (-D_{M_1,U_2} \, \text{COS(A)SIN}^2(B)$$

$$- D_{M_2,U_2} \, \text{COS(A)COS(B)SIN(B)} + D_{M_1,U_1} \, \text{COS(A)COS(B)SIN(B)}$$

$$- D_{M_1,U_3} \, \text{SIN(A)SIN(B)} + D_{M_2,U_1} \, \text{COS(A)COS}^2(B)$$

$$- D_{M_2,U_3} \, \text{SIN(A)COS(B))DEL}(U_1)$$

$$\text{MDU}_3 = (D_{M_2,U_2} \, \text{SIN}^2(A)\text{SIN}^2(B) - D_{M_2,U_1} \, \text{SIN}^2(A)\text{COS(B)SIN(B)}$$

$$- D_{M_1,U_2} \, \text{SIN}^2(A)\text{COS(B)SIN(B)} - D_{M_3,U_2} \, \text{COS(A)SIN(A)SIN(B)}$$

$$- D_{M_2,U_3} \, \text{COS(A)SIN(A)SIN(B)} + D_{M_1,U_1} \, \text{SIN}^2(A)\text{COS}^2(B)$$

$$+ D_{M_3,U_1} \, \text{COS(A)SIN(A)COS(B)} + D_{M_1,U_3} \, \text{COS(A)SIN(A)COS(B)}$$

$$+ D_{M_3,U_3} \, \text{COS}^2(A))\text{DEL}(U_3) + (-D_{M_2,U_1} \, \text{SIN(A)SIN}^2(B)$$

$$- D_{M_2,U_2} \, \text{SIN(A)COS(B)SIN(B)} + D_{M_1,U_1} \, \text{SIN(A)COS(B)SIN(B)}$$

$$+ D_{M_3,U_1} \, \text{COS(A)SIN(B)} + D_{M_1,U_2} \, \text{SIN(A)COS}^2(B)$$

$$+ D_{M_3,U_2} \, \text{COS(A)COS(B))DEL}(U_2) + (D_{M_2,U_2} \, \text{COS(A)SIN(A)SIN}^2(B)$$

$$- D_{M_2,U_1} \, \text{COS(A)SIN(A)COS(B)SIN(B)}$$

$$- D_{M_1,U_2} \cos(A)\sin(A)\cos(B)\sin(B) + D_{M_2,U_3} \sin^2(A)\sin(B)$$

$$- D_{M_3,U_2} \cos^2(A)\sin(B) + D_{M_1,U_1} \cos(A)\sin(A)\cos^2(B)$$

$$- D_{M_1,U_3} \sin^2(A)\cos(B) + D_{M_3,U_1} \cos^2(A)\cos(B)$$

$$- D_{M_3,U_3} \cos(A)\sin(A))\mathrm{DEL}(U_1)$$

6.24 MOMENTS PRODUCED BY ANGULAR VELOCITY PERTURBATIONS

The program used in section 6.23 can, with suitable notational changes, be used to formulate the moments produced by angular velocity perturbations. However, whereas in the preceding application the required moments were obtained by multiplying the transformed aerodynamic stability derivatives by linear velocity increments, in the present case the transformed derivatives must be multiplied by angular velocity increments. In view of these similarities, the following program and displayed moments will be presented without further comment, except to point out that the aerodynamic stability derivatives of the ith moment with respect to the jth angular velocity component are denoted by D_{M_i,P_j}. The corresponding transformed derivatives are denoted by TD_{M_i,P_j}, and the resulting moments by MDP_i.

```
(C17) TDP[I,J]:=0$

(C18) FOR I:1 THRU 3 DO FOR J:1 THRU 3 DO
 FOR R:1 THRU 3 DO FOR N:1 THRU 3 DO
 TDP[I,J]:TDP[I,J]+DIFF(Y[I],X[R])*DIFF(Y[J],X[N])*D[M[R],P[N]]$

(C19) MDP[I]:=0$

(C21) FOR I:1 THRU 3 DO FOR J:1 THRU 3 DO
 MDP[I]:MDP[I]+TDP[I,J]*DEL(P[J])$

(C22) FOR I:1 THRU 3 DO DISPLAY(MDP[I])$
```

$$MDP_1 = (D_{M_2,P_2} \cos(A)\sin(A)\sin^2(B)$$

$$- D_{M_2,P_1} \cos(A)\sin(A)\cos(B)\sin(B)$$

$$- D_{M_1,P_2} \cos(A)\sin(A)\cos(B)\sin(B) + D_{M_3,P_2} \sin^2(A)\sin(B)$$

$$- D_{M_2,P_3} \cos^2(A)\sin(B) + D_{M_1,P_1} \cos(A)\sin(A)\cos^2(B)$$

$$- D_{M_3,P_1} \sin^2(A)\cos(B) + D_{M_1,P_3} \cos^2(A)\cos(B)$$

$$- D_{M_3,P_3} \cos(A)\sin(A))DEL(P_3) + (-D_{M_2,P_1} \cos(A)\sin^2(B)$$

$$- D_{M_2,P_2} \cos(A)\cos(B)\sin(B) + D_{M_1,P_1} \cos(A)\cos(B)\sin(B)$$

$$- D_{M_3,P_1} \sin(A)\sin(B) + D_{M_1,P_2} \cos(A)\cos^2(B)$$

$$- D_{M_3,P_2} \sin(A)\cos(B))DEL(P_2) + (D_{M_2,P_2} \cos^2(A)\sin^2(B)$$

$$- D_{M_2,P_1} \cos^2(A)\cos(B)\sin(B) - D_{M_1,P_2} \cos^2(A)\cos(B)\sin(B)$$

$$+ D_{M_3,P_2} \cos(A)\sin(A)\sin(B) + D_{M_2,P_3} \cos(A)\sin(A)\sin(B)$$

$$+ D_{M_1,P_1} \cos^2(A)\cos^2(B) - D_{M_3,P_1} \cos(A)\sin(A)\cos(B)$$

$$- D_{M_1,P_3} \cos(A)\sin(A)\cos(B) + D_{M_3,P_3} \sin^2(A))DEL(P_1)$$

$$MDP_2 = (-D_{M_1,P_2} \, SIN(A)SIN^2(B) - D_{M_2,P_2} \, SIN(A)COS(B)SIN(B)$$

$$+ \, D_{M_1,P_1} \, SIN(A)COS(B)SIN(B) + D_{M_1,P_3} \, COS(A)SIN(B)$$

$$+ \, D_{M_2,P_1} \, SIN(A)COS^2(B) + D_{M_2,P_3} \, COS(A)COS(B))DEL(P_3)$$

$$+ \, (D_{M_1,P_1} \, SIN^2(B) + D_{M_2,P_1} \, COS(B)SIN(B) + D_{M_1,P_2} \, COS(B)SIN(B)$$

$$+ \, D_{M_2,P_2} \, COS^2(B))DEL(P_2) + (-D_{M_1,P_2} \, COS(A)SIN^2(B)$$

$$- \, D_{M_2,P_2} \, COS(A)COS(B)SIN(B) + D_{M_1,P_1} \, COS(A)COS(B)SIN(B)$$

$$- \, D_{M_1,P_3} \, SIN(A)SIN(B) + D_{M_2,P_1} \, COS(A)COS^2(B)$$

$$- \, D_{M_2,P_3} \, SIN(A)COS(B))DEL(P_1)$$

$$MDP_3 = (D_{M_2,P_2} \, SIN^2(A)SIN^2(B) - D_{M_2,P_1} \, SIN^2(A)COS(B)SIN(B)$$

$$- \, D_{M_1,P_2} \, SIN^2(A)COS(B)SIN(B) - D_{M_3,P_2} \, COS(A)SIN(A)SIN(B)$$

$$- \, D_{M_2,P_3} \, COS(A)SIN(A)SIN(B) + D_{M_1,P_1} \, SIN^2(A)COS^2(B)$$

$$+ \, D_{M_3,P_1} \, COS(A)SIN(A)COS(B) + D_{M_1,P_3} \, COS(A)SIN(A)COS(B)$$

$$+ \, D_{M_3,P_3} \, COS^2(A))DEL(P_3) + (-D_{M_2,P_1} \, SIN(A)SIN^2(B)$$

$$- \, D_{M_2,P_2} \, SIN(A)COS(B)SIN(B) + D_{M_1,P_1} \, SIN(A)COS(B)SIN(B)$$

$$+ \ D_{M_3,P_1} \ \cos(A)\sin(B) \ + \ D_{M_1,P_2} \ \sin(A)\cos^2(B)$$

$$+ \ D_{M_3,P_2} \ \cos(A)\cos(B))\text{DEL}(P_2) \ + \ (D_{M_2,P_2} \ \cos(A)\sin(A)\sin^2(B)$$

$$- \ D_{M_2,P_1} \ \cos(A)\sin(A)\cos(B)\sin(B)$$

$$- \ D_{M_1,P_2} \ \cos(A)\sin(A)\cos(B)\sin(B) \ + \ D_{M_2,P_3} \ \sin^2(A)\sin(B)$$

$$- \ D_{M_3,P_2} \ \cos^2(A)\sin(B) \ + \ D_{M_1,P_1} \ \cos(A)\sin(A)\cos^2(B)$$

$$- \ D_{M_1,P_3} \ \sin^2(A)\cos(B) \ + \ D_{M_3,P_1} \ \cos^2(A)\cos(B)$$

$$- \ D_{M_3,P_3} \ \cos(A)\sin(A))\text{DEL}(P_1)$$

The same procedure may be used to formulate the aerodynamic moments produced by linear and angular accelerations. These moments will not be included here, since the cases considered so far are sufficient to demonstrate the facility with which symbolic mathematical computation can be used to formulate and transform aerodynamic moments.

6.25 INERTIA MOMENTS

The formulation of inertia moments involves the determination of the product of an angular velocity matrix, a matrix of inertia coefficients, and a column vector of angular velocity components. This product is the matrix equivalent of the familiar vector product $\bar{\omega} \times \bar{H}$, where $\bar{\omega}$ is the angular velocity vector and $\bar{H}$ is the angular momentum vector. By adding to the components of this vector, a vector which represents the rate of change of angular momentum relative to the moving body axes, the inertial moments relative to these axes are obtained (ref. 9). The rate of change of angular momentum relative to the moving body axes may be expressed as the product of the inertia matrix and a column vector of angular acceleration components. The required matrices may be entered and multiplied as follows: The first matrix to be entered is the inertia matrix, with elements $J_{i,j}$. It is entered by typing the statement ENTERMATRIX(3,3) and responding to the system's request for elements.

```
(C23) ENTERMATRIX(3,3);

ROW 1 COLUMN 1   J[1,1];

ROW 1 COLUMN 2   J[1,2];

ROW 1 COLUMN 3   J[1,3];

ROW 2 COLUMN 1   J[2,1];

ROW 2 COLUMN 2   J[2,2];

ROW 2 COLUMN 3   J[2,3];

ROW 3 COLUMN 1   J[3,1];

ROW 3 COLUMN 2   J[3,2];

ROW 3 COLUMN 3   J[3,3];

MATRIX-ENTERED
```

$$(D23) \qquad \begin{bmatrix} J_{1,1} & J_{1,2} & J_{1,3} \\ J_{2,1} & J_{2,2} & J_{2,3} \\ J_{3,1} & J_{3,2} & J_{3,3} \end{bmatrix}$$

A statement of the fact that the ith component of the angular velocity vector is a function of time requires the use of the DEPENDENCIES function. The use of this function permits the system to differentiate the components P_i with respect to time, and to enter the resulting acceleration components in the form of a column vector as follows:

```
(C24) DEPENDENCIES(P(I,T))$

(C25) ENTERMATRIX(3,1);

ROW 1 COLUMN 1  DIFF(P[1],T);

ROW 2 COLUMN 1  DIFF(P[2],T);

ROW 3 COLUMN 1  DIFF(P[3],T);


MATRIX-ENTERED
```

$$(D25) \qquad \begin{bmatrix} \dfrac{d}{dT}\,P_1 \\[2ex] \dfrac{d}{dT}\,P_2 \\[2ex] \dfrac{d}{dT}\,P_3 \end{bmatrix}$$

The angular velocity matrix and a column vector of angular velocity components are entered next

```
(C26) ENTERMATRIX(3,3);

ROW 1 COLUMN 1  0;

ROW 1 COLUMN 2  -P[3];

ROW 1 COLUMN 3  P[2];

ROW 2 COLUMN 1  P[3];

ROW 2 COLUMN 2  0;
```

```
ROW 2 COLUMN 3  -P[1];

ROW 3 COLUMN 1  -P[2];

ROW 3 COLUMN 2  P[1];

ROW 3 COLUMN 3  0;

MATRIX-ENTERED
```

$$
(D26) \qquad \begin{bmatrix} 0 & -P_3 & P_2 \\ P_3 & 0 & -P_1 \\ -P_2 & P_1 & 0 \end{bmatrix}
$$

```
(C27) ENTERMATRIX(3,1);

ROW 1 COLUMN 1  P[1];

ROW 2 COLUMN 1  P[2];

ROW 3 COLUMN 1  P[3];

MATRIX-ENTERED
```

$$
(D27) \qquad \begin{bmatrix} P_1 \\ P_2 \\ P_3 \end{bmatrix}
$$

These four matrices are now combined to yield a column vector of inertia moments relative to aircraft body axes.

```
(C28) ((D23).(D25)+(D26).(D23).(D27));
```

(D28) MATRIX([$J_{1,3}$ ($\frac{d}{dT}$ P_3) + $J_{1,2}$ ($\frac{d}{dT}$ P_2) + $J_{1,1}$ ($\frac{d}{dT}$ P_1)

+ $P_2(P_3J_{3,3}$ + $P_2J_{3,2}$ + $P_1J_{3,1})$ - $P_3(J_{2,3}P_3$ + $P_2J_{2,2}$ + $P_1J_{2,1})$],

[$J_{2,3}$ ($\frac{d}{dT}$ P_3) + $J_{2,2}$ ($\frac{d}{dT}$ P_2) + $J_{2,1}$ ($\frac{d}{dT}$ P_1) - $P_1(P_3J_{3,3}$ + $P_2J_{3,2}$

+ $P_1J_{3,1})$ + $P_1J_{3,1})$ + $P_3(J_{1,3}P_3$ + $J_{1,2}P_2$ + $P_1J_{1,1})$], [$J_{3,3}$ ($\frac{d}{dT}$ P_3)

+ $J_{3,2}$ ($\frac{d}{dT}$ P_2) + $J_{3,1}$ ($\frac{d}{dT}$ P_1) + $P_1(J_{2,3}P_3$ + $P_2J_{2,2}$ + $P_1J_{2,1})$

- $P_2(J_{1,3}P_3$ + $J_{1,2}P_2$ + $P_1J_{1,1})$])

The next two programming steps enable the system to express these inertia moments in conventional functional form.

(C29) FOR I:1 THRU 3 DO ROW[I]:FIRST(ROW((D28),I))\$

(C30) FOR I:1 THRU 3 DO (IM[I]:ROW[I][1],DISPLAY(IM[I]))\$

IM_1 = $J_{1,3}$ ($\frac{d}{dT}$ P_3) + $J_{1,2}$ ($\frac{d}{dT}$ P_2) + $J_{1,1}$ ($\frac{d}{dT}$ P_1) + $P_2(P_3J_{3,3}$

+ $P_2J_{3,2}$ + $P_1J_{3,1})$ - $P_3(J_{2,3}P_3$ + $P_2J_{2,2}$ + $P_1J_{2,1})$

IM_2 = $J_{2,3}$ ($\frac{d}{dT}$ P_3) + $J_{2,2}$ ($\frac{d}{dT}$ P_2) + $J_{2,1}$ ($\frac{d}{dT}$ P_1) - $P_1(P_3J_{3,3}$

+ $P_2J_{3,2}$ + $P_1J_{3,1})$ + $P_3(J_{1,3}P_3$ + $J_{1,2}P_2$ + $P_1J_{1,1})$

IM_3 = $J_{3,3}$ ($\frac{d}{dT}$ P_3) + $J_{3,2}$ ($\frac{d}{dT}$ P_2) + $J_{3,1}$ ($\frac{d}{dT}$ P_1) + $P_1(J_{2,3}P_3$

+ $P_2J_{2,2}$ + $P_1J_{2,1})$ - $P_2(J_{1,3}P_3$ + $J_{1,2}P_2$ + $P_1J_{1,1})$

6.26 RESULTANT MOMENTS

It only remains to request MACSYMA to combine the aerodynamic and inertia moments which have been formulated in preceding sections and to display the results. The ith component of the resultant moment will be denoted by TM_i, where TM_i is the ith component of the moment due to thrust. The two programming steps and the formulated equations follow.

```
(C31) FOR I:1 THRU 3 DO TM[I]:IM[I]-SM[I]-CM[I]-MDU[I]-MDP[I]$
```

```
(C32) FOR I:1 THRU 3 DO DISPLAY(TM[I])$
```

$$
\begin{aligned}
TM_1 = &-(-D_{M_2,C_K} \cos(A)\sin(B) + D_{M_1,C_K} \cos(A)\cos(B) \\
&- D_{M_3,C_K} \sin(A))DEL(C_K) - (D_{M_2,U_2} \cos(A)\sin(A)\sin^2(B) \\
&- D_{M_2,U_1} \cos(A)\sin(A)\cos(B)\sin(B) \\
&- D_{M_1,U_2} \cos(A)\sin(A)\cos(B)\sin(B) + D_{M_3,U_2} \sin^2(A)\sin(B) \\
&- D_{M_2,U_3} \cos^2(A)\sin(B) + D_{M_1,U_1} \cos(A)\sin(A)\cos^2(B) \\
&- D_{M_3,U_1} \sin^2(A)\cos(B) + D_{M_1,U_3} \cos^2(A)\cos(B) \\
&- D_{M_3,U_3} \cos(A)\sin(A))DEL(U_3) - (D_{M_2,P_2} \cos(A)\sin(A)\sin^2(B) \\
&- D_{M_2,P_1} \cos(A)\sin(A)\cos(B)\sin(B) \\
&- D_{M_1,P_2} \cos(A)\sin(A)\cos(B)\sin(B) + D_{M_3,P_2} \sin^2(A)\sin(B) \\
&- D_{M_2,P_3} \cos^2(A)\sin(B) + D_{M_1,P_1} \cos(A)\sin(A)\cos^2(B)
\end{aligned}
$$

6.26

$$- D_{M_3,P_1} \, SIN^2(A)COS(B) + D_{M_1,P_3} \, COS^2(A)COS(B)$$

$$- D_{M_3,P_3} \, COS(A)SIN(A))DEL(P_3) - (-D_{M_2,U_1} \, COS(A)SIN^2(B)$$

$$- D_{M_2,U_2} \, COS(A)COS(B)SIN(B) + D_{M_1,U_1} \, COS(A)COS(B)SIN(B)$$

$$- D_{M_3,U_1} \, SIN(A)SIN(B) + D_{M_1,U_2} \, COS(A)COS^2(B)$$

$$- D_{M_3,U_2} \, SIN(A)COS(B))DEL(U_2) - (-D_{M_2,P_1} \, COS(A)SIN^2(B)$$

$$- D_{M_2,P_2} \, COS(A)COS(B)SIN(B) + D_{M_1,P_1} \, COS(A)COS(B)SIN(B)$$

$$- D_{M_3,P_1} \, SIN(A)SIN(B) + D_{M_1,P_2} \, COS(A)COS^2(B)$$

$$- D_{M_3,P_2} \, SIN(A)COS(B))DEL(P_2) - (D_{M_2,U_2} \, COS^2(A)SIN^2(B)$$

$$- D_{M_2,U_1} \, COS^2(A)COS(B)SIN(B) - D_{M_1,U_2} \, COS^2(A)COS(B)SIN(B)$$

$$+ D_{M_3,U_2} \, COS(A)SIN(A)SIN(B) + D_{M_2,U_3} \, COS(A)SIN(A)SIN(B)$$

$$+ D_{M_1,U_1} \, COS^2(A)COS^2(B) - D_{M_3,U_1} \, COS(A)SIN(A)COS(B)$$

$$- D_{M_1,U_3} \, COS(A)SIN(A)COS(B) + D_{M_3,U_3} \, SIN^2(A))DEL(U_1)$$

$$- (D_{M_2,P_2} \, COS^2(A)SIN^2(B) - D_{M_2,P_1} \, COS^2(A)COS(B)SIN(B)$$

$$- D_{M_1,P_2} \, COS^2(A)COS(B)SIN(B) + D_{M_3,P_2} \, COS(A)SIN(A)SIN(B)$$

376

$$+ \, D_{M_2,P_3} \, \cos(A)\sin(A)\sin(B) + D_{M_1,P_1} \, \cos^2(A)\cos^2(B)$$

$$- \, D_{M_3,P_1} \, \cos(A)\sin(A)\cos(B) - D_{M_1,P_3} \, \cos(A)\sin(A)\cos(B)$$

$$+ \, D_{M_3,P_3} \, \sin^2(A))\mathrm{DEL}(P_1) + S_{M_2} \, \cos(A)\sin(B) - S_{M_1} \, \cos(A)\cos(B)$$

$$+ \, S_{M_3} \, \sin(A) + J_{1,3}\left(\frac{d}{dT} \, P_3\right) + J_{1,2}\left(\frac{d}{dT} \, P_2\right) + J_{1,1}\left(\frac{d}{dT} \, P_1\right)$$

$$+ \, P_2(P_3 J_{3,3} + P_2 J_{3,2} + P_1 J_{3,1}) - P_3(J_{2,3}P_3 + P_2 J_{2,2} + P_1 J_{2,1})$$

$$TM_2 = -(D_{M_1,C_K} \, \sin(B) + D_{M_2,C_K} \, \cos(B))\mathrm{DEL}(C_K)$$

$$- \, (-D_{M_1,U_2} \, \sin(A)\sin^2(B) - D_{M_2,U_2} \, \sin(A)\cos(B)\sin(B)$$

$$+ \, D_{M_1,U_1} \, \sin(A)\cos(B)\sin(B) + D_{M_1,U_3} \, \cos(A)\sin(B)$$

$$+ \, D_{M_2,U_1} \, \sin(A)\cos^2(B) + D_{M_2,U_3} \, \cos(A)\cos(B))\mathrm{DEL}(U_3)$$

$$- \, (-D_{M_1,P_2} \, \sin(A)\sin^2(B) - D_{M_2,P_2} \, \sin(A)\cos(B)\sin(B)$$

$$+ \, D_{M_1,P_1} \, \sin(A)\cos(B)\sin(B) + D_{M_1,P_3} \, \cos(A)\sin(B)$$

$$+ \, D_{M_2,P_1} \, \sin(A)\cos^2(B) + D_{M_2,P_3} \, \cos(A)\cos(B))\mathrm{DEL}(P_3)$$

$$- \, (D_{M_1,U_1} \, \sin^2(B) + D_{M_2,U_1} \, \cos(B)\sin(B) + D_{M_1,U_2} \, \cos(B)\sin(B)$$

$$+ \, D_{M_2,U_2} \, \cos^2(B))\mathrm{DEL}(U_2) - (D_{M_1,P_1} \, \sin^2(B) + D_{M_2,P_1} \, \cos(B)\sin(B)$$

$$+ \; D_{M_1,P_2} \; COS(B)SIN(B) \; + \; D_{M_2,P_2} \; COS^2(B))DEL(P_2)$$

$$- \; (-D_{M_1,U_2} \; COS(A)SIN^2(B) \; - \; D_{M_2,U_2} \; COS(A)COS(B)SIN(B)$$

$$+ \; D_{M_1,U_1} \; COS(A)COS(B)SIN(B) \; - \; D_{M_1,U_3} \; SIN(A)SIN(B)$$

$$+ \; D_{M_2,U_1} \; COS(A)COS^2(B) \; - \; D_{M_2,U_3} \; SIN(A)COS(B))DEL(U_1)$$

$$- \; (-D_{M_1,P_2} \; COS(A)SIN^2(B) \; - \; D_{M_2,P_2} \; COS(A)COS(B)SIN(B)$$

$$+ \; D_{M_1,P_1} \; COS(A)COS(B)SIN(B) \; - \; D_{M_1,P_3} \; SIN(A)SIN(B)$$

$$+ \; D_{M_2,P_1} \; COS(A)COS^2(B) \; - \; D_{M_2,P_3} \; SIN(A)COS(B))DEL(P_1) \; - \; S_{M_1} \; SIN(B)$$

$$- \; S_{M_2} \; COS(B) \; + \; J_{2,3}(\frac{d}{dT} \; P_3) \; + \; J_{2,2}(\frac{d}{dT} \; P_2) \; + \; J_{2,1}(\frac{d}{dT} \; P_1)$$

$$- \; P_1(P_3J_{3,3} \; + \; P_2J_{3,2} \; + \; P_1J_{3,1}) \; + \; P_3(J_{1,3}P_3 \; + \; J_{1,2}P_2 \; + \; P_1J_{1,1})$$

$$TM_3 \; = \; -(-D_{M_2,C_K} \; SIN(A)SIN(B) \; + \; D_{M_1,C_K} \; SIN(A)COS(B)$$

$$+ \; D_{M_3,C_K} \; COS(A))DEL(C_K) \; - \; (D_{M_2,U_2} \; SIN^2(A)SIN^2(B)$$

$$- \; D_{M_2,U_1} \; SIN^2(A)COS(B)SIN(B) \; - \; D_{M_1,U_2} \; SIN^2(A)COS(B)SIN(B)$$

$$- \; D_{M_3,U_2} \; COS(A)SIN(A)SIN(B) \; - \; D_{M_2,U_3} \; COS(A)SIN(A)SIN(B)$$

$$+ \; D_{M_1,U_1} \; SIN^2(A)COS^2(B) \; + \; D_{M_3,U_1} \; COS(A)SIN(A)COS(B)$$

$$+ D_{M_1,U_3} \cos(A)\sin(A)\cos(B) + D_{M_3,U_3} \cos^2(A))DEL(U_3)$$

$$- (D_{M_2,P_2} \sin^2(A)\sin^2(B) - D_{M_2,P_1} \sin^2(A)\cos(B)\sin(B)$$

$$- D_{M_1,P_2} \sin^2(A)\cos(B)\sin(B) - D_{M_3,P_2} \cos(A)\sin(A)\sin(B)$$

$$- D_{M_2,P_3} \cos(A)\sin(A)\sin(B) + D_{M_1,P_1} \sin^2(A)\cos^2(B)$$

$$+ D_{M_3,P_1} \cos(A)\sin(A)\cos(B) + D_{M_1,P_3} \cos(A)\sin(A)\cos(B)$$

$$+ D_{M_3,P_3} \cos^2(A))DEL(P_3) - (-D_{M_2,U_1} \sin(A)\sin^2(B)$$

$$- D_{M_2,U_2} \sin(A)\cos(B)\sin(B) + D_{M_1,U_1} \sin(A)\cos(B)\sin(B)$$

$$+ D_{M_3,U_1} \cos(A)\sin(B) + D_{M_1,U_2} \sin(A)\cos^2(B)$$

$$+ D_{M_3,U_2} \cos(A)\cos(B))DEL(U_2) - (-D_{M_2,P_1} \sin(A)\sin^2(B)$$

$$- D_{M_2,P_2} \sin(A)\cos(B)\sin(B) + D_{M_1,P_1} \sin(A)\cos(B)\sin(B)$$

$$+ D_{M_3,P_1} \cos(A)\sin(B) + D_{M_1,P_2} \sin(A)\cos^2(B)$$

$$+ D_{M_3,P_2} \cos(A)\cos(B))DEL(P_2) - (D_{M_2,U_2} \cos(A)\sin(A)\sin^2(B)$$

$$- D_{M_2,U_1} \cos(A)\sin(A)\cos(B)\sin(B)$$

$$- D_{M_1,U_2} \cos(A)\sin(A)\cos(B)\sin(B) + D_{M_2,U_3} \sin^2(A)\sin(B)$$

$$- D_{M_3,U_2} \cos^2(A)\sin(B) + D_{M_1,U_1} \cos(A)\sin(A)\cos^2(B)$$

$$- D_{M_1,U_3} \sin^2(A)\cos(B) + D_{M_3,U_1} \cos^2(A)\cos(B)$$

$$- D_{M_3,U_3} \cos(A)\sin(A))DEL(U_1) - (D_{M_2,P_2} \cos(A)\sin(A)\sin^2(B)$$

$$- D_{M_2,P_1} \cos(A)\sin(A)\cos(B)\sin(B)$$

$$- D_{M_1,P_2} \cos(A)\sin(A)\cos(B)\sin(B) + D_{M_2,P_3} \sin^2(A)\sin(B)$$

$$- D_{M_3,P_2} \cos^2(A)\sin(B) + D_{M_1,P_1} \cos(A)\sin(A)\cos^2(B)$$

$$- D_{M_1,P_3} \sin^2(A)\cos(B) + D_{M_3,P_1} \cos^2(A)\cos(B)$$

$$- D_{M_3,P_3} \cos(A)\sin(A))DEL(P_1) + S_{M_2} \sin(A)\sin(B)$$

$$- S_{M_1} \sin(A)\cos(B) - S_{M_3} \cos(A) + J_{3,3}(\frac{d}{dT} P_3) + J_{3,2}(\frac{d}{dT} P_2)$$

$$+ J_{3,1}(\frac{d}{dT} P_1) + P_1(J_{2,3}P_3 + P_2J_{2,2} + P_1J_{2,1})$$

$$- P_2(J_{1,3}P_3 + J_{1,2}P_2 + P_1J_{1,1})$$

6.27 SPECIAL FORMS OF THE MOMENT EQUATION

As in the case of the force equations, the investigator sometimes wishes to modify the moment equations to determine how the vehicle will respond if the motion is restricted in some way (ref. 8). For the case of zero sideslip, MACSYMA goes through the equations, makes the appropriate substitutions, and displays the modified results. The zero sideslip condition requires that $\sin(B) = 0$ and $\cos(B) = 1$. The substitution and display statements required to implement this assumption and the modified moment equations assume the following form:

```
(C33) FOR I:1 THRU 3 DO
 TM[I]:SUBST([SIN(B)=0,COS(B)=1],TM[I])$

(C34) FOR I:1 THRU 3 DO DISPLAY(TM[I])$
```

$$TM_1 = -(D_{M_1,C_K} \cos(A) - D_{M_3,C_K} \sin(A))DEL(C_K)$$

$$- (-D_{M_3,U_1} \sin^2(A) - D_{M_3,U_3} \cos(A)\sin(A) + D_{M_1,U_1} \cos(A)\sin(A)$$

$$+ D_{M_1,U_3} \cos^2(A))DEL(U_3) - (-D_{M_3,P_1} \sin^2(A) - D_{M_3,P_3} \cos(A)\sin(A)$$

$$+ D_{M_1,P_1} \cos(A)\sin(A) + D_{M_1,P_3} \cos^2(A))DEL(P_3) - (D_{M_1,U_2} \cos(A)$$

$$- D_{M_3,U_2} \sin(A))DEL(U_2) - (D_{M_1,P_2} \cos(A) - D_{M_3,P_2} \sin(A))DEL(P_2)$$

$$- (D_{M_3,U_3} \sin^2(A) - D_{M_3,U_1} \cos(A)\sin(A) - D_{M_1,U_3} \cos(A)\sin(A)$$

$$+ D_{M_1,U_1} \cos^2(A))DEL(U_1) - (D_{M_3,P_3} \sin^2(A) - D_{M_3,P_1} \cos(A)\sin(A)$$

$$- D_{M_1,P_3} \cos(A)\sin(A) + D_{M_1,P_1} \cos^2(A))DEL(P_1) + S_{M_3} \sin(A)$$

$$- S_{M_1} \cos(A) + J_{1,3}(\frac{d}{dT} P_3) + J_{1,2}(\frac{d}{dT} P_2) + J_{1,1}(\frac{d}{dT} P_1)$$

$$+ P_2(P_3 J_{3,3} + P_2 J_{3,2} + P_1 J_{3,1}) - P_3(J_{2,3} P_3 + P_2 J_{2,2} + P_1 J_{2,1})$$

$$TM_2 = -D_{M_2,C_K} DEL(C_K) - (D_{M_2,U_1} \sin(A) + D_{M_2,U_3} \cos(A))DEL(U_3)$$

$$- (D_{M_2,P_1} \sin(A) + D_{M_2,P_3} \cos(A))DEL(P_3) - D_{M_2,U_2} DEL(U_2)$$

$$- D_{M_2,P_2} \; DEL(P_2) - (D_{M_2,U_1} \; COS(A) - D_{M_2,U_3} \; SIN(A))DEL(U_1)$$

$$- (D_{M_2,P_1} \; COS(A) - D_{M_2,P_3} \; SIN(A))DEL(P_1) + J_{2,3}(\frac{d}{dT} P_3)$$

$$+ J_{2,2}(\frac{d}{dT} P_2) - S_{M_2} + J_{2,1}(\frac{d}{dT} P_1) - P_1(P_3 J_{3,3} + P_2 J_{3,2} + P_1 J_{3,1})$$

$$+ P_3(J_{1,3} P_3 + J_{1,2} P_2 + P_1 J_{1,1})$$

$$TM_3 = -(D_{M_1,C_K} \; SIN(A) + D_{M_3,C_K} \; COS(A))DEL(C_K) - (D_{M_1,U_1} \; SIN^2(A)$$

$$+ D_{M_3,U_1} \; COS(A)SIN(A) + D_{M_1,U_3} \; COS(A)SIN(A)$$

$$+ D_{M_3,U_3} \; COS^2(A))DEL(U_3) - (D_{M_1,P_1} \; SIN^2(A) + D_{M_3,P_1} \; COS(A)SIN(A)$$

$$+ D_{M_1,P_3} \; COS(A)SIN(A) + D_{M_3,P_3} \; COS^2(A))DEL(P_3) - (D_{M_1,U_2} \; SIN(A)$$

$$+ D_{M_3,U_2} \; COS(A))DEL(U_2) - (D_{M_1,P_2} \; SIN(A) + D_{M_3,P_2} \; COS(A))DEL(P_2)$$

$$- (-D_{M_1,U_3} \; SIN^2(A) - D_{M_3,U_3} \; COS(A)SIN(A) + D_{M_1,U_1} \; COS(A)SIN(A)$$

$$+ D_{M_3,U_1} \; COS^2(A))DEL(U_1) - (-D_{M_1,P_3} \; SIN^2(A) - D_{M_3,P_3} \; COS(A)SIN(A)$$

$$+ D_{M_1,P_1} \; COS(A)SIN(A) + D_{M_3,P_1} \; COS^2(A))DEL(P_1) - S_{M_1} \; SIN(A)$$

$$- S_{M_3} \; COS(A) + J_{3,3}(\frac{d}{dT} P_3) + J_{3,2}(\frac{d}{dT} P_2) + J_{3,1}(\frac{d}{dT} P_1)$$

$$+ P_1(J_{2,3} P_3 + P_2 J_{2,2} + P_1 J_{2,1}) - P_2(J_{1,3} P_3 + J_{1,2} P_2 + P_1 J_{1,1})$$

In addition to the zero sideslip condition, the investigator frequently wishes to determine vehicle response when the angle of attack is limited to small values. For this condition MACSYMA would implement the assumption that SIN(A) = A. Moreover, if the angle of attack were sufficiently small, the program would request MACSYMA to make the additional substitution COS(A) = 1.

In this case, the required substitution and display statements give rise to the following modified moment equations:

```
(C35) FOR I:1 THRU 3 DO TM[I]:SUBST([SIN(A)=A,COS(A)=1],TM[I])$
```

```
(C36) FOR I:1 THRU 3 DO DISPLAY(TM[I])$
```

$$TM_1 = -(D_{M_1,C_K} - D_{M_3,C_K} A)DEL(C_K) - (-D_{M_3,U_1} A^2 - D_{M_3,U_3} A$$

$$+ D_{M_1,U_1} A + D_{M_1,U_3})DEL(U_3) - (-D_{M_3,P_1} A^2 - D_{M_3,P_3} A + D_{M_1,P_1} A$$

$$+ D_{M_1,P_3})DEL(P_3) - (D_{M_1,U_2} - D_{M_3,U_2} A)DEL(U_2) - (D_{M_1,P_2}$$

$$- D_{M_3,P_2} A)DEL(P_2) - (D_{M_3,U_3} A^2 - D_{M_3,U_1} A - D_{M_1,U_3} A$$

$$+ D_{M_1,U_1})DEL(U_1) - (D_{M_3,P_3} A - D_{M_3,P_1} A - D_{M_1,P_3} A$$

$$+ D_{M_1,P_1})DEL(P_1) + S_{M_3} A + J_{1,3}(\frac{d}{dT} P_3) + J_{1,2}(\frac{d}{dT} P_2) + J_{1,1}(\frac{d}{dT} P_1)$$

$$- S_{M_1} + P_2(P_3 J_{3,3} + P_2 J_{3,2} + P_1 J_{3,1}) - P_3(J_{2,3} P_3 + P_2 J_{2,2}$$

$$+ P_1 J_{2,1})$$

$$TM_2 = -D_{M_2,C_K} DEL(C_K) - (D_{M_2,U_1} A + D_{M_2,U_3})DEL(U_3) - (D_{M_2,P_1} A$$

$$+ D_{M_2,P_3})DEL(P_3) - D_{M_2,U_2} DEL(U_2) - D_{M_2,P_2} DEL(P_2) - (D_{M_2,U_1}$$

$$- D_{M_2,U_3} A)DEL(U_1) - (D_{M_2,P_1} - D_{M_2,P_3} A)DEL(P_1) + J_{2,3}(\frac{d}{dT} P_3)$$

$$+ J_{2,2}(\frac{d}{dT} P_2) - S_{M_2} + J_{2,1}(\frac{d}{dT} P_1) - P_1(P_3 J_{3,3} + P_2 J_{3,2} + P_1 J_{3,1})$$

$$+ P_3(J_{1,3} P_3 + J_{1,2} P_2 + P_1 J_{1,1})$$

$$TM_3 = -(D_{M_1,C_K} A + D_{M_3,C_K})DEL(C_K) - (D_{M_1,U_1} A^2 + D_{M_3,U_1} A$$

$$+ D_{M_1,U_3} A + D_{M_3,U_3})DEL(U_3) - (D_{M_1,P_1} A^2 + D_{M_3,P_1} A + D_{M_1,P_3} A$$

$$+ D_{M_3,P_3})DEL(P_3) - (D_{M_1,U_2} A + D_{M_3,U_2})DEL(U_2) - (D_{M_1,P_2} A$$

$$+ D_{M_3,P_2})DEL(P_2) - (-D_{M_1,U_3} A^2 - D_{M_3,U_3} A + D_{M_1,U_1} A$$

$$+ D_{M_3,U_1})DEL(U_1) - (-D_{M_1,P_3} A^2 - D_{M_3,P_3} A + D_{M_1,P_1} A$$

$$+ D_{M_3,P_1})DEL(P_1) - S_{M_1} A + J_{3,3}(\frac{d}{dT} P_3) - S_{M_3} + J_{3,2}(\frac{d}{dT} P_2)$$

$$+ J_{3,1}(\frac{d}{dT} P_1) + P_1(J_{2,3} P_3 + P_2 J_{2,2} + P_1 J_{2,1})$$

$$- P_2(J_{1,3} P_3 + J_{1,2} P_2 + P_1 J_{1,1})$$

Examination of these equations reveals the existence of terms such as A^2. If it is assumed that second-order terms in A are negligible, a program statement instructing MACSYMA to make the substitution $A^2 = 0$ would simplify the moment equations as follows:

```
(C37) FOR I:1 THRU 3 DO
 TM[I]:SUBST([A**2=0],TM[I])$

(C39) FOR I:1 THRU 3 DO DISPLAY(TM[I])$
```

$$TM_1 = -(D_{M_1,C_K} - D_{M_3,C_K}A)DEL(C_K) - (-D_{M_3,U_3}A + D_{M_1,U_1}A$$

$$+ D_{M_1,U_3})DEL(U_3) - (-D_{M_3,P_3}A + D_{M_1,P_1}A + D_{M_1,P_3})DEL(P_3)$$

$$- (D_{M_1,U_2} - D_{M_3,U_2}A)DEL(U_2) - (D_{M_1,P_2} - D_{M_3,P_2}A)DEL(P_2)$$

$$- (-D_{M_3,U_1}A - D_{M_1,U_3}A + D_{M_1,U_1})DEL(U_1) - (-D_{M_3,P_1}A - D_{M_1,P_3}A$$

$$+ D_{M_1,P_1})DEL(P_1) + S_{M_3}A + J_{1,3}(\frac{d}{dT}P_3) + J_{1,2}(\frac{d}{dT}P_2)$$

$$+ J_{1,1}(\frac{d}{dT}P_1) - S_{M_1} + P_2(P_3J_{3,3} + P_2J_{3,2} + P_1J_{3,1})$$

$$- P_3(J_{2,3}P_3 + P_2J_{2,2} + P_1J_{2,1})$$

$$TM_2 = -D_{M_2,C_K}DEL(C_K) - (D_{M_2,U_1}A + D_{M_2,U_3})DEL(U_3) - (D_{M_2,P_1}A$$

$$+ D_{M_2,P_3})DEL(P_3) - D_{M_2,U_2}DEL(U_2) - D_{M_2,P_2}DEL(P_2) - (D_{M_2,U_1}$$

$$- D_{M_2,U_3}A)DEL(U_1) - (D_{M_2,P_1} - D_{M_2,P_3}A)DEL(P_1) + J_{2,3}(\frac{d}{dT}P_3)$$

$$+ J_{2,2}(\frac{d}{dT}P_2) - S_{M_2} + J_{2,1}(\frac{d}{dT}P_1) - P_1(P_3J_{3,3} + P_2J_{3,2} + P_1J_{3,1})$$

$$+ P_3(J_{1,3}P_3 + J_{1,2}P_2 + P_1J_{1,1})$$

$$TM_3 = -(D_{M_1,C_K}A + D_{M_3,C_K})DEL(C_K) - (D_{M_3,U_1}A + D_{M_1,U_3}A$$

$$+ D_{M_3,U_3})DEL(U_3) - (D_{M_3,P_1}A + D_{M_1,P_3}A + D_{M_3,P_3})DEL(P_3)$$

$$- (D_{M_1,U_2} A + D_{M_3,U_2})DEL(U_2) - (D_{M_1,P_2} A + D_{M_3,P_2})DEL(P_2)$$

$$- (-D_{M_3,U_3} A + D_{M_1,U_1} A + D_{M_3,U_1})DEL(U_1) - (-D_{M_3,P_3} A + D_{M_1,P_1} A$$

$$+ D_{M_3,P_1})DEL(P_1) - S_{M_1} A + J_{3,3}(\frac{d}{dT} P_3) - S_{M_3} + J_{3,2}(\frac{d}{dT} P_2)$$

$$+ J_{3,1}(\frac{d}{dT} P_1) + P_1(J_{2,3}P_3 + P_2 J_{2,2} + P_1 J_{2,1})$$

$$- P_2(J_{1,3}P_3 + J_{1,2}P_2 + P_1 J_{1,1})$$

An additional simplification is possible if the assumption that angular velocity perturbations are negligible is a valid one. Implementation of the assumption that $DEL(P_i) = 0$ yields the following greatly simplified equations:

```
(C40) FOR I:1 THRU 3 DO FOR J:1 THRU 3 DO
TM[I]:SUBST([DEL(P[J])=0],TM[I])$
```

```
(C41) FOR I:1 THRU 3 DO DISPLAY(TM[I])$
```

$$TM_1 = -(D_{M_1,C_K} - D_{M_3,C_K} A)DEL(C_K) - (-D_{M_3,U_3} A + D_{M_1,U_1} A$$

$$+ D_{M_1,U_3})DEL(U_3) - (D_{M_1,U_2} - D_{M_3,U_2} A)DEL(U_2) - (-D_{M_3,U_1} A$$

$$- D_{M_1,U_3} A + D_{M_1,U_1})DEL(U_1) + S_{M_3} A + J_{1,3}(\frac{d}{dT} P_3) + J_{1,2}(\frac{d}{dT} P_2)$$

$$+ J_{1,1}(\frac{d}{dT} P_1) - S_{M_1} + P_2(P_3 J_{3,3} + P_2 J_{3,2} + P_1 J_{3,1})$$

$$- P_3(J_{2,3}P_3 + P_2 J_{2,2} + P_1 J_{2,1})$$

$$TM_2 = -D_{M_2,C_K} DEL(C_K) - (D_{M_2,U_1} A + D_{M_2,U_3})DEL(U_3)$$

$$- D_{M_2,U_2} \, \text{DEL}(U_2) - (D_{M_2,U_1} - D_{M_2,U_3} \, A) \text{DEL}(U_1) + J_{2,3}(\tfrac{d}{dT} P_3)$$

$$+ J_{2,2}(\tfrac{d}{dT} P_2) - S_{M_2} + J_{2,1}(\tfrac{d}{dT} P_1) - P_1(P_3 J_{3,3} + P_2 J_{3,2} + P_1 J_{3,1})$$

$$+ P_3(J_{1,3} P_3 + J_{1,2} P_2 + P_1 J_{1,1})$$

$$TM_3 = -(D_{M_1,C_K} A + D_{M_3,C_K}) \text{DEL}(C_K) - (D_{M_3,U_1} A + D_{M_1,U_3} A$$

$$+ D_{M_3,U_3}) \text{DEL}(U_3) - (D_{M_1,U_2} A + D_{M_3,U_2}) \text{DEL}(U_2) - (-D_{M_3,U_3} A$$

$$+ D_{M_1,U_1} A + D_{M_3,U_1}) \text{DEL}(U_1) - S_{M_1} A + J_{3,3}(\tfrac{d}{dT} P_3) - S_{M_3}$$

$$+ J_{3,2}(\tfrac{d}{dT} P_2) + J_{3,1}(\tfrac{d}{dT} P_1) + P_1(J_{2,3} P_3 + P_2 J_{2,2} + P_1 J_{2,1})$$

$$- P_2(J_{1,3} P_3 + J_{1,2} P_2 + P_1 J_{1,1})$$

6.28 THRUST MOMENTS

As indicated in section 6.19, the thrust moments TM_i appearing on the left-hand side of these equations are the resultant of the moments produced by a number of thrust generating systems. The equations (6.19.1) relate the thrust axes coordinates $X_n{}^i$ to the coordinate system $Y_n{}^i$, which has the same origin as the thrust axes but is parallel to the body axes system.

To facilitate the formulation, equations (6.19.1) are entered here.

```
(C1) Y[1,N]:X[1,N]*COS(K[N])*COS(P[N])$

(C2) Y[2,N]:X[1,N]*COS(K[N])*SIN(P[N])$

(C3) Y[3,N]:-X[1,N]*SIN(K[N])$
```

The point of application of the nth thrust vector relative to the body axes system, with origin at the center of gravity, has components $(L_{1,n}, L_{2,n}, L_{3,n})$. The components of the nth thrust vector in this coordinate system are given by equations (6.19.2). The product of the position matrix with elements $(L_{1,n}, L_{2,n}, L_{3,n})$ and a column vector of thrust components can be processed as follows.

First enter the (3,3) position matrix, element by element, as requested by MACSYMA. Next enter the (3,1) column vector of thrust components in the same manner. When the matrices are entered the displayed form of each matrix assumes the conventional textbook form

```
(C4) ENTERMATRIX(3,3);

ROW 1 COLUMN 1   0;

ROW 1 COLUMN 2   -L[3,N];

ROW 1 COLUMN 3   L[2,N];

ROW 2 COLUMN 1   L[3,N];

ROW 2 COLUMN 2   0;

ROW 2 COLUMN 3   -L[1,N];

ROW 3 COLUMN 1   -L[2,N];

ROW 3 COLUMN 2   L[1,N];

ROW 3 COLUMN 3   0;

MATRIX-ENTERED
```

$$\text{(D4)} \qquad \begin{bmatrix} 0 & -L_{3,N} & L_{2,N} \\ L_{3,N} & 0 & -L_{1,N} \\ -L_{2,N} & L_{1,N} & 0 \end{bmatrix}$$

```
(C5) ENTERMATRIX(3,1);
```

```
ROW 1 COLUMN 1  T[N]*DIFF(Y[1,N],X[1,N]);

ROW 2 COLUMN 1  T[N]*DIFF(Y[2,N],X[1,N]);

ROW 3 COLUMN 1  T[N]*DIFF(Y[3,N],X[1,N]);
```

MATRIX-ENTERED

(D5)

$$\begin{bmatrix} T_N \cos(K_N) \cos(P_N) \\ T_N \cos(K_N) \sin(P_N) \\ -T_N \sin(K_N) \end{bmatrix}$$

By requesting the system to multiply these two matrices, the following product matrix is obtained:

(C6) (D4).(D5);

(D6)

$$\begin{bmatrix} -L_{3,N} \, T_N \cos(K_N) \sin(P_N) - L_{2,N} \, T_N \sin(K_N) \\ L_{3,N} \, T_N \cos(K_N) \cos(P_N) + L_{1,N} \, T_N \sin(K_N) \\ L_{1,N} \, T_N \cos(K_N) \sin(P_N) - L_{2,N} \, T_N \cos(K_N) \cos(P_N) \end{bmatrix}$$

In order to express this column vector of thrust moments in conventional functional form, the following two programming steps are required:

(C7) FOR I:1 .THRU 3 DO ROW[I]:FIRST(ROW((D6),I));

(C8) FOR I:1 THRU 3 DO (TM[I]:ROW[I][1],DISPLAY(TM[I]))$

$$TM_1 = -L_{3,N} \, T_N \cos(K_N) \sin(P_N) - L_{2,N} \, T_N \sin(K_N)$$

$$TM_2 = L_{3,N} \, T_N \cos(K_N) \cos(P_N) + L_{1,N} \, T_N \sin(K_N)$$

$$TM_3 = L_{1,N} \, T_N \cos(K_N) \sin(P_N) - L_{2,N} \, T_N \cos(K_N) \cos(P_N)$$

These equations give the moments produced by the nth thrust vector. When the number of thrust generating systems is known, these equations can be summed on n to obtain the total thrust moments.

6.29 SPATIAL ORIENTATION IN TERMS OF THE DIRECTION COSINES

The differential equations for the direction cosines can be obtained by first entering a (3,1) column vector of direction cosines, with elements D_{I1}, D_{I2}, and D_{I3}, where I can assume the values 1,2,3, and by premultiplying this vector by the angular velocity matrix. This operation is equivalent to the vector cross product of the angular velocity vector and the unit vectors $\hat{I}, \hat{J}$, and $\hat{K}$ (ref. 9). The programming steps and the displayed output are

```
(C1) ENTERMATRIX(3,1);

ROW 1 COLUMN 1   D[I,1];

ROW 2 COLUMN 1   D[I,2];

ROW 3 COLUMN 1   D[I,3];

MATRIX-ENTERED
```

$$(D1) \qquad \begin{bmatrix} D_{I,1} \\ D_{I,2} \\ D_{I,3} \end{bmatrix}$$

```
(C2) ENTERMATRIX(3,3);

ROW 1 COLUMN 1   0;

ROW 1 COLUMN 2   -P[3];

ROW 1 COLUMN 3   P[2];

ROW 2 COLUMN 1   P[3];
```

```
ROW 2 COLUMN 2   0;

ROW 2 COLUMN 3   -P[1];

ROW 3 COLUMN 1   -P[2];

ROW 3 COLUMN 2   P[1];

ROW 3 COLUMN 3   0;

MATRIX-ENTERED
```

(D2)

$$\begin{bmatrix} 0 & -P_3 & P_2 \\ P_3 & 0 & -P_1 \\ -P_2 & P_1 & 0 \end{bmatrix}$$

The product of these two matrices is

```
(C3) (D2).(D1);
```

(D3)

$$\begin{bmatrix} P_2 D_{I,3} - P_3 D_{I,2} \\ P_3 D_{I,1} - P_1 D_{I,3} \\ P_1 D_{I,2} - P_2 D_{I,1} \end{bmatrix}$$

The individual terms of this column vector can be evaluated for $I = 1,2,3$ by executing the following program statement:

```
(C4) FOR I:1 THRU 3 DO FOR J:1 THRU 3 DO
EV(C[I,J]:ROW((D3),J))$
```

The evaluated terms can be printed out by using the now familiar display statement

```
(C5) FOR I:1 THRU 3 DO FOR J:1 THRU 3 DO DISPLAY(C[I,J])$
```

$$C_{1,1} = [D_{1,3} P_2 - D_{1,2} P_3]$$

$$C_{1,2} = [D_{1,1}P_3 - P_1 D_{1,3}]$$

$$C_{1,3} = [P_1 D_{1,2} - D_{1,1}P_2]$$

$$C_{2,1} = [P_2 D_{2,3} - D_{2,2}P_3]$$

$$C_{2,2} = [D_{2,1}P_3 - P_1 D_{2,3}]$$

$$C_{2,3} = [P_1 D_{2,2} - P_2 D_{2,1}]$$

$$C_{3,1} = [P_2 D_{3,3} - P_3 D_{3,2}]$$

$$C_{3,2} = [P_3 D_{3,1} - P_1 D_{3,3}]$$

$$C_{3,3} = [P_1 D_{3,2} - P_2 D_{3,1}]$$

The dependence of the direction cosines on the indices I and J and the time T can be shown by using the **DEPENDENCIES** statement. The use of this statement facilitates the formulation of the differential coefficients

```
(C6) DEPENDENCIES(D(I,J,T))$
```

It only remains to request that the differential coefficients of the direction cosines DC_{IJ} with respect to the time T be added to the coefficients C_{IJ} and displayed as follows:

```
(C7) FOR I:1 THRU 3 DO FOR J:1 THRU 3 DO
DC[I,J]:C[I,J]+DIFF(D[I,J],T)$
```

```
(C8) FOR I:1 THRU 3 DO FOR J:1 THRU 3 DO DISPLAY(DC[I,J])$
```

$$DC_{1,1} = [D_{1,3}P_2 - D_{1,2}P_3] + \frac{d}{dT}D_{1,1} = 0$$

$$DC_{1,2} = [D_{1,1}P_3 - P_1 D_{1,3}] + \frac{d}{dT}D_{1,2} = 0$$

$$DC_{1,3} = [P_1 D_{1,2} - D_{1,1}P_2] + \frac{d}{dT}D_{1,3} = 0$$

$$DC_{2,1} = [P_2 D_{2,3} - D_{2,2}P_3] + \frac{d}{dT}D_{2,1} = 0$$

$$DC_{2,2} = [D_{2,1}P_3 - P_1 D_{2,3}] + \frac{d}{dT}D_{2,2} = 0$$

$$DC_{2,3} = [P_1 D_{2,2} - P_2 D_{2,1}] + \frac{d}{dT}D_{2,3} = 0$$

$$DC_{3,1} = [P_2 D_{3,3} - P_3 D_{3,2}] + \frac{d}{dT}D_{3,1} = 0$$

$$DC_{3,2} = [P_3 D_{3,1} - P_1 D_{3,3}] + \frac{d}{dT}D_{3,2} = 0$$

$$DC_{3,3} = [P_1 D_{3,2} - P_2 D_{3,1}] + \frac{d}{dT}D_{3,3} = 0$$

This concludes the formulation of the simplified aeronautical model considered. The formulation gave rise to 18 equations: 3 force equations; 3 moment equations; 9 direction cosine equations to determine the spatial orientation of the vehicle; and 3 equations to determine the geographical location of the vehicle relative to an Earth-fixed reference frame. It is seen that the technique of symbolic mathematical computation, as implemented by the MACSYMA system, can be used to facilitate the formulation of complex mathematical models of physical systems and reduce the errors to which human operators are prone. The versatility and simplicity of the system make it attractive to programmers and nonprogrammers alike. Moreover, as already noted, the capability of working interactively enhances the utility of the system by permitting the user to modify the formulation as he proceeds.

6.30

6.30 REFERENCES

1. Mathlab Group, Project MAC, M.I.T.: MACSYMA Reference Manual. Massachusetts Institute of Technology, 1957.

2. McConnell, Albert J.: Applications of Tensor Analysis. Dover Publications, Inc., 1957.

3. Spain, Barry: Tensor Calculus. Third ed., revised. Oliver and Boyd, Edinburgh, 1960.

4. Tolman, Richard C.: Relativity, Thermodynamics and Cosmology. Clarendon Press, Oxford, 1958.

5. Gainer, Thomas G.; and Hoffman, Sherwood: Summary of Transformation Equations and Equations of Motion Used in Free-Flight and Wind-Tunnel Data Reduction and Analysis. NASA SP-3070, 1972.

6. Sokolnikoff, Ivan S.: Tensor Analysis; Theory and Applications. John Wiley & Sons, Inc., 1960.

7. Tobak, Murray: On the Use of the Indicial Function Concept in the Analysis of Unsteady Motions of Wings and Wing-Tail Combinations. NACA Report 1188, 1954.

8. Etkin, Bernard: Dynamics of Atmospheric Flight. John Wiley & Sons, Inc., 1972.

9. Kolk, W. Richard: Modern Flight Dynamics. Prentice-Hall, Inc., 1961.

3597 6

394